Lecture Notes in Mechanical Engineering

[illegible]

[illegible]

[illegible]

[illegible]

[illegible]

[illegible]

Lecture Notes in Mechanical Engineering (LNME) publishes the latest developments in Mechanical Engineering - quickly, informally and with high quality. Original research reported in proceedings and post-proceedings represents the core of LNME. Volumes published in LNME embrace all aspects, subfields and new challenges of mechanical engineering. Topics in the series include:

- Engineering Design
- Machinery and Machine Elements
- Mechanical Structures and Stress Analysis
- Automotive Engineering
- Engine Technology
- Aerospace Technology and Astronautics
- Nanotechnology and Microengineering
- Control, Robotics, Mechatronics
- MEMS
- Theoretical and Applied Mechanics
- Dynamical Systems, Control
- Fluid Mechanics
- Engineering Thermodynamics, Heat and Mass Transfer
- Manufacturing
- Precision Engineering, Instrumentation, Measurement
- Materials Engineering
- Tribology and Surface Technology

To submit a proposal or request further information, please contact the Springer Editor in your country:

China: Li Shen at li.shen@springer.com
India: Dr. Akash Chakraborty at akash.chakraborty@springernature.com
Rest of Asia, Australia, New Zealand: Swati Meherishi at swati.meherishi@springer.com
All other countries: Dr. Leontina Di Cecco at Leontina.dicecco@springer.com

To submit a proposal for a monograph, please check our Springer Tracts in Mechanical Engineering at http://www.springer.com/series/11693 or contact Leontina.dicecco@springer.com

Indexed by SCOPUS. The books of the series are submitted for indexing to Web of Science.

More information about this series at http://www.springer.com/series/11236

Abdelmajid Benamara ·
Mohamed Haddar · Benameur Tarek ·
Mezlini Salah · Chaari Fakher
Editors

Advances in Mechanical Engineering and Mechanics

Selected Papers from the 4th Tunisian Congress on Mechanics, CoTuMe 2018, Hammamet, Tunisia, October 13–15, 2018

Springer

Editors
Abdelmajid Benamara
National School of Engineers of Monastir
Monastir, Tunisia

Mohamed Haddar
National School of Engineers of Sfax
Sfax, Tunisia

Benameur Tarek
National School of Engineers of Monastir
Monastir, Tunisia

Mezlini Salah
National School of Engineers of Monastir
Monastir, Tunisia

Chaari Fakher
National School of Engineers of Sfax
Sfax, Tunisia

ISSN 2195-4356 ISSN 2195-4364 (electronic)
Lecture Notes in Mechanical Engineering
ISBN 978-3-030-19780-3 ISBN 978-3-030-19781-0 (eBook)
https://doi.org/10.1007/978-3-030-19781-0

This Springer imprint is published by the registered company Springer Nature Switzerland AG
The registered company address is: Gewerbestrasse 11, 6330 Cham, Switzerland

Preface

In the early 2005, the Tunisian Association for Mechanics (ATM) was founded to promote activities related to the large field of mechanics and has made impressive progress. The most important scientific event of ATM is the TUnisian COngress on MEchanics (COTUME), which is organized every three years.

The first was held in Hammamet, Tunisia, 17–19 March 2008, then in Sousse, 19–21 March 2012, and then in Sousse, 24–26 March 2014; such meetings are now being settled as a comprehensive series entitled COTUME. This year's COTUME was held in Hammamet from 13 to 15 October 2018.

Major topics included design methodology, dynamics and vibration of structures, manufacturing processes, structure modelling and computational mechanics, advanced materials and mechanical behaviour, contact mechanics, reliability and risk, fluid mechanics, heat transfer and robotics.

We were delighted by the enthusiastic response to this year's meeting. About 120 papers were presented, and more than 200 participants discussed, during the three days of the congress, latest advances in the field of mechanics, advanced materials, manufacturing processes and opportunities for Tunisian laboratories within the framework of European research programmes.

The proceedings contains 32 peer-reviewed conference papers which are selected from 126 papers submitted to COTUME 2018. The chapters included in this book cover a broad overview of the state of the art in the field and a useful resource for academic researchers and industry specialists active in the field of mechanical engineering and have been classified into the four following parts:

1. Structure modelling and computation
2. Design methodology and manufacturing process
3. Materials: Mechanical behaviour and structure
4. Fluid mechanics, energy, mass and heat transfer

The organizers of the congress were honoured by the presence of international associations and keynote speakers, who are experts in the covered research topics, namely

Prof. Arquis Eric, President of the French Association of Mechanics (AFM), I2M Bordeaux, France,

Prof. El Had Khalid, President of the Moroccan Society for Mechanical Science (SMSM), Morocco,

Prof. Rossignol Philippe, European Research Programmes Advisor, France,

Prof. Bouguecha Anas, LA2MP, ENIG, Gafsa University and IFTM, Leibniz Universität Hannover, Garbsen, Germany,

Prof. BenSalah Nizar, LMMP, ENSIT University of Tunis, Tunisia,

Prof Hamdi M.A, ISI group, Université de Technologie de Compiègne, UTC, France,

Prof. Bouraoui Chokri, LMS, ENISo, Sousse University, Tunisia, and

Prof. Chrigui Mouldi, UR-MMS, ENIG, Gabes University and IEPPT, Technical University of Darmstadt, Germany.

We also would like to take this opportunity to thank all members of the organizing committee, the scientific review committee and the doctoral school of Science and Technology for Engineers at ENIM, as well as the generous sponsors such as JUSTECH for helping make COTUME 2018 a successful event.

Tunisian Association of Mechanics, ATM
BenAmara A. (President)
Dogui A. (First Honorary President)

Organizing Committee
Bradai C. (President)

Scientific Committee
Benameur T.
Mezlini S.

Organization

Editors

Benamara Abdelmajid	National School of Engineers of Monastir, Monastir, Tunisia
Benameur Tarek	National School of Engineers of Monastir, Monastir, Tunisia
Haddar Mohamed	National School of Engineers of Sfax, Sfax, Tunisia
Mezlini Salah	National School of Engineers of Monastir, Monastir, Tunisia
Chaari Fakher	National School of Engineers of Sfax, Sfax, Tunisia

Scientific Committee

Abbes M. S.	ENIS, Tunisia
Abdennadher M.	IPEIS, Tunisia
Abid S.	IPEIS, Tunisia
Affi Z.	ENIM, Tunisia
Aifaoui N.	IPEIM, Tunisia
Akrout A.	ENIT, Tunisia
Akrout M.	ENIS, Tunisia
Aloui A.	ISSATG, Tunisia
Aloui F.	UVHC, France
Ammar L.	ENIT, Tunisia

Amri R.	ISETR, Tunisia
Ayadi A.	ENIS, Tunisia
Baccar M.	ENIS, Tunisia
Bahloul R.	ISSATS, Tunisia
Baili M.	ENIT, France
Barkallah M.	ENIS, Tunisia
Bel Hadj Salah H.	ENIM, Tunisia
Bellagi A.	ENIM, Tunisia
Ben Abdallah J.	ENIT, Tunisia
Ben Amara A.	ENIM, Tunisia
Ben Ameur T.	ENIM, Tunisia
Ben Bacha H.	ENIS, Tunisia
Ben Bettaieb M.	ENSM, France
Ben Chaabene A.	ISETM, Tunisia
Ben Cheikh Larbi A.	ENSIT, Tunisia
Ben Daly H.	ENISO, Canada
Ben Dhia H.	ECP, France
Ben Fredj N.	ENSIT, Tunisia
Ben Jeddou A.	SUPMECA, France
Ben Kahla N.	ISSATS, Tunisia
Ben Nejma F.	LIRIS, France
Ben Ouezdou F.	IPEIM, Tunisia
Ben Salah N.	ENSIT, Tunisia
Ben Salem W.	IPEIM, Tunisia
Ben Slama R.	ISSATG, Tunisia
Ben Slima Ghzaiel S.	ENIT, Tunisia
Ben Tahar M.	UTC, France
Ben Yahya	ENSIT, Tunisia
Ben Zineb T.	Univ. Lorraine, France
Bensalah W.	IPEIM, Tunisia
Ben Salem S.	IPEIN, Tunisia
Bessrour J.	ENIT, Tunisia
Bettaieb H.	ENIM, Tunisia
Bouaziz Z.	ENIS, Tunisia
Bouazizi M. L.	IPEIN, Tunisia
Bouguecha A.	ENIG, Tunisia
Bouhafs M.	ENIT, Tunisia
Bouraoui C.	ISIG, Tunisia
Bouraoui T.	ENIM, Tunisia
Bouzid Sai W.	ENIS, Tunisia
Bradai C.	ENIS, Tunisia
Braham C.	ENSAM, France
Briki J.	ENIT, France
Chaari F.	ENIS, Tunisia
Chafra M.	EPT, Tunisia

Chatti S.	ISSATS, Tunisia
Chatti S.	IUL, Germany
Chebbeh A.	ISETN, Tunisia
Chelbi A.	ENSIT, Tunisia
Chouchane M.	ENIM, Tunisia
Choura S.	ENIS, Tunisia
Chrigui M.	ENIG, Tunisia
Chtourou H.	IPEIS, Tunisia
Dammak F.	ENIS, Tunisia
Dammak M.	IPEIS, Tunisia
Dhouib K.	ENSIT, Tunisia
Doghri I.	UC-Louvain, Belgium
Dogui A.	ENIM, Tunisia
El Borgi S.	EPT, Tunisia
Elleuch K.	ENIS, Tunisia
Elleuch R.	IPEIS, Tunisia
Fakhfakh T.	ENIS, Tunisia
Fathallah R.	ENISO, Tunisia
Gamaoun F.	ENISO, Tunisia
Ghanem F.	ENSIT, Tunisia
Ghorbel F.	RICE, USA
Guedri K.	UQU, KSA
Guedri M.	IPEIN, Tunisia
Haddar M.	ENIS, Tunisia
Hadj Taieb E.	ENIS, Tunisia
Hajlaoui K.	IMISU, KSA
Halouani F.	ENIS, Tunisia
Hambli R.	UNIV-ORL, France
Hamdi H.	ENISE, France
Hamdi M. A.	ESI, France
Hammami L.	ENIS, Tunisia
Hassine T.	ENISO, Tunisia
Hassis H.	ENIT, Tunisia
Hbaieb M.	ISETS, Tunisia
Hor A.	ISAE, France
Jaoua M.	ENIT, Tunisia
Jemni A.	ENIM, Tunisia
Kairaouani L.	ENIT, Tunisia
Kallel Kamoun I.	ENIS, Tunisia
Kammoun M.	ENIS, Tunisia
Karkoub M.	PI, UAE
Karra C.	IPEIS, Tunisia
Kchaou H.	IPEIS, Tunisia
Kchaou M.	ISAMS, Tunisia
Khabou M. T.	ENIS, Tunisia

Khalfallah A.	ISSATS, Tunisia
Kharrat M.	FSG, Tunisia
Khir T.	ISSATG, Tunisia
Khmiri J.	ENIM, Tunisia
Krichen A.	ISSATG, Tunisia
Lafhaj Z.	ECL, France
Larbi W.	CNAM, France
Lili T.	FST, Tunisia
Louati J.	ENIS, Tunisia
Louhichi B.	ISSATS, Tunisia
Maalej A.	ENIS, Tunisia
Maalej Y.	ENIT, Tunisia
Maatar M.	ENIS, Tunisia
Mahjoub N.	IPEIM, Tunisia
Makhlouf K.	ENSIT, Tunisia
Masmoudi F.	ENIS, Tunisia
Masmoudi N.	ENIS, Tunisia
Mbarek M.	ISETM, Tunisia
Mbarek T.	IGM, Germany
Mehdi K.	IPEIM, Tunisia
Mezlini S.	ENIM, Tunisia
Mhiri H.	ENIM, Tunisia
Mlika A.	ENISO, Tunisia
Mzali F.	ENIM, Tunisia
Nahdi E.	ESSTEB, Tunisia
Najar F.	IPEIM, Tunisia
Nasri R.	ENIT, Tunisia
Romdhane L.	ENISO, Tunisia
Saanouni K.	UTT, France
Safi M.	ENIT, Tunisia
Sahlaoui H.	ENSIT, Tunisia
Sai K.	ENIS, Tunisia
Said R.	IPEIM, Tunisia
Sassi M.	MASDAR, UAE
Sghaier S.	ISSATK, Tunisia
Sidhom H.	ENSIT, Tunisia
Sidhom N.	ENSIT, Tunisia
Smaoui H.	ENIT, Tunisia
Soula M.	ENSIT, Tunisia
Terres M. A.	ISSATS, Tunisia
Timoumi Y.	ENIM, Tunisia
Tounsi N.	CNRC, Tunisia
Tourki Z.	ENISO, Tunisia
Trigui M.	IPEIM, Tunisia
Triki E.	ENIT, Tunisia

Tsomarev O.	ENIT, Tunisia
Walha L.	ENIS, Tunisia
Wali M.	ENIS, Tunisia
Zairi F.	Polytech-Lille, France
Zghal A.	ENSIT, Tunisia
Zidi M.	ENIM, Tunisia
Zouari B.	ENIS, Tunisia

Organizing Committee

Ayadi Walid	ISET-Rades
Ben Moussa Naoufel	LMMP-ENSIT
Bouaziz Mohamed	CFDTP-ENIS
Chaabane Makram	LGM-ENIM
Chaieb Iheb	LMMP-ENSIT
Chaouachi Fraj	UMSSDT-ENSIT
Ennetta Ridha	M2EM-ENIG
Frija Mounir	UGPMM-ENIS
Ghiss Moncef	LMS-ENISo
Kchaou Mohamed	LASEM-ENIS
Khir Tahar	M2EM-ENIG
Mars Jamel	UGPMM-ENIS
Masmoudi Neila	LASEM-ENIS
Mezlini Salah	LGM-ENIM
Yangui Majdi	LA2MP-ENIS
Zemzemi Farhat	LMS-ENISo
Znaidi Amna	LRMAI-ENIT

About this Book

This book offers a selection of original peer-reviewed papers presented at the 4th TUnisian COngress on MEchanics (COTUME 2018), held in Hammamet, Tunisia, from 13 to 15 October 2018. It reports on fundamental research studies and innovative industrial applications mainly related to structure modelling and computation, design methodology and manufacturing processes, mechanical behaviour of materials, fluid mechanics, energy, mass and heat transfer. It covers a broad overview of the state of the art in the field and a useful resource for academic researchers, postgraduates and industrial specialists in mechanical engineering. The COTUME 2018 was organized by the Tunisian Association of Mechanics (ATM) and honoured by the active participation of the President of the Mediterranean Network of Engineering School, General Director of Scientific Research in Tunisia, the French Association of Mechanics (AFM), the European Research Council and the Moroccan Society for Mechanical Science (SMSM).

Contents

Materials: Mechanical Behaviour and Structure

Fluid Mechanics and Energy, Mass and Heat Transfer

Structure Modelling and Computation

Femoral Postoperative Bone Adaptation – Numerical Calculation and Clinical Validation with DEXA Investigations

Anas Bouguecha[1](✉), Bernd-Arno Behrens[2], and Matthias Lerch[3]

[1] ENIGa, Laboratory of Mechanics, Modeling and Manufacturing (LA2MP) of the National School of Engineers of Sfax, Sfax, Tunisia
Anas.bouguecha@gmx.de
[2] Institute of Forming Technology and Machines, Leibniz Universitaet Hannover, Hannover, Germany
behrens@ifum.uni-hannover.de
[3] Clinic of Orthopaedic Surgery, Hannover Medical School, Hannover, Germany
Matthias.Lerch@diakovere.de

Abstract. Nowadays the finite element method (FEM) presents an established tool in the field of biomedical engineering and is applied for the numerical simulation of several biomechanical phenomena. Nevertheless, the accuracy of the simulation results essentially depends, on the material modeling. To point this up, FE computations of the strain-adaptive bone adaptation (remodeling) after total hip arthroplasty (THA) are presented in order to predict the secondary stability of the prostheses. Reliable calculation results are possible here only, if prerequisites are fulfilled. These include especially an accurate mathematical model for the description of the postoperative changes in the apparent bone density (ABD). In addition, clinical studies play a crucial role in the validation of these complex computations. In this context corresponding DEXA (dual-energy x-ray absorptiometry) investigations are presented in this work.

Keywords: FEM · Bone remodeling · THA · DEXA

1 Introduction

The physiological function of the hip joint can be disturbed by osteoarthritis (oa) and diseases leading to oa such as hip dysplasia (Fig. 1a), avascular necrosis of the femoral head or by trauma such as the femoral neck fracture (Fig. 1a). All these conditions will cause severe pain which will be first treated conservatively. However, the conservative treatment of hip pain (of the hip joint) is often limited and the patient needs surgery on the hip joint.

A. Benamara et al. (Eds.): CoTuMe 2018, LNME, pp. 3–15, 2019.
https://doi.org/10.1007/978-3-030-19781-0_1

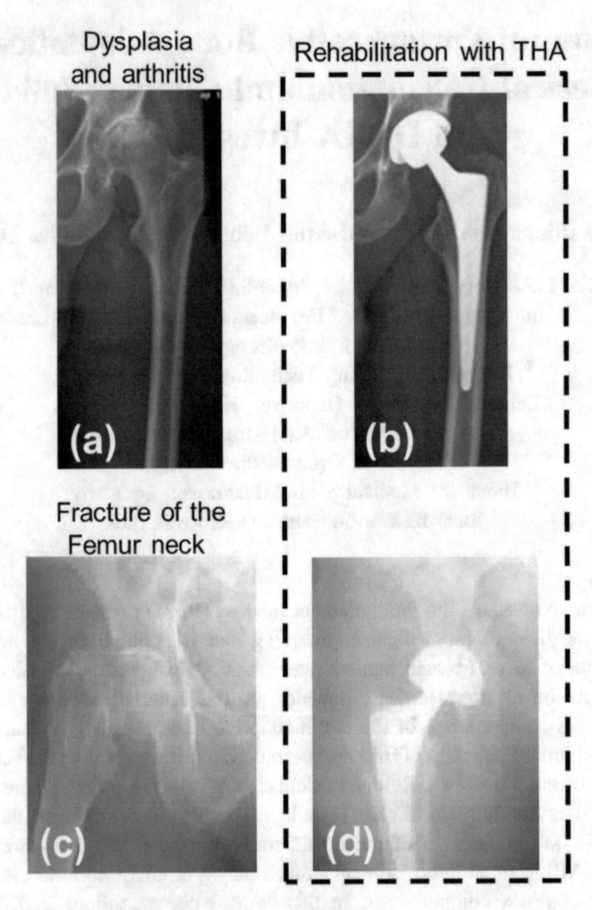

Fig. 1. Rehabilitation with THA (b, d) for the treatment of osteoarthritis or dysplasia (a) as well as the fracture of the femur neck (c) (according to Stöckle et al. 2005; Gabler 2010)

For the treatment of advanced degenerative (Fig. 1c) or traumatic injuries (Fig. 1d) of hip joints, the use of hip implants has been proven (Adam and Kohn 2000).

Nevertheless, the aseptic loosening of the prosthesis, which is caused inter alia by bone loss processes in the periprosthetic bone, remains a problem of hip arthroplasty. These degradation processes often result from a change in the physiological load distribution in the joint treated with THA (Kuiper and Huiskes 1997) and are usually associated with prosthetic migration.

There is a considerable variety of THA available today. These are usually modular. Figure 2 shows an example of this modular design for the AnCA Fit (Cremascoli Ortho, Italy) prosthesis system. Here, the femoral component consists of the prosthetic head, the cone and the shaft. The pan and the inlay together form the pan component. In particular for the femoral components, many different designs exist.

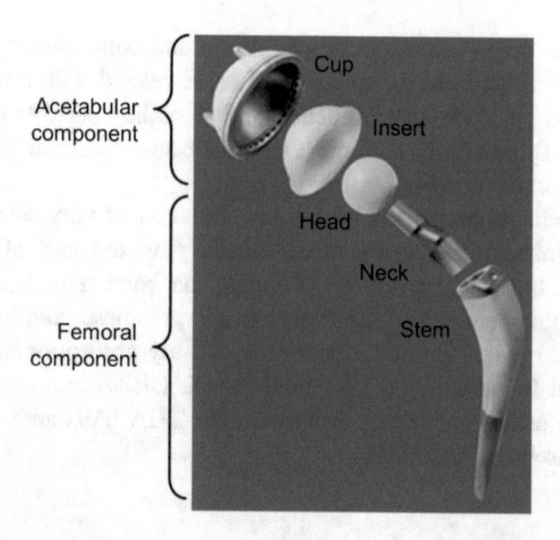

Fig. 2. Modular design of a THA (according to Toni et al. 2001)

But all these stems have the common problem of stress shielding due to alterations of load transfer (Aamodt et al. 2001; Goetzen et al. 2005). The femoral loading is changed from externally to internally, thus severely disturbing the bone stress gradient that reaches from the metaphysis to the diaphysis of the femur. The more the design of the femoral stem enables the natural physiologic load transfer, the less change to the bone is to be expected. However, one of the most important reasons for stress shielding is non-physiologic load transfer (Bryan et al. 1996; Goetzen et al. 2005; Pritchett 1995; Sumner et al. 1998).

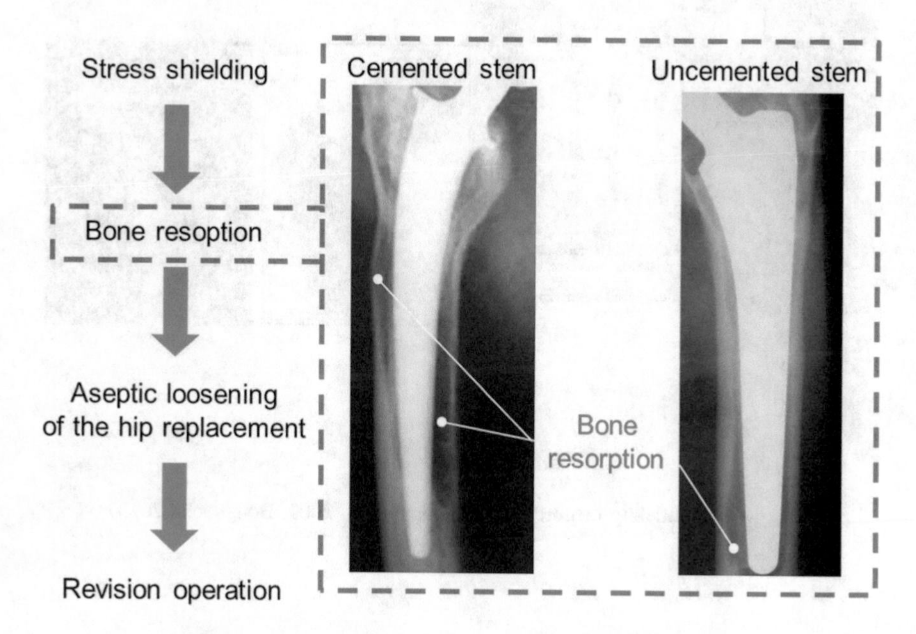

Fig. 3. Problem of THA (according to Bouguecha 2013)

This stress shielding causes a bone resorption and consequently the aseptic loosening of the implant and is in the majority of cases related with a revision operation. This is a very strong problem that influences the secondary stability of the implant and can be detected until now only in clinical studies. Bone resorption can happen around an uncemented or cemented implant (Fig. 3).

Therefore, a reliable prediction of this phenomenon of bone adaptation (remodeling) in the preclinical phase is essential. That's why the aim of this work is the development of a numerical model to calculate the bone remodeling for preclinical evaluation of the biomechanical performance of hip prostheses and its validation with a clinical study. Thus we performed a dual-energy x-ray absorptiometry (DEXA) study as it is well suited to validate the numerical model. DEXA is a very well-established method for the evaluation of bone adaptation after THA (Albanese et al. 2006; Lerch et al. 2012a; Panisello et al. 2006).

2 Materials and Methods

2.1 Modelling

In the first step, the FE model for the physiological femur was established.

From CT data the CAD surface model of the femur was created using reverse engineering, and a meshed solid model was generated. This model was meshed using 4-noded tetrahedrons (Fig. 4).

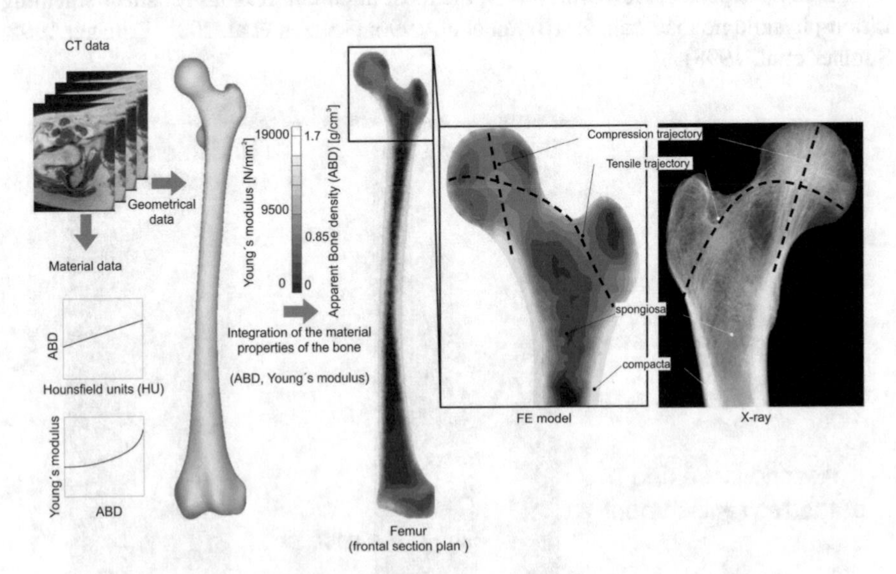

Fig. 4. FE modeling (according to Behrens et al. 2009, Bouguecha 2013)

Then the material data will be defined. From the HU the apparent bone density (ABD) (Rho et al. 1995) and the YOUNG's modulus as a function of the ABD (Carter and Hayas 1977) are determined.

$$\rho = 0.114 + 916 \cdot 10^{-6} \cdot HU \quad (1)$$

$$\mathrm{E} = 3790 \cdot \rho^3 \quad (2)$$

On the right side you see the distribution of the ABD in the frontal section of the femur. The comparison of an x-ray image of this FE model shows a very good conformity.

The implantation of the implant is done virtually based on the OP instructions of the manufacturer as well as the radiographs from the patient data base of the Clinic of Orthopaedic Surgery (in the DIAKOVERE Annastift) of the Medical School in Hanover.

In this work the focus will be on the cement less short stem prosthesis the META (from Aesculap B.Braun). This implant is anchored per press-fit technique and has a coating with pure titanium Plasmapore in the proximal part of the stem (Fig. 5).

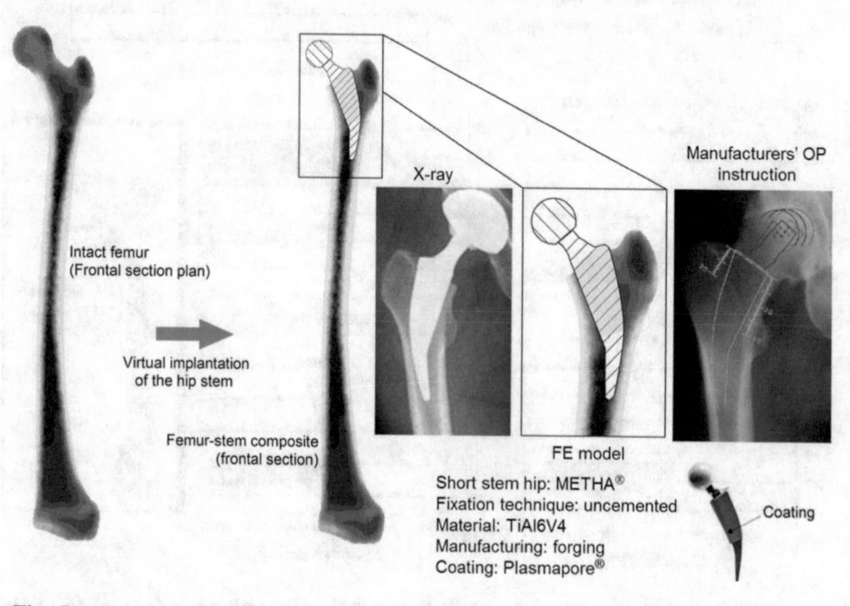

Fig. 5. Virtual implantation of the stem (according to Lerch 1012a; Bouguecha 2013)

For each simulation boundary conditions are required. In this application physiological boundary conditions are necessary, which conform to the patient activities after THA.

Investigations of Morlock et al. 2001 show that the most frequent dynamic activity of people with a hip implant is walking. This presents about 10% of the whole activities. Lying sitting and standing are not very important due to the lower stresses acting on the hip joint.

So the loads during a gait cycle are needed to use them in the boundary conditions of the bone remodeling tool.

As former numerical studies have shown the influence of the muscle forces on the load distribution and on the computation of bone adaptation, a reduced muscle system according to Heller et al. 2005 was used. It consists of abductors (M. gluteus minimus, M. gluteus maximus and M. gluteus medius), the M. tensor fascia latae, the M. vastus medialis and the M. vastus lateralis. The acting points of the hip resulting force and the different forces in the used reduced muscle system are illustrated in Fig. 6. The progress of these forces during the gait cycle is taken from the investigations of Bergmann et al. 2001 and Duda et al. 1997.

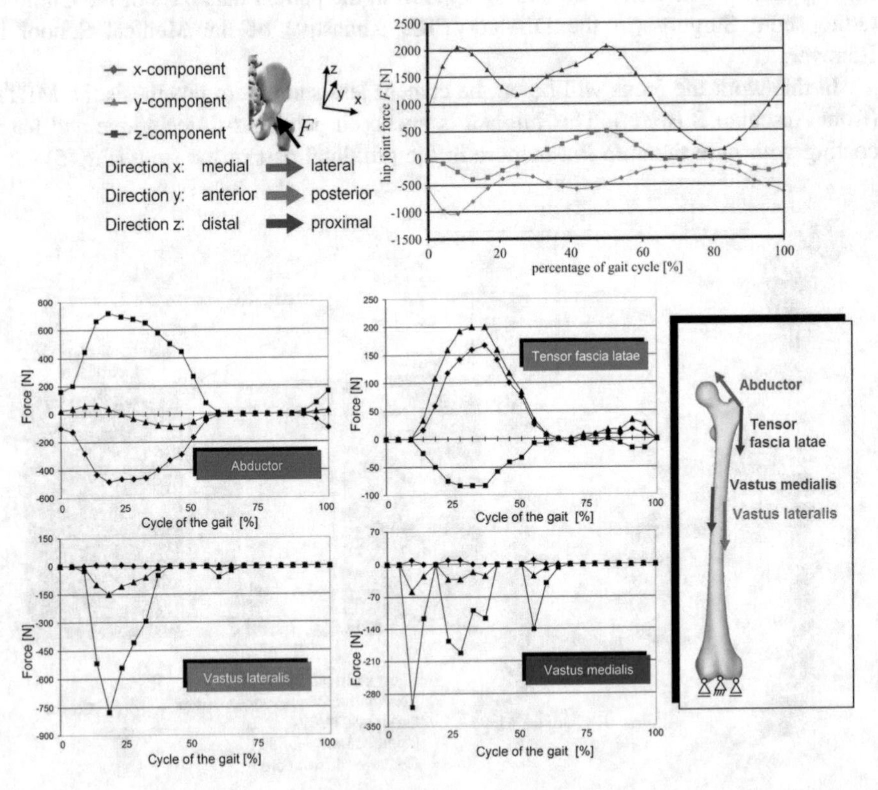

Fig. 6. Evolution of hip contact and muscle forces during the gait cycle (according to Heller et al. 2005; Bergmann et al. 2001; Duda et al. 1997)

2.2 Bone Adaptation Law

For the determination of the postoperative bone adaptation first a reference is needed. This is the distribution of the load in the intact bone. This calculation will be done in a single run (Fig. 7).

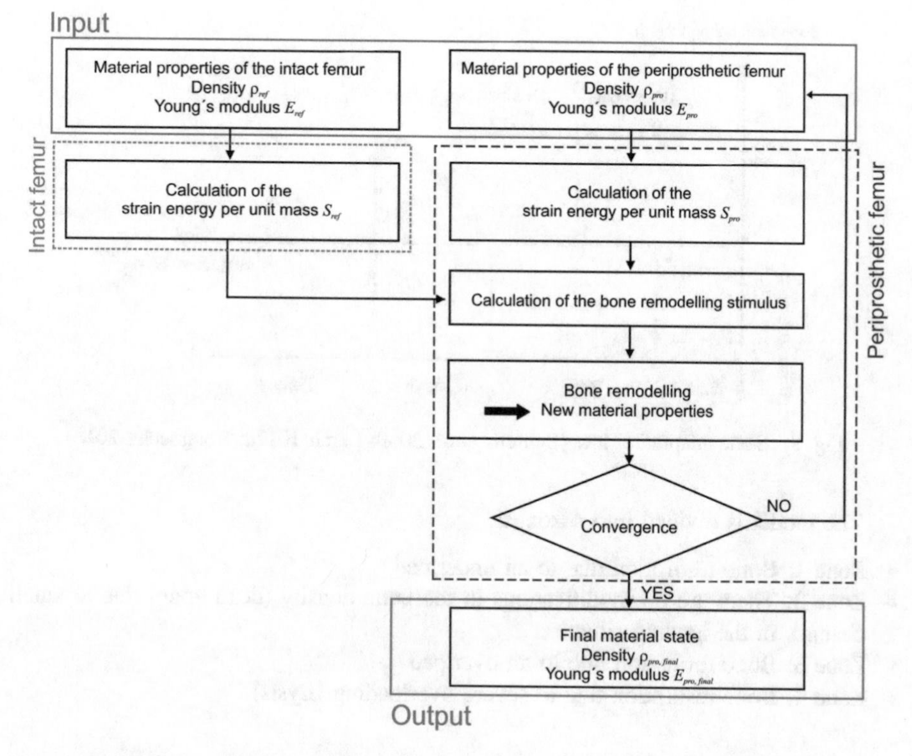

Fig. 7. Numerical computation method of the postoperative bone adaptation (Behrens et al. 2008; Behrens et al. 2009; Bouguecha 2013)

Then the load in the bone after implantation of the stem is computed. We determine the bone remodeling stimulus as a ratio of the elastic strain energy in the periprosthetic bone to the one in the intact femur (Eqs. 3 and 4).

$$\aleph = \frac{S_{pro}}{S_{ref}} \quad (3)$$

$$S = \frac{D}{\rho} = \frac{\frac{1}{2} \cdot \sigma^T \cdot \varepsilon}{\rho} \quad (4)$$

With a bone adaptation law the new material properties can be calculated. This is an iterative loop that will be executed until the convergence condition is achieved (Fig. 7).

The basis of this simulation is the bone adaptation law, which has been developed at the IFUM in Hannover. This mathematical model gives the evolution of the bone adaptation rate as a function of the stimulus (Fig. 8).

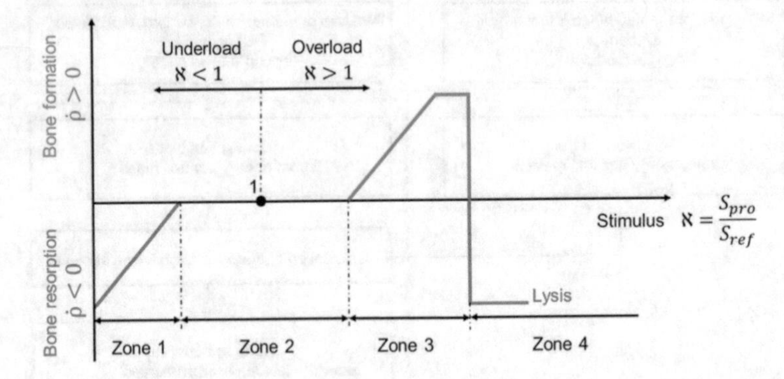

Fig. 8. Bone adaptation law (Behrens et al. 2009; Lerch 1012a; Bouguecha 2013)

The model is divided into 4 zones:

- Zone 1: Bone resorption due to an underload
- Zone 2: There are no modifications in the bone density (dead zone) due to small changes in the loading situation
- Zone 3: Bone formation due to an overload
- Zone 4: Bone resorption due to severe overloading (Lysis)

2.3 DEXA Investigations

By means of the dual-energy x-ray absorptiometry (DEXA) method the bone mineral density (BMD) can be assessed by a fan beam going through the objects tissue. The system detects the radiation after passing the object and then calculates the bone mineral content (BMC). BMD is calculated dividing the measured BMC by the detected scan area. DEXA is usually performed in the spine or hip in humans, both areas of major fracture risk in the elderly or in people with osteoporosis (Lucas et al. 2017). DEXA is considered the most reliable tool for the evaluation of bone remodeling after THA using different stem designs (Albanese et al. 2006; Lerch et al. 2012a, b, c; Lerch et al. 2013b; Panisello et al. 2006; Stukenborg-Colsman et al. 2012). The analysis of the 7 periprosthetic Gruen zones (Regions of Interest, ROI) is the most commonly used protocol to evaluate bone adaptation after the implantation of conventional femoral stems (Stukenborg-Colsman et al. 2012; Boden et al. 2006; Panisello et al. 2006; Aldinger et al. 2003; Kobayashi et al. 2000).

The complete DEXA study protocol can be viewed in our previous publications (Lerch et al. a, b, c; Lerch et al. 2013b; Stukenborg-Colsman et al. 2012).

In brief: From July 2008 to January 2009, a consecutive series of 25 patients (9 (36%) female and 16 (64%) male) with unilateral short stem (Metha® BBraun, AESCULAP AG, Tuttlingen, Germany) implantations were included in a prospective study. The number of patients was calculated by a power-analysis performed by our institute for biometry. This was required by our institute's institutional review board committee to obtain approval for this study (Ethic Committee No. 4226). Inclusion criteria were the indication for unilateral implantation of the implant due to osteoarthritis of the hip and an age between 35 and 70 years. Patients with a body mass index (BMI) over 30, with a history of previous surgery on the same hip, femoral fracture, metabolic bone disease, use of steroids or other drugs affecting bone metabolism, intraoperative cracks, or severe osteoarthritis of the contralateral hip were excluded from the study. Mean age was 58.9 years (range 38–69 years), mean preoperative BMI was 24.6 (range 20.6–27.4) The patients underwent DEXA examinations preoperatively and after 1 week, 6 months, 12 months and 2 years after implantation. Bone mineral density (BMD, g/cm^2) data collected one week after surgery served as baseline value for the following DEXA examinations. All patients had full weight bearing postoperatively. DEXA scans were performed using a HOLOGIC Discovery A S/N 80600 device (Hologic Inc., Waltham, MA). The BMD of the operated hip was measured using the "metal-removal hip" scanning mode. Conventional Gruen's zones were adapted to the short stem design in Metha cases (Falez et al. 2008; Falez et al. 2015; Lerch et al. 2012a, c; Lerch et al. 2013a; Roth et al. 2005; Speirs et al. 2007). The Shapiro-Wilk-tests did not show a normal distribution for the DEXA measurements; the Wilcoxon signed-ranks test was used to statistically compare the density changes. $P < 0.005$ was considered significant. Data analysis was performed using SPSS (11.05 SPSS Inc. Chicago, Illinois).

3 Results

In the numerical FE calculation of the postoperative bone adaptation after THA, convergence was reached after 40 computation steps. The ABD distribution in this increment represents the medical long-term situation, while the initial state (step 1) corresponds to the medical situation directly after TRA. Figure 9 illustrates the ABD distribution in the frontal section of the proximal part of the composite. It clearly shows that the ABD decreases considerably in the bone close to the prosthesis due to the change in the physiological load situation.

Figure 9 shows a heavy decrease of the bone density in the distal and lateral proximal part of the femur. The progress of the bone mass in the femur is presented in Fig. 9.

In Table 1 the calculation and the results of the DEXA investigations are compared. Especially, in the last column of this Table the difference between measurement and simulation results in each ROI is presented. In Table 1 a very good conformity in all areas is shown except from ROI 7.

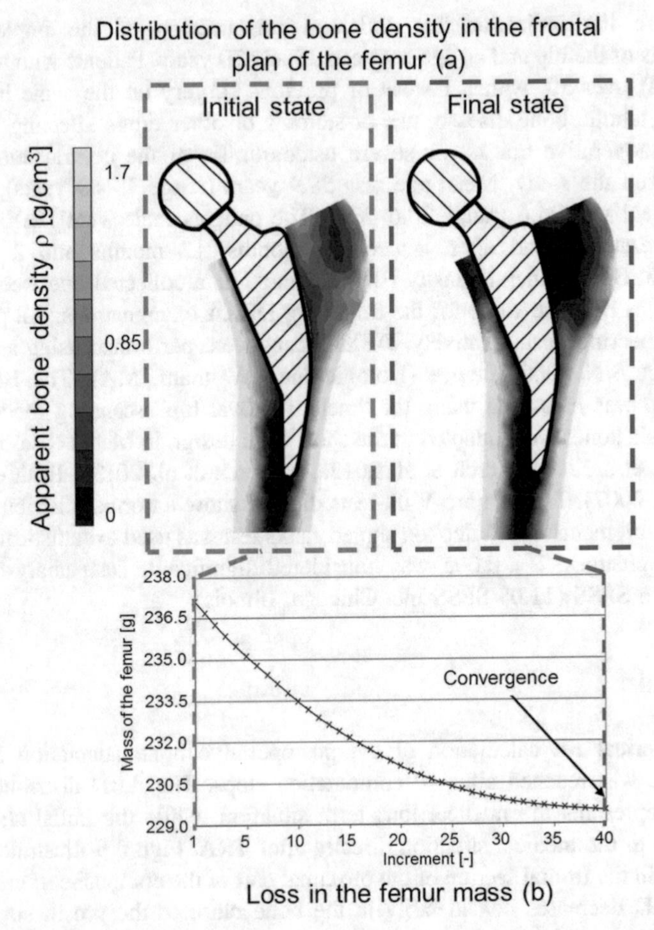

Fig. 9. Femoral postoperative bone remodeling after THA with the Metha stem (Lerch 1012a; Bouguecha 2013)

Table 1. Change in bone mass after implantation of the Metha short stem (*: Lerch et al. 2012a, c).

Regions of interest	Simulation [%]	DEXA* [%]	Deviation [%]
ROI 1	−16.7	−7.7	9.0
ROI 2	−8.8	−1.4	7.4
ROI 3	−3.1	−8.8	5.7
ROI 4	0	−5.3	5.3
ROI 5	−0.6	−6.0	5.4
ROI 6	−6.8	10.6	17.4
ROI 7	−52	5.7	57.7

4 Conclusion and Discussion

The established numerical model shows good accordance with the clinically collected data thus our model is well suited to predict periprosthetic bone mineral changes after THA. The deviation between the numerical and the DEXA data was pronounced in the very proximal regions (Table 1).

We can find a higher ratio of spongiosa in these areas representing more biological activity in reaction to the implant. This makes the proximal areas less predictable. Further studies are needed to assess the biological impact on bone adaptation in these areas. However, probably the validation itself needs to be further assessed as we detected a reduction of scanned bone surface in ROI7 (calcar region). This led to an increase in BMD in ROI7 between the 1st- and 2nd year after surgery due to strong bone resorption in the calcar. The system identifies tissue as bone and measures the detected area. If this tissue is not dense enough, it is not included into the BMD measurement. Thus, the Hologic system only detects the residual bone nearby the minor trochanter where stress transfer increases bone mass. This might lead to a false increase in BMD in ROI7.

If the changing of the bone surface, a very distinct bone resorption is calculated. This corresponds to a mass loss of about 32%. And thus a small difference between the simulation and the clinical values of bone degradation of only 20% is the result.

We already reported the limitations of our model and the validation (Lerch et al. 2012a, b, c; Lerch et al. 2013a, b; Stukenborg-Colsman et al. 2012): The individual bone remodeling variations are unclear, especially among different ages and bone morphologies. It is likely that some individuals have better implant match than others.

For the simulation only no biological considerations were taken into account. The implant coating seems to have an influence on the periprosthetic bone changes. A representative study showed that hydroxyapatite-tricalcium phosphate coated femoral implants caused significantly less femoral bone loss than the uncoated stems (Tanzer et al., 2001). We have to take into account that the dicalcium phosphate dihydrate layer has an unquantified effect on periprosthetic bone remodeling that was not considered in our simulation. We assumed that the proximal portion of the stem has full bonding to the bone. Probably, this does not correspond to the physiological facts, as different periprosthetic tissue differentiations are known to affect bone formation around the implant (Puthumanapully and Browne 2011).

Acknowledgements. This work has been done at the Institute of Forming Technology and Machines (IFUM) of the Leibniz Universitaet Hannover (LUH) as well as in the Clinic for Orthopaedic Surgery (in the DIAKOVERE Annastift) of the Hannover Medical School (MHH).

This study is part of the investigations within the subproject D6 of the collaborative research center SFB 599 "Sustainable bioresorbable and permanent implants of metal and ceramic materials". The authors would like to thank the German Research foundation (DFG) for the financial support of the project and also the company.

Furthermore, the authors would like to thank AESCULAP AG for providing CAD data of the short stem METHA.

References

Adam, F., Kohn, D.: Computergestützte Entwicklung eines anatomischen Hüftprothesenschaftes. Magazin Forschung **1**, 41–48 (2000)

Albanese, C.V., Rendine, M., De Palma, F., Impagliazzo, A., Falez, F., Postacchini, F., Villani, C., Passariello, R., Santori, F.S.: Bone remodelling in THA: a comparative DXA scan study between conventional implants and a new stemless femoral component. A preliminary report. Hip. Int. **16**(Suppl 3), 9–15 (2006)

Aldinger, P.R., Sabo, D., Pritsch, M., Thomsen, M., Mau, H., Ewerbeck, V., Breusch, S.J.: Pattern of periprosthetic bone remodeling around stable uncemented tapered hip stems: a prospective 84-month follow-up study and a median 156-month cross-sectional study with DXA. Calcif. Tissue Int. **73**, 115–121 (2003)

Bergmann, G., Deuretbacher, G., Heller, M., Graichen, F., Rohlmann, A., Strauss, J., Duda, G.N.: Hip contact forces and gait patterns from routine activities. J. Biomech. **34**, 859–871 (2001)

Behrens, B.A., Bouguecha, A., Nolte, I., Meyer-Lindenberg, A., Stukenborg-Colsman, C., Pressel, T.: Strain adaptive bone remodelling: influence of the implantation technique. Stud. Health Technol. Inf. **133**, 33–44 (2008). Medicine Meets Engineering

Behrens, B.-A., Nolte, I., Wefstaedt, P., Stukenborg-Colsman, C., Bouguecha, A.: Numerical investigations on the strain-adaptive bone remodelling in the periprosthetic femur: influence of the boundary conditions. BioMed. Eng. OnLine **8**, 7 (2009)

Boden, H.S., Skoldenberg, O.G., Salemyr, M.O., Lundberg, H.J., Adolphson, P.Y.: Continuous bone loss around a tapered uncemented femoral stem: a long-term evaluation with DEXA. Acta Orthop. **77**, 877–885 (2006)

Bouguecha, A.: Importance of material and friction modeling in the numerical simulation based on the FEM, Habilitation thesis, Faculty of Mechanics, Gottfried Wilhelm Leibniz Universität Hannover (2013)

Brodner, W., Bitzan, P., Lomoschitz, F., Krepler, P., Jankovsky, R., Lehr, S., Kainberger, F., Gottsauner-Wolf, F.: Changes in bone mineral density in the proximal femur after cementless total hip arthroplasty. A five-year longitudinal study. J. Bone Joint Surg. Br. **86**, 20–26 (2004)

Carter, D.R., Hayas, W.C.: The compressive behaviour of bone as a two-phase porous structure. J. Bone & Joint Surgery **59-A**(7), 954–962 (1977)

Duda, G.N., Schneider, E., Chao, E.Y.S.: Internal forces and moments in the femur during walking. J. Biomech. **30**, 933–941 (1997)

Falez, F., Casella, F., Panegrossi, G., Favetti, F., Barresi, C.: Perspectives on metaphyseal conservative stems. J. Orthop Traumatol. **9**, 49–54 (2008)

Falez, F., Casella, F., Papalia, M.: Current concepts, classification, and results in short stem hip arthroplasty. Orthopedics **38**, S6–S13 (2015)

Gabler, C.-A.M.: Die Wiederherstellung des Hüftgelenkdreh-zentrums bei der endoprothetischen Versorgung der Dysplasiecoxarthrose über eine Pfannendachrekonstruktion mittels Kopfspan und Schraubpfanne. Dissertation an der Medizinischen Fakultät der Ludwig-Maximilians-Universität zu München (2010)

Gotze, C., Ehrenbrink, J., Ehrenbrink, H.: Is there a bone-preserving bone remodelling in short-stem prosthesis? DEXA analysis with the Nanos total hip arthroplasty. Z. Orthop. Unfall. **148**, 398–405 (2010)

Heller, M.O., Bergmann, G., Kassi, J.-P., Claes, L., Haas, N.P., Duda, G.N.: Determination of muscle loading at the hip joint for use in pre-clinical testing. J Biomechanics **38**, 1155–1163 (2005)

Kobayashi, S., Saito, N., Horiuchi, H., Iorio, R., Takaoka, K.: Poor bone quality or hip structure as risk factors affecting survival of total-hip arthroplasty. Lancet **355**, 1499–1504 (2000)

Kuiper, J.H., Huiskes, R.: The predictive value of stress shielding for quantification of adaptive bone resorption around hip replacements. J. Biomech. Eng. **119**, 228–231 (1997)

Lerch, M., Kurtz, A., Stukenborg-Colsman, C., Nolte, I., Weigel, N., Bouguecha, A., Behrens, B. A.: Bone remodeling after total hip arthroplasty with a short stemmed metaphyseal loading implant: finite element analysis validated by a prospective DEXA investigation. J. Orthop. Res. **30**, 1822–1829 (2012a)

Lerch, M., Kurtz, A., Windhagen, H., Bouguecha, A., Behrens, B.A., Wefstaedt, P., Stukenborg-Colsman, C.M.: The cementless Bicontact((R)) stem in a prospective dual-energy X-ray absorptiometry study 1. Int. Orthop. **36**, 2211–2217 (2012b)

Lerch, M., von der Haar-Tran, A., Windhagen, H., Behrens, B.A., Wefstaedt, P., Stukenborg-Colsman, C.M.: Bone remodelling around the Metha short stem in total hip arthroplasty: a prospective dual-energy X-ray absorptiometry study. Int. Orthop. **36**, 533–538 (2012c)

Lerch, M., Weigel, N., Windhagen, H., Ettinger, M., Thorey, F., Kurtz, A., Stukenborg-Colsman, C., Bouguecha, A.: Finite element model of a novel short stemmed total hip arthroplasty implant developed from cross sectional CT scans. Technol. Health Care **21**, 493–500 (2013a)

Lerch, M., Windhagen, H., Stukenborg-Colsman, C.M., Kurtz, A., Behrens, B.A., Almohallami, A., Bouguecha, A.: Numeric simulation of bone remodelling patterns after implantation of a cementless straight stem. Int. Orthop. **37**, 2351–2356 (2013b)

Morlock, M., Schneider, E., Bluhm, A., Vollmer, M., Bergmann, G., Müller, V., Honl, M.: Duration and frequency of everyday activities in total hip patients. J. Biomechanics **34**(7), 873–881 (2001)

Panisello, J.J., Herrero, L., Herrera, A., Canales, V., Martinez, A., Cuenca, J.: Bone remodelling after total hip arthroplasty using an uncemented anatomic femoral stem: a three-year prospective study using bone densitometry. J. Orthop Surg. (Hong Kong) **14**, 32–37 (2006)

Puthumanapully, P.K., Browne, M.: Tissue differentiation around a short stemmed metaphyseal loading implant employing a modified mechanoregulatory algorithm: a finite element study. J. Orthop. Res. **29**, 787–794 (2011)

Reiter, T.J., Böhm, H.J., Krach, W., Rammerstorfer, F.G.: Some applications of the finite-element method in biomechanical stress analyses. Int. J. Comput. Appl. Technol. **7**(3–6), 233–241 (1994)

Rho, J.Y., Hobatho, M.C., Ashman, R.B.: Relations of mechanical properties to density and CT numbers in human bone. Med. Eng. Phys. **17**, 347–355 (1995)

Roth, A., Richartz, G., Sander, K., Sachse, A., Fuhrmann, R., Wagner, A., Venbrocks, R.A.: Periprosthetic bone loss after total hip endoprosthesis. Dependence on the type of prosthesis and preoperative bone configuration. Orthopade **34**, 334–344 (2005)

Speirs, A.D., Heller, M.O., Taylor, W.R., Duda, G.N., Perka, C.: Influence of changes in stem positioning on femoral loading after THR using a short-stemmed hip implant. Clin. Biomech. (Bristol Avon) **22**, 431–439 (2007)

Stöckle, U., Lucke, M., Haas, N.P.: Der Oberschenkelhalsbruch. Deutsches Ärzteblatt, Heft 49 (2005)

Stukenborg-Colsman, C.M., Haar-Tran, A., Windhagen, H., Bouguecha, A., Wefstaedt, P., Lerch, M.: Bone remodelling around a cementless straight THA stem: a prospective dual-energy X-ray absorptiometry study 2. Hip. Int. **22**, 166–171 (2012)

Tanzer, M., Kantor, S., Rosenthall, L., Bobyn, J.D.: Femoral remodeling after porous-coated total hip arthroplasty with and without hydroxyapatite-tricalcium phosphate coating: a prospective randomized trial. J. Arthroplasty **16**, 552–558 (2001)

Toni, A., Paterni, S., Sudanese, A., Guerra, E., Traina, F., Giardina, F., Antonietti, B., Giunti, A.: Anatomic cementless total hip arthroplasty with ceramic bearings and modular necks: 3 to 5 years follow-up. HIP Int. **11–1**, 1–17 (2001)

Intelligent Neural Network Control for Active Heavy Truck Suspension

Anis Hamza(✉) and Noureddine Ben Yahia

Mechanical, Production and Energy Laboratory (LMPE),
National School of Engineering of Tunis (ENSIT),
University of Tunis, Tunis, Tunisia
anis7amza@gmail.com, nourdine.benyahia@yahoo.com

Abstract. In this paper, we will explain the interest of developing a new active suspension control approach for trucks HGVs based on Artificial Neural Network ANNs. The suspension system can be categorized into passive, semi-active and active suspension system. In active suspension system control, the required performance is based on the following: Ride comfort, Suspension travels and Road handling. The model is developed with MATLAB Toolbox and estimated and validated using data collected through tests done with a truck. There are several methods for modelling a system. One of them is to use the laws of physics to describe the system and use experimental data or given information about the system to determine the system's parameters. The statement of the problem of this research is to develop a robust controller that can improve the performances of the nonlinear active suspension system of the Heavy Truck and its verifications using graphical and animation output.

Keywords: Heavy Goods Vehicles (HGV) · Suspension system · Vibration · MATLAB Toolbox · Artificial Neural Networks (ANN)

1 Introduction

Freight transport is a vital part of the global economy and the lives of citizens. The share of road transport is largely dominating in Tunisia with 85% of the land freight tonnage and 95% of passenger transport. The situation is similar in most countries. Trucks are therefore becoming more and more important in our country, and pose increasingly serious problems in terms of road safety, comfort and damage to infrastructure. The implementation of effective suspension overload control systems has been a stated goal of research for the past decade, to improve road safety and preserve infrastructure. In a first part, we will detail the different types of the suspension system. In a second part we will discuss the problematic of our research. Finally we will explain the relief used based on artificial intelligence.

The model-free fuzzy logic control (Cherry and Jones 1995; Yeh and Tsao 1994) and neural network control (Hampo and Marko 1992; Huang and Lian 1996) were employed to design the controllers of vehicle active suspension systems for releasing the requirement of complicated dynamic model. Rao and Prahlad (1997) proposed a tuneable fuzzy controller for an active suspension system. Huang and Lian (1996)

A. Benamara et al. (Eds.): CoTuMe 2018, LNME, pp. 16–23, 2019.
https://doi.org/10.1007/978-3-030-19781-0_2

proposed a fuzzy and neural network hybrid control scheme to compensate the coupling dynamics for improving control performance. However, these approaches need a complicated learning mechanism or a specific performance decision table, which is designed by a trial-and-error process. The Radial Basis Function (RBF) scheme was first proposed by Hardy (1971). It has been used to represent nonlinear correspondence between inputs and outputs of nonlinear control systems. Sanner and Slotine (1992) employed Gaussian basis functions in nonlinear adaptive control, (Lu and Basar 1998) used RBF to develop a neural-network based identification algorithm, and (Liu and Chen 1993) employed the Radial Basis Function Neural Networks (RBFNN) scheme to model some unknown nonlinear functions for deriving a feedback linearization control law. Here, a new sliding mode controller based on RBFNN is developed for direct control purposes and implemented on a quarter-car active suspension system. This control strategy is based on a RBF structure and combines the advantages of adaptive control schemes and sliding modes. The adaptive rule is used to adjust the weighting of the RPF online using the condition of reaching a specified sliding surface. Since this approach has learning capability to establish and regulate continuous RPF weightings, its control implementation can be started with zero initial weighting RBFNN. Ieluzzi et al. (2006) develop control design, development process and overall performance of a semi-active suspension control for a heavy truck.

2 Suspension Systems

Suspension systems are among the most critical components that guarantee driving comfort, good handling and safety in vehicles on the ground. A vehicle is always subject to random excitation due to an uneven road profile. Vehicle suspension systems are designed to absorb energy and mitigate annoying vibrations due to this random excitation. Therefore, the way to develop a completely model-free adaptive control structure has become an interesting area of research. These suspension systems can be classified as passive, semi-active and active suspension systems depending on the external power supply of the system and/or a control bandwidth (Appleyard and Wellstead 1995). The suspension consists of the spring system (Hamza et al. 2015), shock absorbers and links that connect a vehicle to its wheels. In other words, the suspension system is a mechanism that physically separates the body of the car. The main function of the vehicle suspension system is to minimize the vertical acceleration transmitted to the passenger that directly provides comfort (Hamza et al. 2013a).

There are three types of suspension: passive suspension, semi-active suspension and active suspension (see Fig. 1). Active and semi-active suspensions aim to improve comfort (especially for active ones) and to modulate the suspension adjustment on demand. Traditionally, the design of truck (truck) height suspension has been compromised between the three conflicting criteria: road handling, loading and passenger comfort. The suspension system must support the vehicle, provide steering control and provide effective passenger isolation. Good driving comfort requires a flexible suspension (Hamza et al. 2013b).

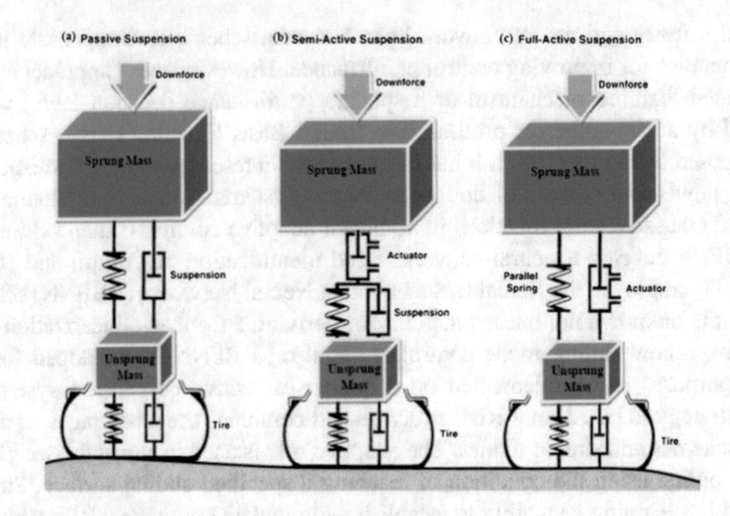

Fig. 1. The suspension system

3 Problem Formulation

The sensitivity of the human body to vibrations is frequency dependent and is high at frequencies where chassis accelerations are increased for higher damping values. Moreover, the sensitivity has also a directional dependence why researchers have differentiated vertical and horizontal vibrations. The human sensitivity to whole-body vibrations in the vertical and horizontal directions have been treated quantitatively in the standard ISO 2631-1 (IOS 1997).

In addition to the comfort of the cabin, we want to increase road safety, the goal of this paper is to enable the design of a controller algorithm that uses the advantages given by an active suspension to retain or adjust the HGVs. The model developed should be accurate enough for the developer to evaluate different controller methods. The model is developed in MATLAB and estimated and validated using data collected through tests done with a truck (See Fig. 2).

Fig. 2. Behaviours of the truck on a curve.

There are several methods for modeling a system. One of them is to use the laws of physics to describe the system and use experimental data or given information about the system to determine the system's parameters. Another way is to use black-box modelling (ANNs) to describe the whole or parts of the system. Black-box modelling is when the relation between input and output is described through a model class and the parameters estimated through statistical methods with the help of experimental data. Thereafter the model is compared against a new data set, to see how well the model describes the true system. This is done without the need of knowing the underlying physical relations. Finally, there is the alternative that is known as grey-box modelling, where the parameters have physical interpretations but are estimated through statistical methods. This can be described as a mix between black-box and physical modeling methods. The statement of the problem of this research is to develop a robust controller that can improve the performances of the nonlinear active suspension system of the Heavy Truck and its verifications using graphical and animation output.

4 Active Suspension System Schema and Model

The dynamic equations of this suspension system are as follows:

$$\begin{cases} M_s\ddot{Z}_s = -k_s(Z_s - Z_u) - \eta_s(\dot{Z}_s - \dot{Z}_u) + F_a - F_f \\ M_u\ddot{Z}_u = k_s(Z_s - Z_u) + \eta_s(\dot{Z}_s - \dot{Z}_u) + k_t(Z_r - Z_u) - F_a + F_f \end{cases} \tag{1}$$

Where Z_r is the road surface position variation, Z_s and Z_u are measured variables representing the sprung mass displacement and the displacement of tire axis, respectively. F_a and F_f are the hydraulic actuating force and the hydraulic friction force, respectively. The relationship between $X_v(t)$ the servo valve spool displacement and $Q(t)$ the hydraulic flow rate, and the continuity equation of the hydraulic cylinder chamber give:

$$Q(t) = K_g(t)X_v(t) - K_c(t)P(t) \tag{2}$$

$$Q(t) = A(\dot{Z}_x(t) - \dot{Z}_u(t)) + C_TP(t) + (\frac{V_T}{4\beta})\dot{P}(t) \tag{3}$$

Where K_g is a time-varying servo valve flow gain, K_c is the servo valve flow pressure coefficient, $P(t)$ is the cylinder differential pressure, A is the cross section area of cylinder, C_T is the total leakage coefficient of the cylinder, V_T is the total compressed volume, and β is the effective bulk modulus of the system.

The relationship between the servo valve spool displacement and the control voltage is described as:

$$X_v(t) = K_vU(t) \tag{4}$$

And K_v is the servo valve gain. Then, the time derivative of the actuating force of this hydraulic suspension system can be derived:

$$\dot{F}_a(t) = A\dot{P}(t) = A(\frac{V_T}{4\beta})\left[K_g(t)K_vU(t) - C_TP(t) - A(\dot{Z}_x(t) - \dot{Z}_u(t))\right] \tag{5}$$

The dynamic equations of this suspension system can be rewritten as follows:

$$\begin{aligned} M_sV_T\ddot{Z}_s = &-4\beta C_Tk_sZ_s - \left[k_s + 4\beta(C_T\eta_s + A^2)\right]V_T\dot{Z}_s \\ &-(\eta_s + 4\beta C_TM_s)\ddot{Z}_s + 4\beta AK_g(t)K_vU(t) - (4\beta C_TF_f + V_T\dot{F}_f) \\ &+\left[\eta_sV_T\ddot{Z}_u + (k_sV_T + 4\beta A^2 + 4\beta\eta_sC_T)\dot{Z}_u + 4\beta C_Tk_sZ_u\right] \end{aligned} \tag{6}$$

The dynamic equation of this hydraulic servo control system comprises a coupling with several outputs, a variable characteristic in time and non-linear. It is difficult to estimate these system parameters and use this dynamic equation to design a model-based controller. Therefore, the intelligent control scheme by Artificial Neural Networks (ANN) is used to design the active suspension controller.

5 Artificial Neural Networks (ANN)

An adaptive neural network (ANN) control method for a continuous damping control (CDC) damper is used in the Heavy Truck suspension systems. The control objective is to suppress positional oscillation of the sprung mass in the presence of road irregularities. To achieve this, a boundary model is first applied to depict dynamic characteristics of the CDC damper based on experimental data (Fig. 3).

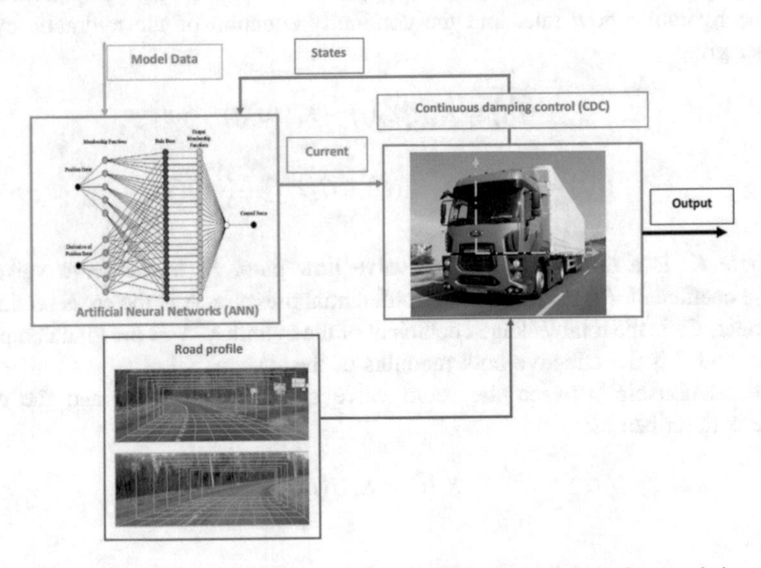

Fig. 3. Damper current control based on HGVs state and their characteristics.

To overcome nonlinearity issues of the model system and uncertainties in the suspension parameters, an adaptive radial basis function neural network (RBFNN) with online learning capability is utilized to approximate unknown dynamics, without the need for prior information related to the suspension system.

In addition, particle swarm optimization (PSO) technique (Fikret Ercan 2009) is adopted to determine and optimize the parameters of the controller. Closed loop stability and asymptotic convergence performance are guaranteed based on Lyapunov stability theory. Finally, simulation results demonstrate that the proposed controller can effectively regulate the chassis vertical position under different road excitations.

Artificial neural networks (ANNs) can be used to design digital controllers that can maintain high dynamic performance of HGVs even with the misalignment problem. ANNs have been proven to be universal approximations of nonlinear dynamical systems (Hunt et al. 1995). ANNs have been developed as distributed network parallel models based on the biological learning of the human brain.

Having been used for many years in pattern recognition applications as well as signal and image processing (Miller et al. 1992), ANNs are currently employed in a wider class of scientific discipline. They are also able to operate in noise environments and have the ability to generalize that allows them to tolerate errors or lack of data. The choice of the architecture of the neuron network is based on the mean squared error (MSE) obtained after learning ANN system. The structure of the network is obtained after study and several tests, by varying the number of entries, of hidden layer and of neurons in each layer. The neurons of the hidden layer have a nonlinear sigmoid transfer function and the output neuron has a linear transfer function (Figs. 4 and 5).

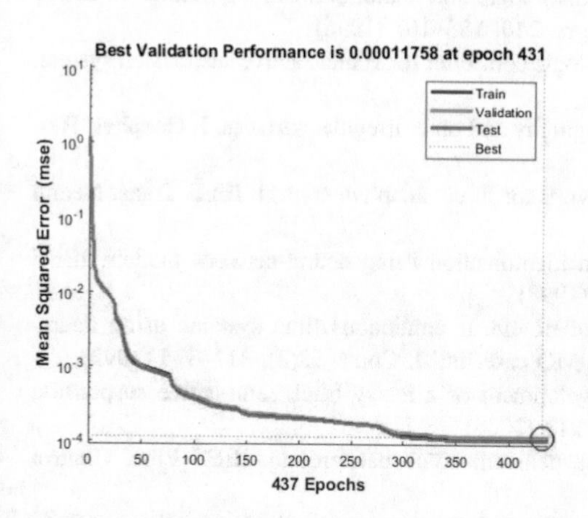

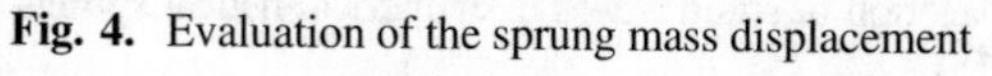

Fig. 4. Evaluation of the sprung mass displacement

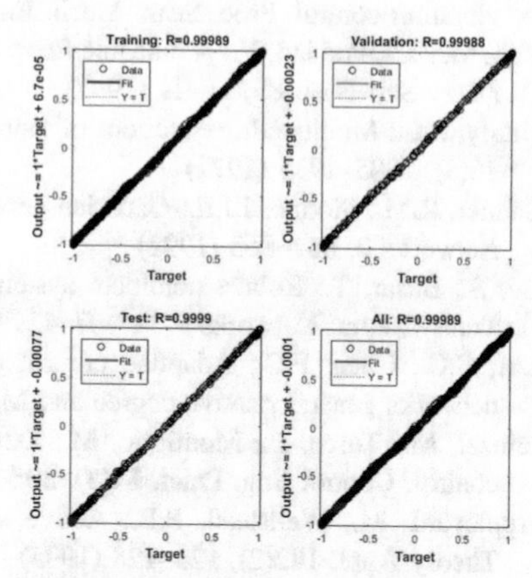

Fig. 5. Regression curve between actual and predicted values

6 Conclusion

In this paper, an adaptive neural network controller for a nonlinear HGVs suspension system using a CDC damper is proposed. An active control design is considered using a boundary model of the CDC damper. The adaptive controller is designed to meet control objectives and RBFNN is used to approximate the nonlinear uncertain part of the suspension system. Finally, the closed loop stability along with asymptotic convergence performance are proved using Lyapunov theory. As prospects, the performance of this controller will be validated by numerical simulations under different road conditions.

Acknowledgements. This work is partially supported by Mr. Abdelmalik Ouamara, Director of Ford Otosan in Algeria & Tunisia.

References

Cherry, A.S., Jones, R.P.: Fuzzy logic control of an automotive suspension systems. IEEE Proc. Control Theory Appl. **142**(2), 149–160 (1995)

Yeh, E.C., Tsao, Y.J.: Fuzzy preview control scheme of active suspension for rough road. Int. J. Veh. Des. **15**, 166–180 (1994)

Hampo, R., Marko K.: Neural network architectures for active suspension control. In: Proceedings of IJCNN-International Joint Conference on Neural Network, pp. 765–770 (1992)

Huang, S.J., Lian, R.J.: A combination of fuzzy logic and neural network algorithms for active vibration control. Proc. Instn. Mech. Engrs. **210**, 153–167 (1996)

Rao, M.V.C., Prahlad, V.: A tuneable fuzzy logic controller for vehicle active suspension system. Fuzzy Sets Syst. **85**, 11–21 (1997)

Hardy, R.L.: Multiquadric equations of topograghy and other irregular surfaces. J. Geophys. Res. **76**(8), 1905–1915 (1971)

Sanner, R.M., Slotine, J.J.E.: Gaussian network for direct adaptive control. IEEE Trans. Neural Networks **3**, 837–863 (1992)

Lu, S., Basar, T.: Robust nonlinear system identification using neural-network models. IEEE Trans. Neural Networks **9**(3), 407–429 (1998)

Liu, C.C., Chen, F.C.: Adaptive control of nonlinear continuous time systems using neural networks general relative degree and MIMO case. Int. J. Contr. **58**(2), 317–335 (1993)

Ieluzzi, M., Turco, P., Montiglio, M.: Development of a heavy truck semi-active suspension control. Control Eng. Pract. **14**(3), 305–312 (2006)

Appleyard, M., Wellstead, P.E.: Active suspension: some background. IEEE Proc. Control Theory Appl. **142**(2), 123–128 (1995)

Hamza, A., Ayadi, S., Hadj-Taieb, E.: Propagation of strain waves in cylindrical helical springs. J. Vib. Control **21**(10), 1914–1929 (2015)

Hamza, A., Ayadi, S., Hadj-Taieb, E.: Resonance phenomenon of strain waves in helical compression springs. Mech. Ind. **14**, 253–265 (2013a)

Hamza, A., Ayadi, S., Hadj-Taieb, E.: The natural frequencies of waves in helical springs. C.R. Mec. **341**, 672–686 (2013b)
International Organization for Standardization: Mechanical Vibration and Shock - Evaluation of Human Exposure to Whole-body Vibration, ISO 2631–1 (1997)
Fikret Ercan, M.: Particle swarm optimization and other metaheuristic methods in hybrid flow shop scheduling problem. In: Lazinica, A. (ed.) Particle Swarm Optimization (2009). ISBN 978-953-7619-48-0
Hunt, K.J., Irwin, G.R., Warwick, K.: Neural Network Engineering in Dynamic Control Systems. Springer, London (1995)
Miller, W.T., Sutton, R.S., Werbos, P.J.: Neural Networks for Control. MIT Press, Cambridge (1992)

Analytical Modeling of the Tool Trajectory with Local Smoothing

Mohamed Essid[1,2](✉), Bassem Gassara[1](✉), Maher Baili[3](✉), Moncef Hbaieb[1,2](✉), Gilles Dessein[3](✉), and Wassila Bouzid[1](✉)

[1] Unité de Génie de Production Mécanique et Matériaux, ENIS, Route Soukra Km 3, 5, B.P. 1173, 3038 Sfax, Tunisia
essid.mh@gmail.com, gassara_bassem@yahoo.fr, mhbaieb@yahoo.fr, wassilabouzid@yahoo.fr
[2] Institut Supérieur des Etudes Technologiques de Sfax, Route de Mahdia Km 2.5, BP 88 A, 3099 El Bustan Sfax, Tunisia
[3] Université de Toulouse, INPT/ENIT, Laboratoire Génie de Production, 47 Avenue d'Azereix, BP 1629, 65016 Tarbes Cedex, France
{maher.baili,gilles.dessein}@enit.fr

Abstract. To improve the quality and the precision of the complex surfaces obtained by high speed milling, it is necessary to ensure the continuity in tangency and curvature of the tool motion. This is possible using one of two re-interpolation methods of the tool path, with local or global smoothing. The present work consists to model the transition element geometry of the locally smoothed tool path based on the sinumerik CNC models. The smoothing element is assimilated to a quintic hermite polynomial and the experimental tests carried out showed a good correlation between the simulated and the smoothed tool path.

Keywords: High-speed milling · Discontinuity · Smoothing · Hermite polynomial

1 Introduction

Milling a free form surface by high-speed milling requires a numerical control (NC) program generated by CAM software. The tool path defined in the NC program is characterized by a succession of linear segments discontinuous in tangency. During high-speed milling of a workpiece, the tool must follow the discontinuous trajectory in tangency defined in the NC program. This mode of machining generates facets and vibrations at the level of the machined surface due to the jerky variation of the kinematic axis motions parameters (Siemens 2004).

Local smoothing of the discontinuities between the successive segments permit to avoid these problems (Siemens 2010).

A. Benamara et al. (Eds.): CoTuMe 2018, LNME, pp. 24–31, 2019.
https://doi.org/10.1007/978-3-030-19781-0_3

The literature interested in the geometric modeling of the smoothing elements can be classified into two categories:

- The real-time machining: the motion setpoint of the machine axes are generated through a program proposed by the user (Sencer et al. 2015; Tulsyan and Altintas 2015; Yang and Yuen 2017; Huang et al. 2018).
- The machining with a computer numerical control (CNC) machines. The Sinumerik CNC can generate two modes of local smoothing: The first involves using circle arc as a transition element (Gassara et al. 2013a, b). The second is to add a spline element in the corner (Siemens 2010).

The geometry of the smoothing element can be defined by three geometrical parameters (Siemens 2010): length transition, contour tolerance or tolerance for each axis of the machine (Fig. 1).

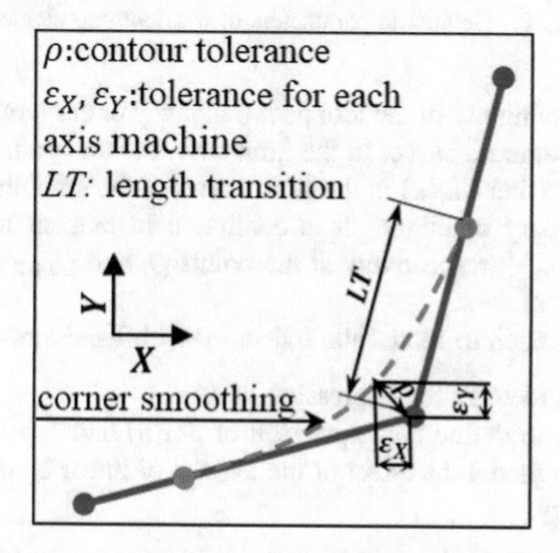

Fig. 1. Parameters defining the geometry of the smoothing element

The optimization of machining parameters and the estimation of the machining time necessitate the simulation of the trajectory. The tool path smoothed with tolerance imposed for each axis was developed by (Pessoles et al. 2010). Until now, the smoothed tool path with contour tolerance programmed (ρ) is unknown.

This paper presents an analytical approach to the geometry of smoothed corner with contour tolerance imposed.

2 Geometric Modeling of the Smoothing Element

The new computer numerical control (CNC) machine tools are capable to interpolate cubic and quintic Hermite polynomial. Quintic Hermite polynomial is used to model the geometry of the transition block. Figure 2 illustrates a tool path composed of two discontinuous linear segments in tangency $([P_iP_{i+1}], [P_{i+1}P_{i+2}])$. The segment $[P_{i+1}Q_{i+1}]$ represents the maximum contour tolerance defined in the machine parameters or in the NC program.

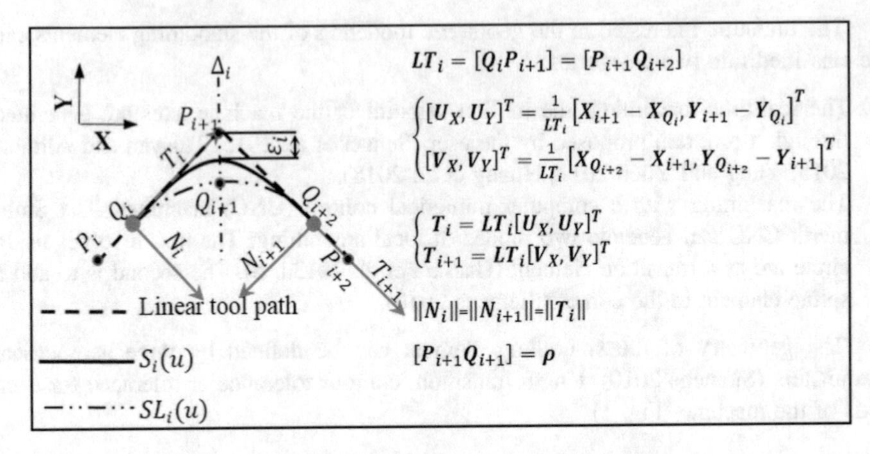

Fig. 2. Geometric parameters of a smoothing element

The type of the continuity of the tool path (tangency or curvature) depends on order of the smoothing parametric curve. In the limit case, the tool path is continued only in tangency. Please note that $SL_i(u)$ is the limit curve of the smoothing element follows this geometric boundary condition. It is continued in tangent to the two segments $[P_iP_{i+1}]$ and $[P_{i+1}P_{i+2}]$, respectively at the points Q_i and Q_{i+2} and passing through the point Q_{i+1} (Fig. 2).

The adopted approach to model the trajectory with local smoothing consists:

- In a first time, to identify the expression of LT_i.
- In a second time, to define the expression of $SL_i(u)$ and $S_i(u)$.
- In a third time, to model the effect of the lengths of linear blocks on the smoothing elements geometry.

The following six hypothesis define the boundary geometric conditions of the rounding contour (see Fig. 2).

- The first geometric derivative $\left(SL_i'(u)\right)$ at two points Q_i and Q_{i+2}, is respectively equal to the two vectors T_i and T_{i+1}.
- The smoothing element $(S_i(u))$ and its limit curve $(SL_i(u))$ are symmetrical with respect to the bisector of the angle formed by the two segments of the tool path Δ_i. This geometric condition can be expressed as follows:

$$SL_i(0.5) = Q_{i+1} \tag{1}$$

- The lengths of the two linear tool path are sufficiently long, that the CNC unit does not reduce the length transition (LT_i).
- Transition block is defined by a quintic Hermite polynomial: $S_i(u)$ is continued in tangency and curvature at the linear tool path.
- $S_i(u)$ and $SL_i(u)$ have the same length transition; therefore they have the same start and ending point.

- N_i and N_{i+1} are two normal vectors to the linear tool path respectively at the two points Q_i and Q_{i+2}.
- The maximum error of the tool path (ε_i) is limited by ρ: the transition curve is bounded by $SL_i(u)$ and the linear tool path (Fig. 1): ($\varepsilon_i \leq \rho$).

According to this boundary conditions, $SL_i(u)$, $TL_i(u)$ and $S_i(u)$ can be expressed as follows:

$$SL_i(u) = \left(1 - 3u^2 + 2u^3\right)Q_i + \left(3u^2 - 2u^3\right)Q_{i+2} + \left(u - 2u^2 + u^3\right)T_i + \left(-u^2 + u^3\right)T_{i+1} \tag{2}$$

$$LT_i = \frac{X_{Q_{i+1}} - X_{i+1}}{0.375(V_X - U_X)} \tag{3}$$

$$\begin{aligned} S_i(u) = {} & \left(1 - 10u^3 + 15u^4 - 6u^5\right)Q_i + \left(u - 6u^3 + 8u^4 - 3u^5\right)wT_i \\ & + \left(\frac{1}{2}u^2 - \frac{3}{2}u^3 + \frac{3}{2}u^4 - \frac{1}{2}u^5\right)N_i + \left(\frac{1}{2}u^3 - u^4 + \frac{1}{2}u^5\right)N_{i+1} \\ & + \left(-4u^3 + 7u^4 - 3u^5\right)wT_{i+1} + \left(10u^3 - 15u^4 + 6u^5\right)Q_{i+1} \end{aligned} \tag{4}$$

To satisfy the last boundary condition, the weight (w) of two tangential vectors (T_i, T_{i+1}) must be greater than or equal to one. w is determined through an experimental test.

If the tool path is described by a small length segments (Milling free form surface). The CNC reduces the length transitions of the smoothing elements in order to ensure the continuity of the tool motion. The geometric parameters of this boundary condition are represented in Fig. 3.

Denoted P_i^* the set of control points of the tool path with local smoothing: P_i^* and P_{i+1}^* are the start and ending points of $S_i(u)$ respectively.

Considering a tool path composed of three linear segments (Fig. 3), the tool path continuity is assured only if the sum of LT_i and LT_{i+1} deduced from Eq. 3 is smaller than the length of the common segment (Fig. 3a).

LTL_i are the lengths transition of the smoothing blocks ensuring continuity in tangency and curvature of the tool motion (Fig. 3b). Along a tool path composed with N linear blocks, LTL_i must satisfy the following geometric boundary condition:

$$LTL_i = \min\left[LT_i, \frac{L_i}{2}, \frac{L_{i+1}}{2}\right] \tag{5}$$

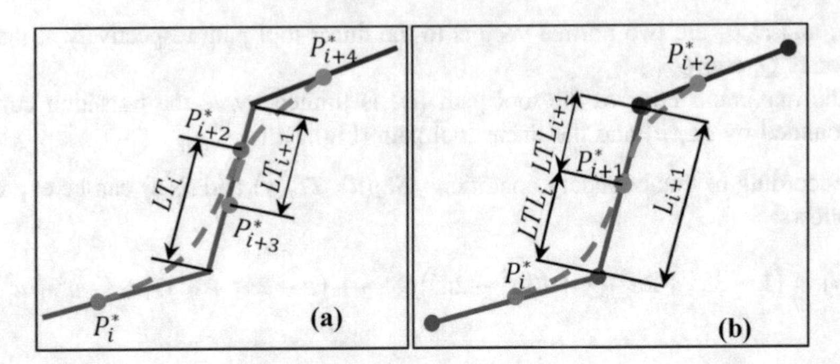

Fig. 3. Modeling of the length transition limit. (a) Without reducing the length transition. (b) With reducing the length transition

3 Experimental Tests and Results

In order to identify the value of the parameter w (weight of two tangent vectors of the smoothing element) and to validate the analytical models of the tool path with local smoothing, a set of experimental tests were carried out using a 3-axis machining of the SPINNER MVC 850 type with a SINUMERIK 840D CNC. The unit NC allows the recording of the commanded position data (position setpoint) and actual position data for each axis machine every 2 ms. The results are saved as an XML file in a flash drive. The experimental tests were performed with 0.01 mm contour tolerance.

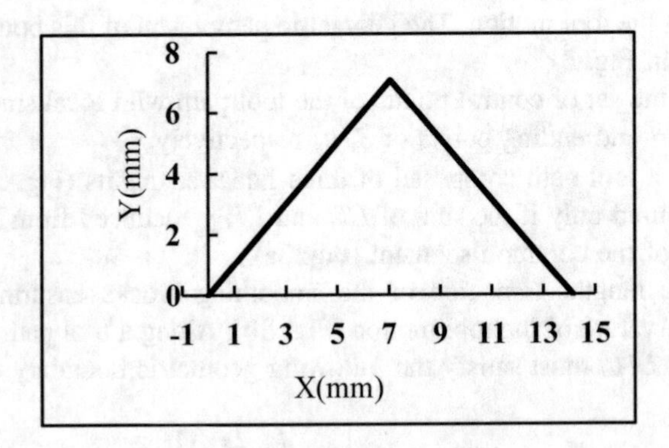

Fig. 4. Linear tool path test

Figure 4 shows the reference tool path, chosen as a test case; the chosen shape of the tool path is composed of two equidistant linear segments equal to 10 mm.

Along the tool path, there is always an offset between the actual and setpoint position of the machine axes, this offset depends on numerical parameters defined in the CNC (Siemens 2010). In order to avoid the offset effect on the searched value of w, we compare the simulated smoothing elements $(w = 1, 2, 3)$ to the axes motion setpoints.

Figure 5 illustrates the simulation results of the smoothed tool path and the set-points of the axes machine with details at the discontinuity point.

One can note that:

- The tool path setpoints are between the two parametric curves simulated with $w = (1, 3)$.
- Simulated tool path with $w = 2$ and tool motion setpoints are merged
- The transition length of the simulated smoothing element are equal to that measured.

We adopt now that the weight of two tangential vectors $w = 2$.

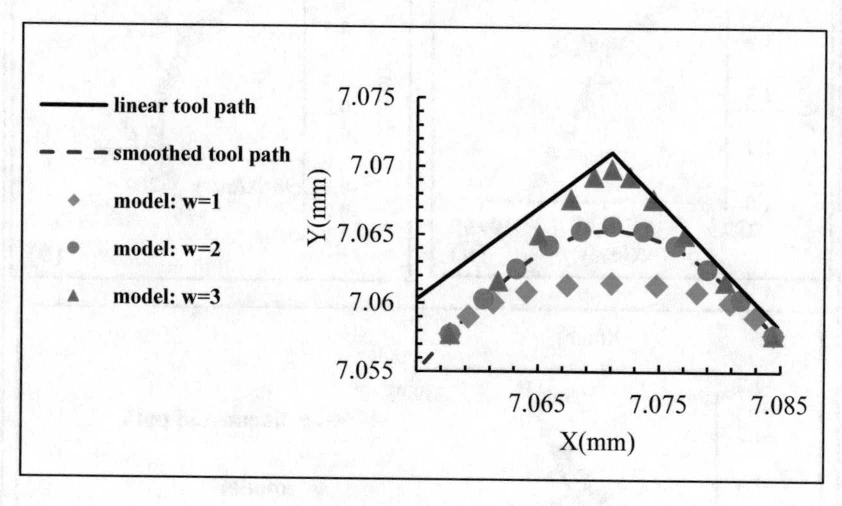

Fig. 5. Identification of the w

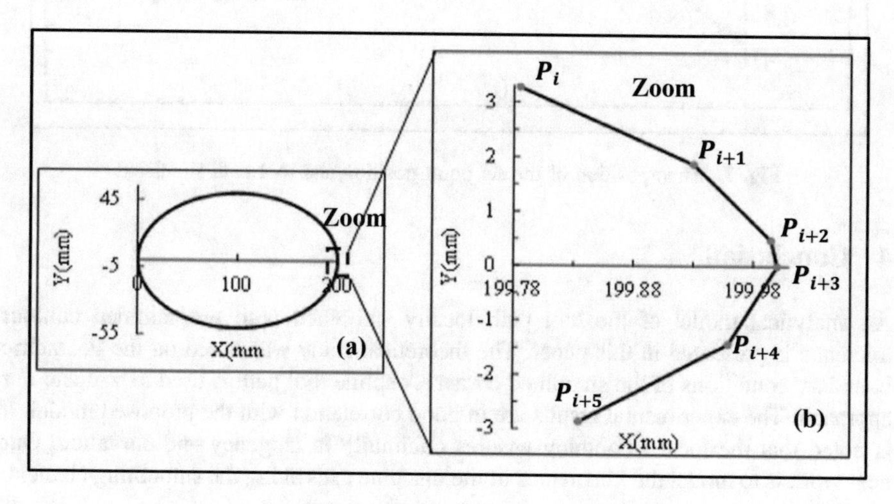

Fig. 6. Model validation path

The second test was performed with an elliptical trajectory (Fig. 6a). The comparative study between simulated and measured trajectory at the junction points defined in Fig. 6b, shows a high accuracy of the adopted model (Fig. 7). The tool trajectory along the segment $[P_{i+2}P_{i+3}]$ is completely transformed into a polynomial motion, the length transition of the two corners smoothed at two points P_{i+2} and P_{i+3} satisfies the boundary condition defined in Eq. 5 (Fig. 7b).

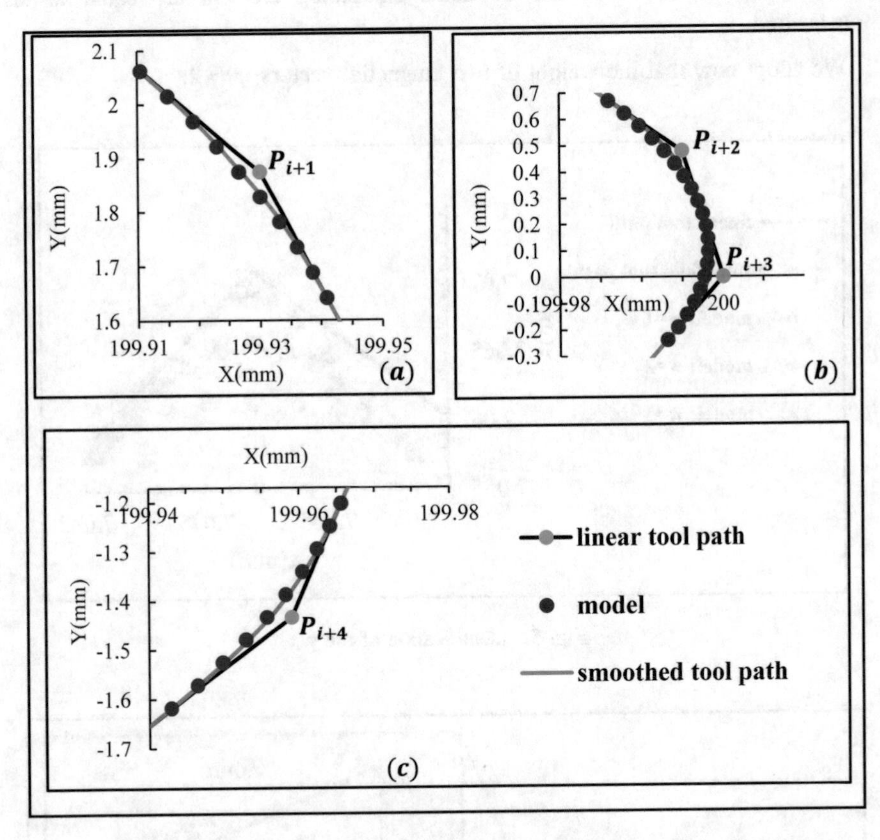

Fig. 7. Superposition of the set point position and tool path simulated

4 Conclusion

An analytical model of the tool path locally smoothed with programmed contour tolerance is presented in this paper. The theoretical study was based on the geometric boundary conditions of the smoothed corner. A spline tool path is used to validate our approach. The experimental results are in good correlation with the proposed model. It is noted that the local smoothing ensures continuity in tangency and curvature. Our next work is to model the kinematics of the machine axes along the smoothing element.

Acknowledgements. This work is carried out thanks to the support and funding allocated to the Unit of Mechanical and Materials Production Engineering (UGPMM/UR17ES43) by the Tunisian Ministry of higher Education and Scientific Research.

References

Gassara, B., Baili, M., Dessein, G., et al.: Feed rate modeling in circular-circular interpolation discontinuity for high-speed milling. Int. J. Adv. Manuf. Technol. **65**, 1619–1634 (2013a). https://doi.org/10.1007/s00170-012-4284-z

Gassara, B., Dessein, G., Baili, M., et al.: Analytical and experimental study of feed rate in high-speed milling. Mach. Sci. Technol. **17**, 181–208 (2013b). https://doi.org/10.1080/10910344.2013.780537

Huang, J., Du, X., Zhu, L.M.: Real-time local smoothing for five-axis linear toolpath considering smoothing error constraints. Int. J. Mach. Tools Manuf. **124**, 67–79 (2018). https://doi.org/10.1016/j.ijmachtools.2017.10.001

Pessoles, X., Landon, X., Rubio, W.: Kinematic modelling of a 3-axis NC machine tool in linear and circular interpolation. Int. J. Adv. Manuf. Technol. **47**, 639–655 (2010). https://doi.org/10.1007/s00170-009-2236-z

Sencer, B., Ishizaki, K., Shamoto, E.: A curvature optimal sharp corner smoothing algorithm for high-speed feed motion generation of NC systems along linear tool paths. Int. J. Adv. Manuf. Technol. **76**, 1977–1992 (2015). https://doi.org/10.1007/s00170-014-6386-2

Siemens: SINUMERIK 810D/840D Fabrication de pièces complexes en fraisage (2004)

Siemens: SINUMERIK 840D sl/828D Basic Functions (2010)

Tulsyan, S., Altintas, Y.: Local toolpath smoothing for five-axis machine tools. Int. J. Mach. Tools Manuf. **96**, 15–26 (2015). https://doi.org/10.1016/j.ijmachtools.2015.04.014

Yang, J., Yuen, A.: An analytical local corner smoothing algorithm for five-axis CNC machining. Int. J. Mach. Tools Manuf. **123**, 22–35 (2017). https://doi.org/10.1016/j.ijmachtools.2017.07.007

Repairing Cracked Structures Using the IF Process

Ahmed Bahloul(✉) and Chokri Bouraoui

Mechanical Laboratory of Sousse, National Engineering School of Sousse, University of Sousse, Bp. 264 Erriadh, 4023 Sousse, Tunisia
bahloulahmadl@outlook.fr, chokri.bouraoui@enim.rnu.tn

Abstract. This paper aims at investigating the Interference Fit Process as a procedure to repair cracked structures. A cracked SENT specimen is considered in the current work using 2D-FE analysis. Lemaitre-Chaboche model was considered for characterizing the material behavior. FE simulations are carried out at different crack hole CAH diameters and different IF sizes. According to the findings, the highest value of fatigue life improvement is obtained at IF size equal to 0.2 mm and CAH diameter equal to 6 mm.

Keywords: Crack repair · Fatigue life improvement · IF process · FE-analysis

1 Introduction

Generally, all mechanical parts contain cracks. These cracks can be always generated and propagated in mechanical structures either during manufacturing, during design stage or during assembly and operation service. Hence, it is very important to find the accessible methods for retarding, or even arresting the fatigue crack propagation in structural elements before final failure. The main purpose of such methods is to improve the fatigue life of cracked structure when its replacement by a new part is time consuming and costly, which commonly happens in the aircraft industry.

Therefore, looking for a procedure/method to arrest fatigue crack growth before failure is very useful. In this context, Several investigations have dealt with the idea of arresting fatigue crack growth (Domazet 1996) such as: repairing the cracked zone by applying of composite patches (Ayatollahi and Hashemi 2007), indentation (Song and Sheu 2002) (Ruzek et al. 2012), cold expansion hole technique (Ghfiri el al. 2000), drilling holes near the crack tip (Song and Shieh 2004) (Ayatollahi et al. 2015) (Ayatollahi et al. 2014) (Fanni et al. 2015) (Razavi et al. 2016), and applying an overloading step (Carlson et al. 1991).

A. Benamara et al. (Eds.): CoTuMe 2018, LNME, pp. 32–38, 2019.
https://doi.org/10.1007/978-3-030-19781-0_4

The current work presents a study aiming at investigating the Interference Fit Process as a procedure to arrest fatigue crack growth. A 2D FE analysis of a cracked SENT specimen using ABAQUS software is implemented in order to predict the residual stress distribution and the fatigue life improvement after repair.

2 FE Analysis

In the present section, a 2D-finite element analysis using ABAQUS software is carried out. A Single Edge Notch Tension (SENT) specimen, having a and D, the crack length and the hole diameter located at the crack tip is considered. In this study, the FE-analysis includes two load steps. In the first step, the interference fit process was simulated by applying a radial displacement on the nodes situated at the hole edge up to the desired interference fit level resulting in expansion similar to the insertion of pin (Fig. 1). In the second step, a cyclic axial loading ($R = 0.1$) is applied to estimate the fatigue life improvement of the cracked SENT specimen after repair. In order to capture the proper stress-strain behavior arising from the interference fit process, a very fine mesh is implemented near the hole edge as illustrated in Fig. 2.

In order to take into account the mean stress relaxation, the Bauschinger effect and the cyclic hardening during cyclic loading, Lemaitre and Chaboche kinematic hardening model, embedded in the commercial code ABAQUS, is used.

It should be noted that the interference fit process can produce a beneficial compressive residual stress along the hole edge for improving the fatigue life. However, this beneficial effect cannot be appreciable due to the damage associated with the high level of the interference fit size. In order to make a compromise between the beneficial compressive residual stress distribution and the damage associated with the interference fit process, Lemaitre and Chaboche's damage model (Lemaitre and Chaboche 2002) is used in its integrated form. User material subroutine is developed to integrate the Lemaitre and Chaboche's damage model in the commercial FE code ABAQUS.

The SWT model (Smith et al. 1970) coupled with the FE analysis is used to predict the fatigue life improvement of the cracked SENT specimen after repair. The ability of this model for estimating the fatigue life of mechanical parts after repair is discussed in Sect. 3.

It should be noted that all the existing studies have essentially used the Interference Fit Process to improve the fatigue resistance of fastener holes or joints in the mechanical structures. However, in our work, the Interference Fit Process is used as a technique to arrest an existing crack in mechanical structures before it becomes critical.

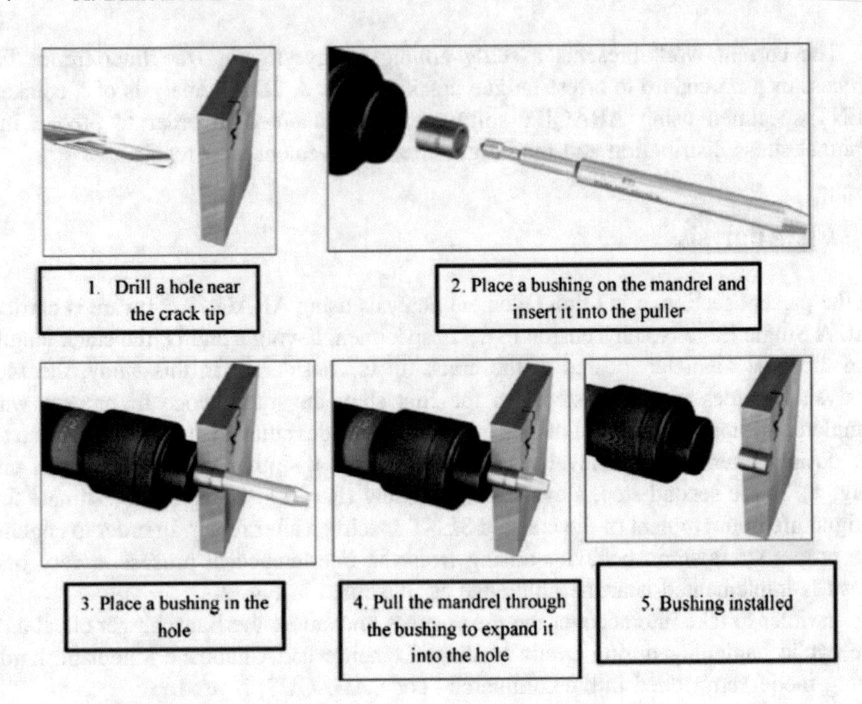

Fig. 1. IF process steps.

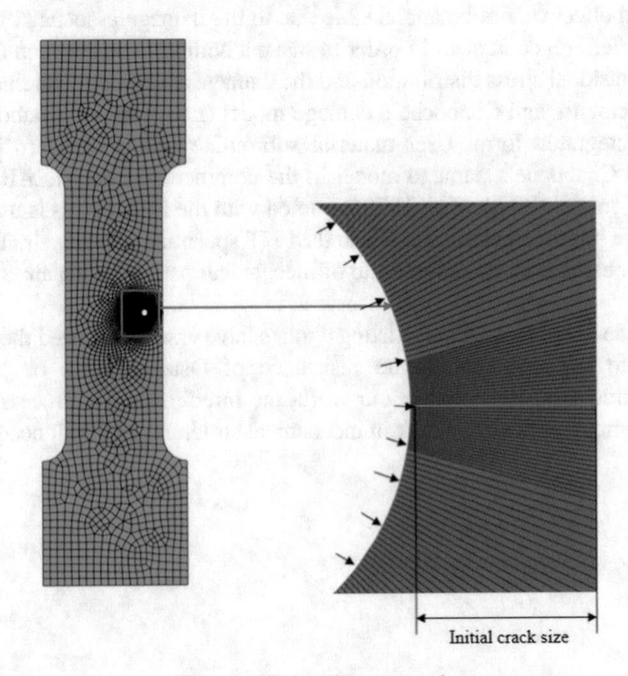

Fig. 2. Finite element mesh.

3 Results and Discussion

(i) The first application is established to validate the SWT model's effectiveness for predicting the fatigue life improvement after repair. Therefore, a holed plate specimen repaired by high interference fit bushing is considered. The interference fit bushing diameter expanded into the hole is equal to 5.05 mm. After the interference fit process simulation, the specimen was applied to a cyclic fatigue loading with zero load ratio (R = 0), for different applied loads. The SWT model is implemented to predict the fatigue life improvement after repair. Figure 3 shows a good agreement between the available experimental data (Chakherlou et al. 2010) and the SWT model for predicting fatigue life improvement after interference fit process.

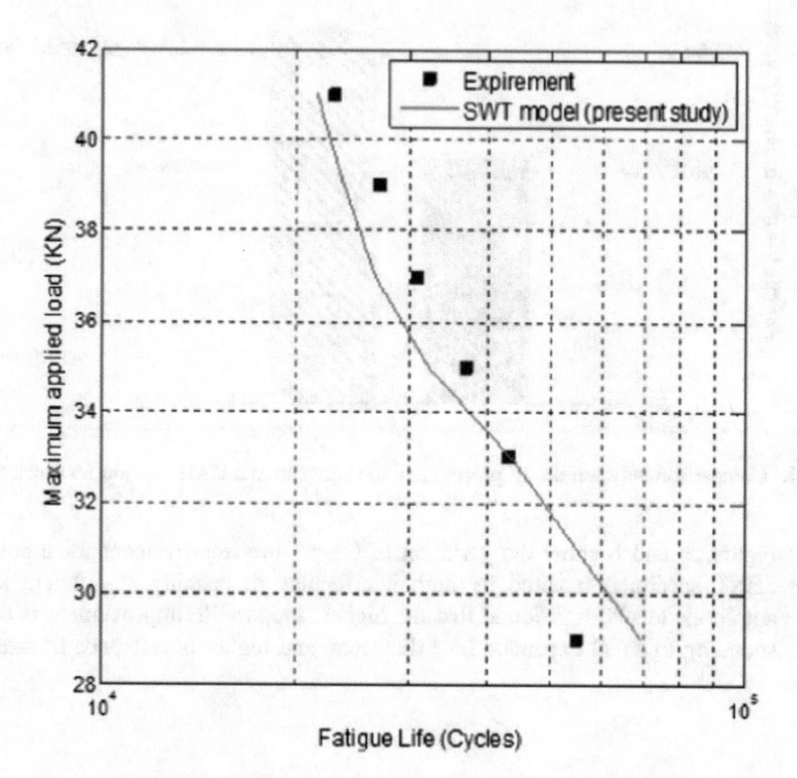

Fig. 3. Comparison between experimental data (Chakherlou et al. 2010) and numerical results.

(ii) Figure 4 shows a comparison between the fatigue life improvement of cracked SENT specimen repaired by crack arrest hole (CAH) method (i.e. Ex = 0%) and by an interference fit bushing (i.e. Ex = 2.5%). It is found that the interference fit process provides significant fatigue life improvement compared to the conventional CAH method, extensively used in several research works (Ghfiri et al. 2000) (Ayatollahi et al. 2015). The effectiveness of the proposed method is attributed to the induced compressive residual stress around the hole which shields it from the effect of cyclic loading.

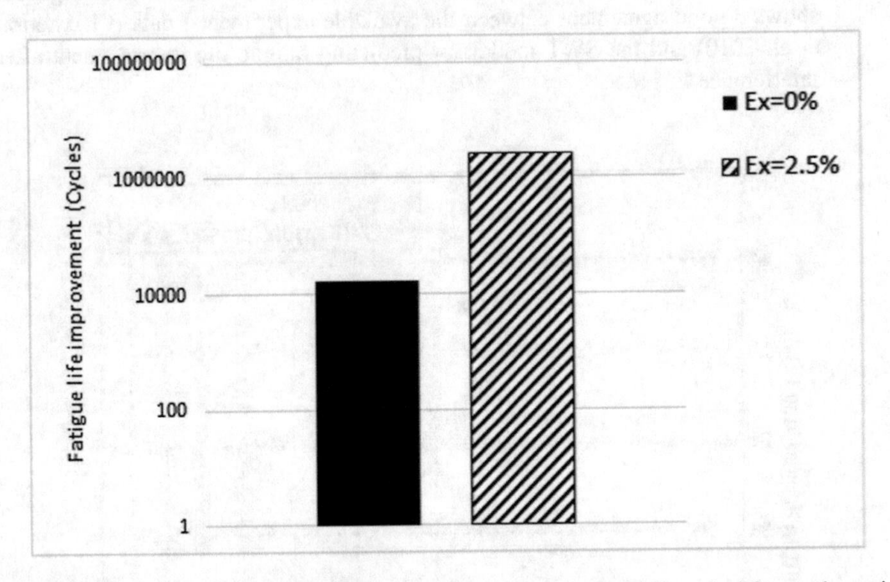

Fig. 4. Comparison between the IF process and the conventional CAH method for crack repair

(iii) Figure 5a and b show the variation of fatigue life improvement for a cracked SENT specimen repaired by high interference fit bushing at different stress amplitude levels. It is found that the highest fatigue life improvement is corresponding to small expanded hole diameters and higher interference fit size.

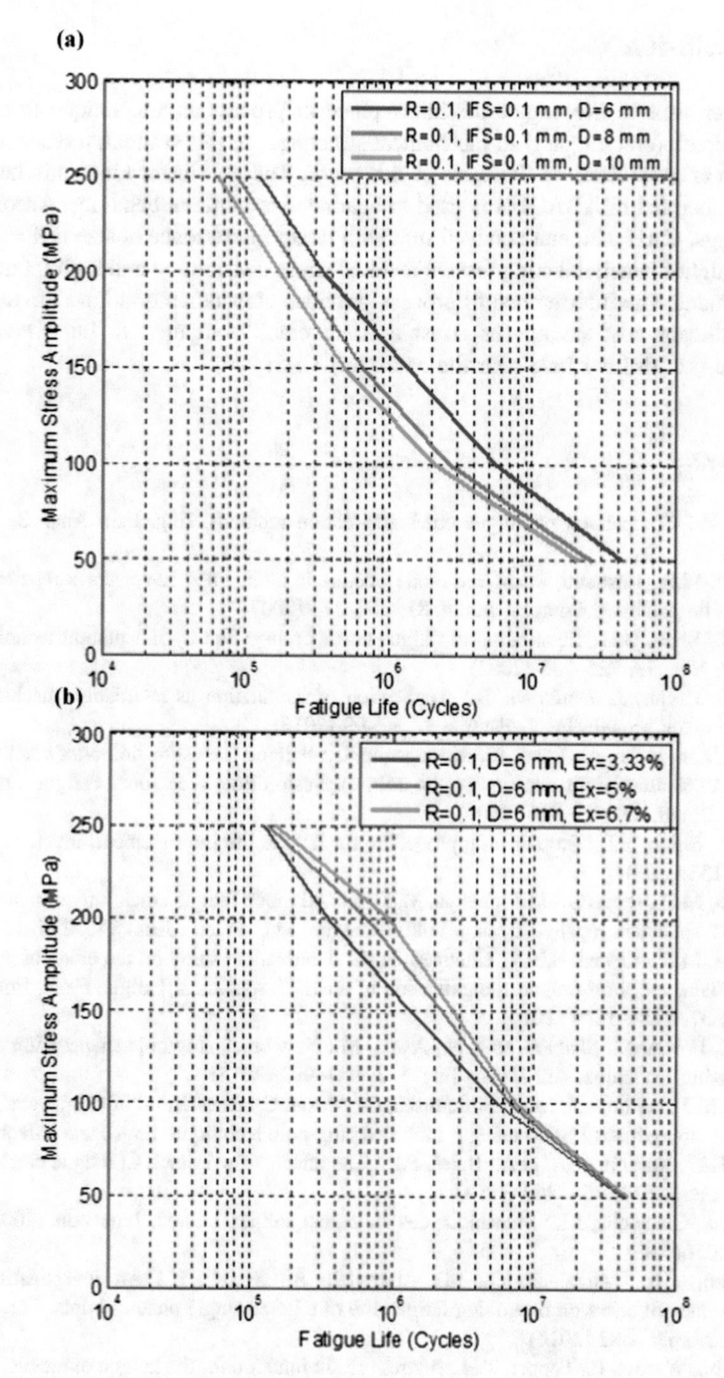

Fig. 5. Fatigue life improvement after repair (a) at different expanded hole diameters and (b) at different Interference Fit sizes.

4 Conclusion

This paper aims to investigate the interference fit process as a technique to arrest an existing crack (crack repair) in mechanical structures. In this context, a cracked SENT specimen is considered on the basis of this work. The non-linear kinematic hardening model embedded in ABAQUS is used to characterize material behavior. According to the findings, CAH diameter equals 6 mm with larger interference fit size IFS = 0.2 mm provide higher beneficial compressive residual stress distribution and higher fatigue life improvement. The interference fit process provides considerable fatigue life extension and significant cost saving for arrest existent crack compared to the conventional technique (i.e. drilling hole near the crack tip).

References

Domazet, Z.: Comparison of fatigue crack retardation methods. Eng. Fail. Anal. **3**, 137–147 (1996)

Ayatollahi, M.R., Hashemi, R.: Mixed mode fracture in an inclined center crack repaired by the composite patching. Compos. Struct. **81**, 264–273 (2007)

Song, P.S., Sheu, G.L.: Retardation of fatigue crack propagation by indentation technique. Int. J. Pres. Ves. **79**, 725–733 (2002)

Ruzek, K., Pavlas, J., Doubrava, R.: Application of indentation as retardation mechanism for fatigue crack growth. Int. J. Fatigue **37**, 92–99 (2012)

Ghfiri, R., Amrouche, A., Imad, A., Mesmacque, G.: Fatigue life estimation after crack repair in 6005 AT-6 aluminium alloy using the cold expansion hole technique. Fatigue Fract. Eng. Mater. Struct. **23**, 911–916 (2000)

Song, P.S., Shieh, Y.L.: Stop drilling procedure for fatigue life improvement. Int. J. Fatigue **26**, 1333–1339 (2004)

Ayatollahi, M.R., Razavi, S.M.J., Yahya, M.Y.: Mixed mode fatigue crack initiation and growth in a CT specimen repaired by stop hole technique. Eng. Fract. Mech. **145**, 115–127 (2015)

Ayatollahi, M.R., Ravazi, S.M.J., Chamani, H.R.: A numerical study on the effect of symmetric crack flank holes on fatigue life extension of a SENT specimen. Fatigue Fract. Eng. Mater. Struct. **37**, 1153–1164 (2014)

Fanni, M., Fouda, N., Shabara, M.A.N., Awad, M.: New crack stop hole shape using structural optimizing technique. Ain Shams Eng. J. **3**, 987–999 (2015)

Razavi, S.M.J., Ayatollahi, M.R., Sommitsch, C., Moser, C.: Retardation of fatigue crack growth in high strength steel S690 using a modified stop-hole technique. Eng. Fract. Mech. (2016)

Carlson, R.L., Kardomateas, G.A., Bates, P.R.: The effects of overloads in fatigue crack growth. Int. J. Fatigue **13**, 453–460 (1991)

Lemaıtre, J., Chaboche, J.L.: Mecanique des materiaux solides, Dunod, 2eme edn. (2002). ISBN 2 10 005662X

Chakherlou, T.N., Mirzajanzadeh, M., Abazadeh, B., Saeedi, K.: An investigation about interference fit effect on improving fatigue life of a holed single plate in joints. Eur. J. Mech. Solid. **29**, 675–682 (2010)

Smith, K.N., Watson, P., Topper, T.H.: A stress-strain function for the fatigue of metals. J. Mater. **15**, 767–778 (1970)

Stochastic Design of Non-linear Electromagnetic Vibration Energy Harvester

Issam Abed[1(✉)] and Mohamed Lamjed Bouazizi[2]

[1] Preparatory Engineering Institute of Nabeul (IPEIN), M'Rezgua, 8000 Nabeul, Tunisia
issam.abed@enit.utm.tn

[2] Mechanical Department, College of Engineering, Prince Sattam Bin Abdulaziz University, Al Kharj, Kingdom of Saudi Arabia
my.bouazizi@psau.edu.sa

Abstract. Perfect design of non-linear electromagnetic vibration energy harvester is disturbed in presence of imperfection. In this paper, we introduced a realistic design of imperfections to study the robustness of the optimized non-linear electromagnetic vibration energy harvester. A generic discrete analytical model is derived and numerically solved. The proposed procedure includes the usage of MANLAB coupled with multi-objective optimization. Thanks to the imperfections, modal interactions and nonlinear coupling, the stochastic design enables for harvesting the vibration energy in the operating power Mean of 373 mW with the frequency Mean of 195 Hz and the bandwidth Mean of 151 Hz.

Keywords: MANLAB · Multi-objective optimization · Stochastic design · Energy harvester · Imperfections

1 Introduction

Harvesting energy has been investigated for more than a decade, resulting from an interesting applications for wireless sensors and electronic devices. Several studies present three main methods for converting mechanical vibrations into electrical energy. These techniques are electromagnetic, piezoelectric and electrostatic. Some authors have presented a hybrid energy harvesting, which associate two or three transduction techniques to produce energy in the same device. Dynamic analysis is really simplified by using the appropriate mathematical equations. Conversely, Imperfections (Qiao et al. 2017) can be due to material defects, manufacturing defaults, structural damage, fatigue, aging, etc., and reflect the authenticity of systems, can disturb the collected energy and modify significantly the dynamic performance of the collected energy. For a bistable generator, stochastic forcing can induce transitions to stable equilibrium positions, resulting in large amplitude of oscillations. For example, Lin and Alphenaar (Lin et al. 2010), and And'o et al. (Andò et al. 2010) established this idea to increase the energy harvesting performance by using a set of bistable cantilevers with repulsive magnets for wide-spectrum vibrations. The purpose of these studies is to permit the harvester to easily shift between two stable positions, which depends on the excitation frequency, amplitude and the extent of non-linearity. To increase the performance of a

A. Benamara et al. (Eds.): CoTuMe 2018, LNME, pp. 39–46, 2019.
https://doi.org/10.1007/978-3-030-19781-0_5

bistable system and to enhance the probability of transition to the potential wells, some authors (Yang et al. 2018) have suggested to exploit the phenomenon of stochastic resonance. The phenomenon of stochastic resonance can occur, when the dynamic system is forced (Wellens et al. 2004). The concept of harvesting energy with multiple degrees of freedom, non-linearity and uncertainty is a complex challenge and the main purpose of this study is to deal with. To explore the vibrational energy of N nonlinear degrees of freedom and to control the modal interactions, previous works (Abed et al. 2016) suggested discrete analytical models joining the Asymptotic Numerical Method (ANM) and the Harmonic Balance method (HBM) method. The core objective of the present study is to extend these works in the presence of imperfections by introducing a constraint of incertitude. For this purpose, uncertainty propagation methods must be used, when imperfections are used as parametric uncertainties modelled by random variables. The objective of using uncertainty parameters in these models is to evaluate robustness of the harvested power against randomness of the uncertain input parameters. The established discrete analytical model leads to a series of coupled algebraic equations. These equations are written according to uncertainty in the generator and numerically solved by using the ANM method coupled with the HBM method (Cochelin et al. 2009). The sampling method Latin Hypercube (LHS) (Helton et al. 2003) is used as a method to propagate uncertainties of the proposed model.

Imperfections effects on the nonlinear dynamics of one, two and three coupled clamped-clamped beams are studied in this paper. Dispersion analysis of the nonlinear bandwidth response, nonlinear frequency responses and maximal nonlinear power is carried out. Additionally, in order to highlight the complexity of multimode solutions in terms of collected power and bandwidth, a detailed analysis through a multi-objective optimization study (Deb et al. 2002) followed by the sensitivity study is performed. The robustness of the harvested power against uncertainties of the selected bandwidth domain and nonlinear frequency is investigated.

2 Mechanical Model

Figure 1 shows a magnetic VEH for N joined clamped-clamped beams of identical rigid length L, elastic length, thickness of the beam h, mass of the magnet m and electromagnetic damping coefficient produced by wire-wound copper coil. The clamped-clamped beams are joined by N moving magnets M, which are attached to the medium of all beams and bottom beam are subject to an external excitation. All coupled beams are positioned between two fixed magnets respectively at the top and the bottom. The magnetic poles are focused on a way that repulsive forces are generated between two adjacent magnets. These forces lead to "nonlinear and linear magnetic stiffness".

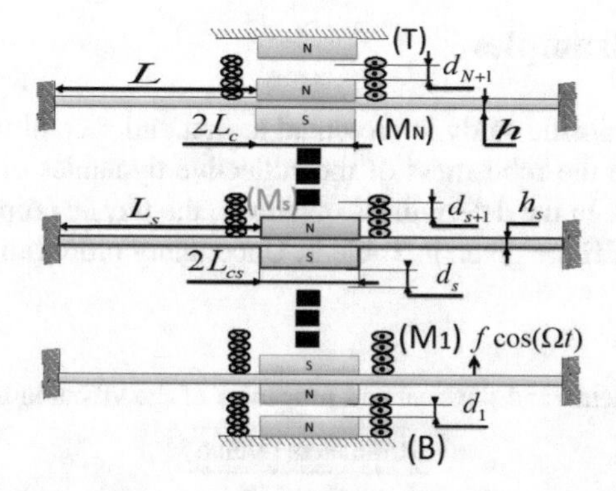

Fig. 1. Periodic non-linear clamped-clamped beams chain with imperfection.

The periodicity of the magnetic VEH is destroyed by existence of p clamped-clamped beams containing parametric uncertainties which can be the beams number s as explained in Fig. 1.

$$\begin{pmatrix} \ddot{\tilde{a}}_i + c_i\dot{\tilde{a}}_i + \omega_i^2\tilde{a}_i + \omega_{i(i-1)}^2(\tilde{a}_i - \tilde{a}_{i-1}) \\ +\omega_{i(i+1)}^2(\tilde{a}_i - \tilde{a}_{i+1}) + \vartheta_i\tilde{a}_1^3 \\ +\vartheta_{i(i-1)}(\tilde{a}_i - \tilde{a}_{i-1})^3 + \vartheta_{i(i+1)}(\tilde{a}_i - \tilde{a}_{i+1})^3 \end{pmatrix} = -\delta_{1i}F_i cos(\Omega t) \quad (1)$$

$$\begin{array}{llll} k_{eq}^L = 12\dfrac{EI}{L^3} & m_{eq} = \dfrac{13}{35}\rho SL + m & c_i = \dfrac{c_{eq}+c_e}{m_{eq}} & \omega_i^2 = \dfrac{k_{eq}^L}{m_{eq}} \\ k_{eq}^{NL} = \dfrac{18}{25}\dfrac{ES}{L^3} & F_{eq} = \dfrac{\rho SL}{2} + m & \omega_{i(i-1)}^2 = \dfrac{k_{i(i-1)}}{m_{eq}} & \omega_{i(i+1)}^2 = \dfrac{k_{i(i+1)}}{m_{eq}} \\ \omega_0 = \sqrt{\dfrac{k_{eq}^L}{m_{eq}}} & c_{eq} = 2\xi\omega_0 m_{eq} & \vartheta_i = \dfrac{k_{eq}^{NL}}{m_{eq}} & \vartheta_{i(i-1)} = \dfrac{\lambda_{i(i-1)}}{m_{eq}} \\ & & \vartheta_{i(i+1)} = \dfrac{\lambda_{i(i+1)}}{m_{eq}} & F_i = \dfrac{F_{eq}f}{m_{eq}} \end{array} \quad (2)$$

Imperfections are supposed to disturb structural input parameters, as rigid lengths, the elastic lengths, thickness of the beam and mass of the magnet, and to vary randomly. A probability modelling on uncertainties is used to estimate the influence of randomness in magnetic VEH input parameter on the coupled clamped-clamped beams chain. Since linear and nonlinear magnetic coupling between beams is very weak. When parameters beam are stochastic, the deterministic displacement is replaced by stochastic parameters $\tilde{a}_i$ in Eq. 1.

Resolving analytically these equations by using Multiple Scale Method is very difficult. To overcome this problem, numerical solving method must be used. In the current work, we used the Asymptotic Numerical Method (ANM) in graphical interactive software named MANLAB.

3 Numerical Examples

In this paper, deterministic study is presented at first and second we present stochastic study to determinate the robustness of the collective dynamics of the VEH structures against uncertainties. In the deterministic example, the physic properties and geometric of the considered VEH are given in Table 1. Uncertainty propagation is achieved using the LHS method.

Table 1. Geometric and physical and properties of the vibration energy harvester.

Parameters	Value
L (mm)	75
L_c (mm)	6
b (mm)	10
h (mm)	0.6
m (g)	5
d_1 (mm)	13
B (T)	0.35
ξ	0.016
c_e (Ns/m)	0.19
$g(\text{ms}^{-2})$	9.8
$f(\text{ms}^{-2})$	5 * g

Effects of uncertainties of the VEH structure are explored through analyses of the frequency response and the dispersion of the collected power. The VEH is composed of two clamped-clamped beams magnetically joined with two moving magnets m, which are attached to the medium of both beam and positioned between two fixed magnets respectively at the top and the bottom.

3.1 Deterministic Study

Figure 2 shows the evolution of the phase, the amplitude and harvested energy for design parameters listed in Table 1. It is presented that the two clamped-clamped beams magnetically coupled vibrate in phase up to an excitation frequency of 50 Hz. Then, a phase between the two signals increases and reaches a maximum (π) up to an excitation frequency of 95 Hz. In the deterministic study, we seek to take benefit of the nonlinear and linear magnetic coupling between two clamped-clamped beams in order to improve the harvested power and the nonlinear bandwidth of the considered device. To do, two objective functions have been defined.

$$g_1 = \frac{BW_{NL}}{f_2} \qquad g_2 = P_{\max} \tag{3}$$

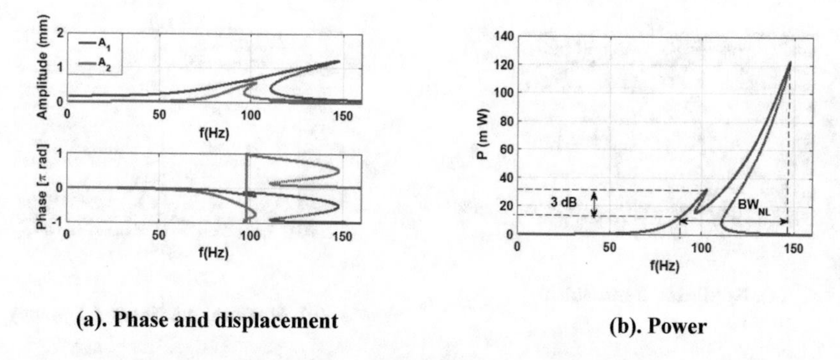

Fig. 2. Forced frequency responses of 2DOF VEH.

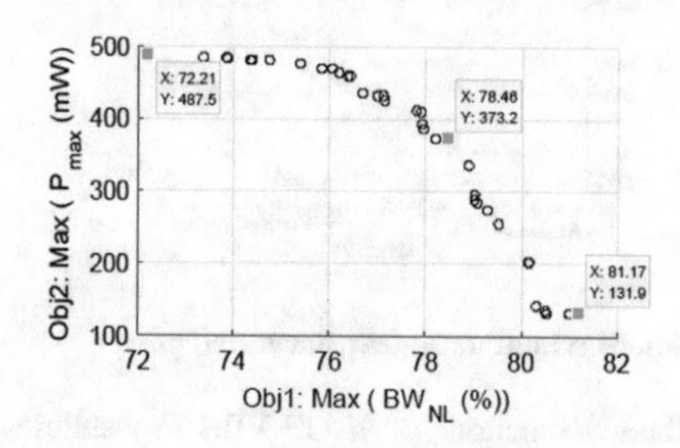

Fig. 3. Periodic non-linear clamped-clamped beams chain with imperfection.

Two cost functions are simultaneously maximized with respect to the seven retained design parameters (c_e, L, h_s, b, L_c, m and d_1). The Chosen solutions to the multi-objective optimization problem by 50 population sizes of 100 generations are exposed in Fig. 3. Three solutions have been selected on the Pareto front.

3.2 Stochastic Study

Stochastic Study for the case of 2DOF VEHs is explained by nonlinear bandwidth response amplitude, the nonlinear frequency responses amplitudes and the maximal nonlinear power P.

This uncertainty is used for two dispersion levels. The uncertainty result of two coupled beams responses depends on the imperfection applied to the input parameters c_e, L, h_s, b, L_c, m and d_1. Figure 4a shows the PDF variation of nonlinear bandwidth Δ_f, Fig. 4b present the PDF variation of the nonlinear frequency f_2 and Fig. 4c shows the PDF variation of the nonlinear power P output parameters for 2DOF VEHs. We remark that the PDF variation of the statistic parameters outputs f_2, Δ_f and P are unequal for everywhere dispersion. For this, we study the statistic parameters of Mean μ, Standard deviation σ, dispersion δ, skewness γ, kurtosis β and confidence intervals CI. The statistic output parameters for the input dispersion variation are given in Table 2.

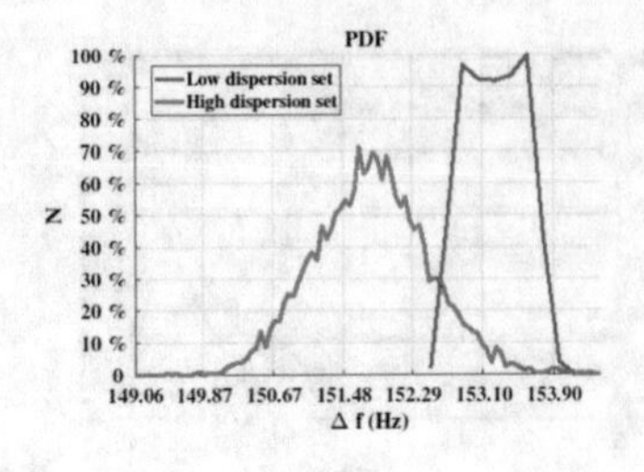

(a). Nonlinear Bandwidth

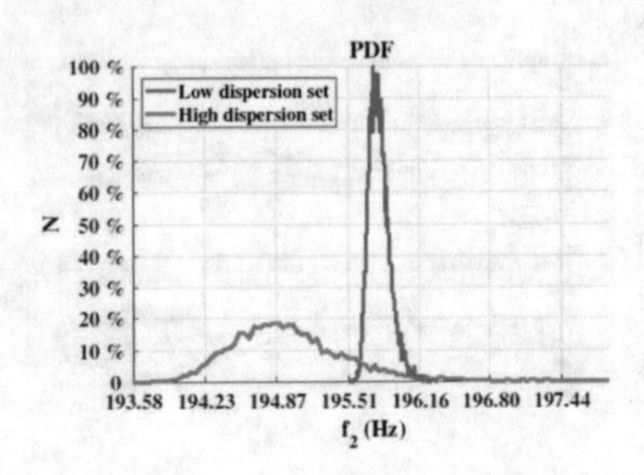

(b). Maximal nonlinear frequency

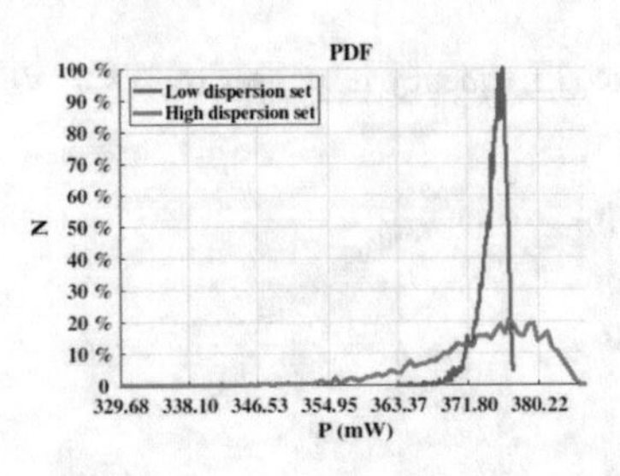

(c). Maximal nonlinear harvested power

Fig. 4. Probability distribution functions of 2DOF VEH Probabilities distribution function of low input dispersion set and for high input dispersion set.

Table 2. Parameters outputs for low and high input dispersion set Design parameters.

Dispersion set	Parameters					
	μ	σ	$\delta(\%)$	γ	β	CI
Low input (Δf (Hz))	153.238	0.37	0.24	10.5	494	[153.23 153.245]
High input (Δf (Hz))	151.8	0.66	0.44	0.11	3.2	[151.79 151.82]
Low input (f_2 (Hz))	195.79	0.095	0.05	0.89	4.6	[195.791 195.795]
High input (f_2 (Hz))	194.96	0.5	0.25	0.74	4.1	[194.95 194.97]
Low input (P (mW))	373.69	1.63	0.44	−1.51	6.57	[374.66 374.73]
High input (P (mW))	373.34	7.24	1.94	−1.15	4.87	[373.19 373.478]

From Table 2, we note that the coefficient skewness of f_2 and Δf are positive and the coefficient skewness of P is negative. Furthermore, when we vary the coefficient dispersion of the input parameters, the coefficient skewness of Δf vary rapidly and the coefficient skewness of f_2 and P varies slightly. Moreover, when we vary the coefficient dispersion of the input parameters, the mean of Δf, f_2 and P varies slightly. When we change the input dispersion set, we note that confidence interval varies as well as other parameters. For this reason, the variation of confidence interval for nonlinear bandwidth Δf, maximal nonlinear frequency f_2 and P with respect to the coefficient dispersion of the input parameters is illustrated in Fig. 5. From Fig. 5a and b, the dispersion level of the input parameters of the thickness and the length of the beam have a small effect on the dispersion and the confidence interval of the bandwidth and the frequency. And from Fig. 5c, the dispersion of the input parameter of the thickness h_s and the length L of the beam has a considerable effect on the dispersion and the confidence interval of the power P. From these results, it is concluded that the dispersion of input parameters such as thickness h_s and length L has a large effect on the confidence interval of the obtained results.

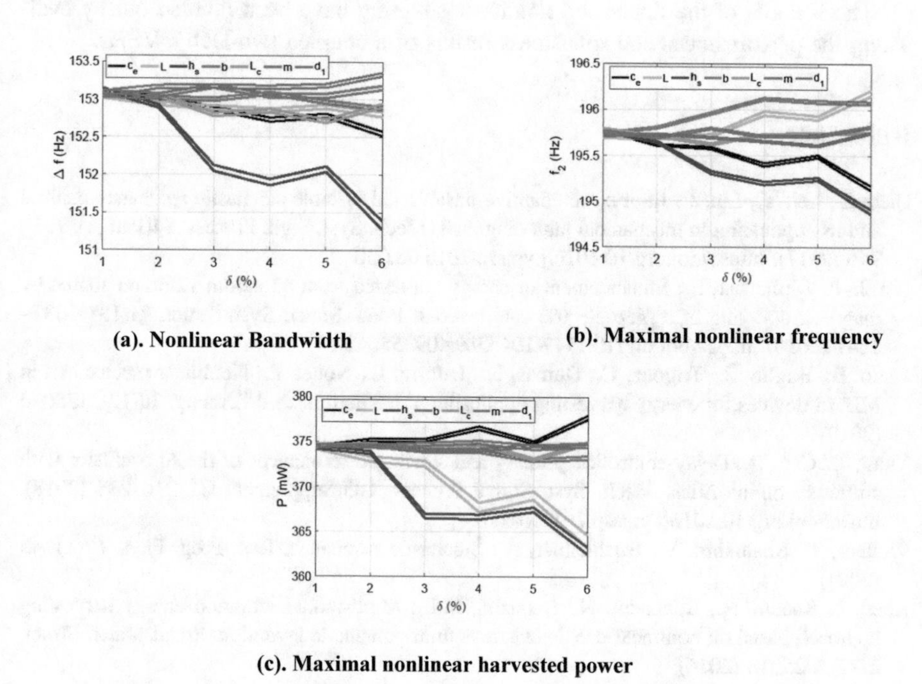

(a). Nonlinear Bandwidth **(b). Maximal nonlinear frequency**

(c). Maximal nonlinear harvested power

Fig. 5. Variation of confidence interval for different input dispersion set.

4 Conclusions

In this paper, we suggest a multi-modal vibration energy harvesting technique based on arrays of a bi-clamped beams coupled magnetically. The equations of motion have been resolved taking into account the mechanic nonlinearity, magnetic nonlinearity and electromagnetic damping. They have been resolved by using the HBM coupled with the ANM method. The case of 2DOFs VEHs have been analytically studied. Multiobjective optimization Procedures are investigated and achieved by using NSGA-II algorithm for the cases of two bi-clamped beams coupled magnetically in order to choice optimal solutions in term of performances.

The VEH permits to study the robustness of the harvested power of a two of bi-clamped beams coupled magnetically and take into account parametric uncertainties. Uncertainties guarantee a realistic modelling on imperfections, which can disturb the perfect collected power of real engineering harvester. In a probabilistic framework, an analysis of dispersion of nonlinear bandwidth, nonlinear frequency and nonlinear power of 2DOFs energy harvesters were achieved. Statistical estimations of the nonlinear frequency, nonlinear bandwidth and nonlinear power responses compute the variability of the thickness, length, width, magnet radius, mass and gap distance. The complexity of the performance responses identified in terms of nonlinear frequency, nonlinear bandwidth and nonlinear power, in deterministic case, was intensified in presence of uncertainty.

The benefits of the linear and nonlinear coupling have been pointed out by evaluating the performances and robustness results of a coupled two-DOFs VEHs.

References

Qiao, Z., Lei, Y., Lin, J., Jia, F.: An adaptive unsaturated bistable stochastic resonance method and its application in mechanical fault diagnosis. Mech. Syst. Sign. Process. **84**(Part A), 731–746 (2017). https://doi.org/10.1016/j.ymssp.2016.08.030

Lin, J.-T., Alphenaar, B.: Enhancement of energy harvested from a random vibration source by magnetic coupling of a piezoelectric cantilever. J. Intell. Mater. Syst. Struct. **21**(13), 1337–1341 (2010). https://doi.org/10.1177/1045389X09355662

Andò, B., Baglio, S., Trigona, C., Dumas, N., Latorre, L., Nouet, P.: Nonlinear mechanism in MEMS devices for energy harvesting applications. J. Micromech. Microeng. **20**(12), 125020 (2010)

Yang, T., Cao, Q.: Delay-controlled primary and stochastic resonances of the sd oscillator with stiffness nonlinearities. Mech. Syst. Signal Process. **103**(Supplement C), 216–235 (2018). https://doi.org/10.1016/j.ymssp.2017.10.002

Wellens, T., Shatokhin, V., Buchleitner, A.: Stochastic resonance. Rep. Prog. Phys. **67**(1), 45 (2004)

Abed, I., Kacem, N., Bouhaddi, N., Bouazizi, M.L.: Multi-modal vibration energy harvesting approach based on nonlinear oscillator arrays under magnetic levitation. Smart Mater. Struct. **25**(2), 025018 (2016)

Cochelin, B., Vergez, C.: A high order purely frequency-based harmonic balance formulation for continuation of periodic solutions. J. Sound Vib. **324**(1), 243–262 (2009). https://doi.org/10.1016/j.jsv.2009.01.054

Helton, J., Davis, F.: Latin hypercube sampling and the propagation of uncertainty in analyses of complex systems. Reliab. Eng. Syst. Saf. **81**(1), 23–69 (2003). https://doi.org/10.1016/S0951-8320(03)00058-9

Deb, K., Pratap, A., Agarwal, S., Meyarivan, T.: A fast and elitist multiobjective genetic algorithm: NSGA-II. IEEE Trans. Evol. Comput. **6**(2), 182–197 (2002). https://doi.org/10.1109/4235.996017

Mechanical Characterization of Coating Materials Based on Nanoindentation Technique

Nadia Chakroun(✉), Aymen Tekaya, Hedi Belhadjsalah, and Tarek Benameur

Mechanical Engineering Laboratory, LR99ES32 ENIM, University of Monastir, 5000 Monastir, Tunisia
lojayna@live.fr, tekaya.aymen@yahoo.fr, hedi.belhadjsalahl@gmail.com, tarektijanibenameur@gmail.com

Abstract. In the present paper, mechanical properties of monolayer coatings were investigated. Particularly, an analytical model was considered to characterize the mechanical properties of TiN thin film deposited on $Zr_{60}Ni_{10}Cu_{20}Al_{10}$ substrate. As a results, the elastic modulus of the TiN thin films is equal to 415.6 GPa. Based on numerical confrontation results, the limit of the considered model was observed for the soft thin film on hard substrate. The main error is caused by a wrong estimation of the contact surface A_c between the indenter tip and the film surface. Hence, the response to the nanoindentation test is quite influenced by the presence of the substrate for thin hard film.

Keywords: TiN thin film · Elastic modulus · Nanoindentation test · Substrate effect

1 Introduction

Development of Thin films technology is stimulated by the industry demand for improving the effectiveness of contact surfaces. Nevertheless, the mechanical properties of thin films are quite different to materials in bulk. As a matter of fact, characterizing the mechanical behavior of thin film systems is quite important in order to understand the performance of these materials in service (Chakroun et al. 2017, 2018). This task is particularly relevant as it is necessary to improve the durability of coated materials (Inui et al. 2016; Kumar and Zeng 2010). The nanoindentation test is the most useful technique for extracting the mechanical properties of materials with low dimensions such as thin films. This method uses an indenter that comes into contact with the surface applying a load. Therefore, using the nanoindentation technique several studies are developed to characterize thin film properties (Doerner and Nix 1986; Oliver and Pharr 1992). In the literature, many models were developed so as to characterize surface coatings (Jung et al. 2004; Liao et al. 2009; Mercier et al. 2011; Bull 2015; Chakroun et al. 2017). These models determine thin film properties taking into account especially the substrate effect.

A. Benamara et al. (Eds.): CoTuMe 2018, LNME, pp. 47–53, 2019.
https://doi.org/10.1007/978-3-030-19781-0_6

In particular, the main objective of this paper is the characterization of thin film materials deposited on a substrate. To that end, we used an analytical model and more specifically the Bec model (Bec et al. 1996). The present study is focused on measuring the intrinsic film properties for experimental TiN/$Zr_{60}Ni_{10}Cu_{20}Al_{10}$ sample (hard film/soft substrate). Furthermore, we evaluated the responses by the considered model using numerical confrontation results.

2 Mechanical Characterization of TiN/$Zr_{60}Ni_{10}Cu_{20}Al_{10}$

2.1 Experimental Details

Monolithic nanocoating of Titanium Nitride (TiN) was deposited on $Zr_{60}Ni_{10}Cu_{20}Al_{10}$ BMG substrate by Radio Frequency (RF) sputtering technique in nitrogen atmosphere at room temperature. In the deposition process, 99.9% pure Titanium (Ti) target was used. A typical base pressure of about 1.33×10^{-6} Pa was in the deposition chamber and the total pressure was 0.66 Pa. The typical deposition rate for titanium was about 0.9 nm/min. A film thickness of 300 nm was obtained. Further preparation details were described elsewhere (Tekaya et al. 2014). Table 1 summarizes the deposition technical parameters of the monolayer TiN. The nanoindentation tests were realized with a Hysitron triboscope apparatus. Before indentation measurements, the system was calibrated by conducting several indents on a fused silica sample under normal loads. Machine compliance, thermal drift and area function were corrected. The nanoindenter depth and load resolution were 0.04 nm and 1 nN, respectively. The indenter was a Berkovich tip with an included angle of about 142.3°. Figure 1 shows experimental nanoindentation curve for TiN film deposited on $Zr_{60}Ni_{10}Cu_{20}Al_{10}$ substrate.

Table 1. Details of deposition parameters for TiN monolayer.

Parameter	Value
N_2 gas pressure (Pa)	0.07
Total pressure (Pa)	0.66
Pre-deposition vacuum (Pa)	1.10^{-6}
Target to substrate distance (mm)	70
Sputtering power density (W/cm^2)	0.63
Deposition time (min)	90
Temperature of substrate holder (°C)	35
Auto bias voltage (V)	−300

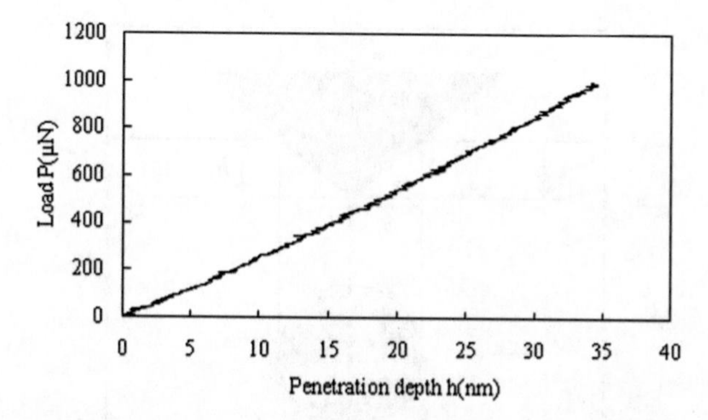

Fig. 1. Nanoindentation curve for $TiN/Zr_{60}Ni_{10}Cu_{20}Al_{10}$

2.2 Analytical Model

In 1996, Bec et al. (Bec et al. 1996) proposed an analytical model for a film deposited on a substrate. The model of Bec et al. models the film/substrate system as a series of two springs. Results obtained on a gold layer (soft film on hard substrate) are presented as an illustration of the method. The model is presented by the following relation:

$$\frac{1}{2a_cE'} = \frac{t}{(\pi a_c^2 + 2ta_c)E_f'} + \frac{1}{2(a_c + \frac{2t}{\pi})E_s'} \tag{1}$$

Where E_s' is the reduced elastic modulus of the substrate and E_f' is the reduced elastic modulus of the film as shown in Fig. 2.

Fig. 2. Film/Substrate system

Furthermore, t and a_c are the thickness and the contact radius of the film, respectively. Figure 3 demonstrates the contact radius a_c.

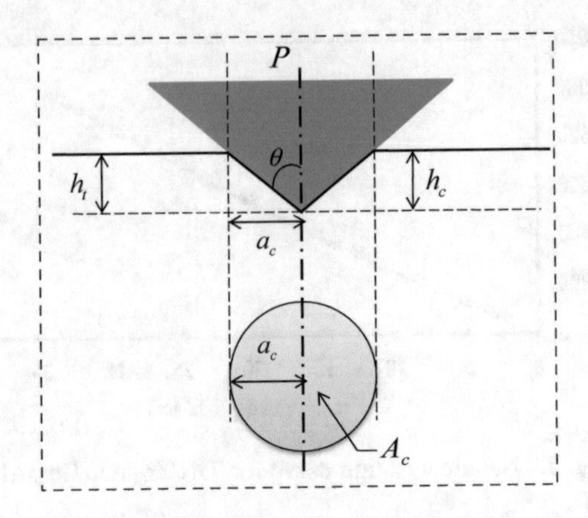

Fig. 3. Nanoindentation details

The analytical model of Bec was considered to characterize the experimental TiN/$Zr_{60}Ni_{10}Cu_{20}Al_{10}$ sample based on the nanoindentation test (Fig. 1). As a results, the elastic modulus of TiN thin film is equal to 415.6 GPa.

3 Results

3.1 Numerical Model

The finite element analyses were carried out using the commercial code ABAQUS 6.12. The indentation problem is considered as an axisymmetric model owing to the conical indenter. The film and the substrate materials are assumed to be isotropic. Four-node axisymmetric elements (CAX4R) with reduced integration are considered, as presented in Fig. 4. Along the axis of symmetry, roller boundary conditions are used. Besides, fixed boundary conditions are applied in the substrate base (Fig. 4). The indenter is a rigid truncated cone with a half angle of 70.3 which makes the same projected area as the Berkovich indenter (Jayaraman et al. 1998). Contact surface is assumed to be frictionless while indentation friction coefficient is a minor factor (Cheng and Cheng 2004).

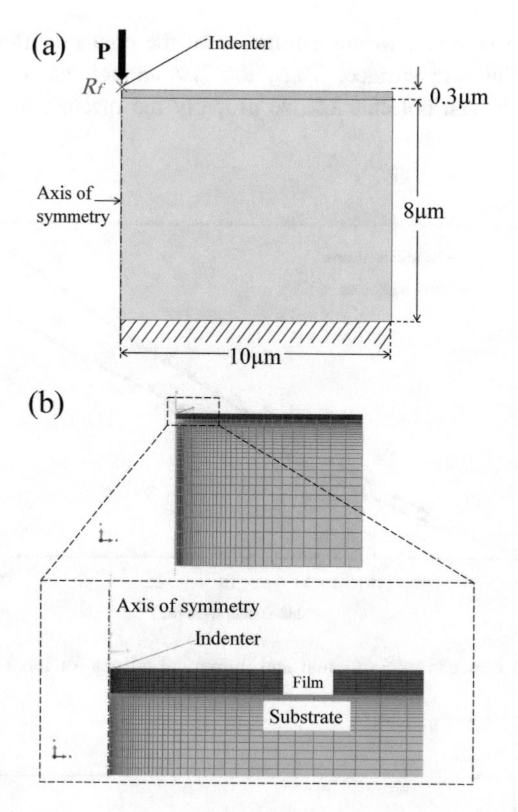

Fig. 4. (a) Schematic of nanoindentation model and (b) Finite element mesh for the film/substrate system.

3.2 Numerical Confrontation

To verify the result found by the model of Bec et al., we performed a numerical nanoindentation test by introducing the value identified by Bec as a modulus of elasticity of the film. Figure 5 presents the experimental nanoindentation curve and the curve obtained by the developed numerical model for the sample $TiN/Zr_{60}Ni_{10}Cu_{20}Al_{10}$. Figure 6 presents the error between the two *P-h* curves presented in Fig. 5. Consequently, the considered model is limited for the case of hard film (TiN) on soft substrate ($Zr_{60}Ni_{10}Cu_{20}Al_{10}$).

The average relative error between the experimental curve for the $TiN/Zr_{60}Ni_{10}Cu_{20}Al_{10}$ sample and the numerical *P-h* curve is equal to 18.2(%). Accordingly, the analytical model considered in the present research is not suitable for the case of hard film deposited on soft substrate such as the $TiN/Zr_{60}Ni_{10}Cu_{20}Al_{10}$ sample. In fact, for soft film on hard substrate, the load applied in the nanoindentation test is essentially absorbed by the film. In this case the response to the nanoindentation test represents mainly the film behavior. In this case, the Bec model characterizes the film properties. However, when the film is harder than the substrate (as the case of $TiN/Zr_{60}Ni_{10}Cu_{20}Al_{10}$), the load applied in the nanoindentation test is transferred to the substrate. This fact causes a wrong estimation of the contact surface A_c between the indenter tip and the film surface. Thus, the Bec model, based essentially on the nanoindentation test, can not characterize properly the mechanical properties of hard film material.

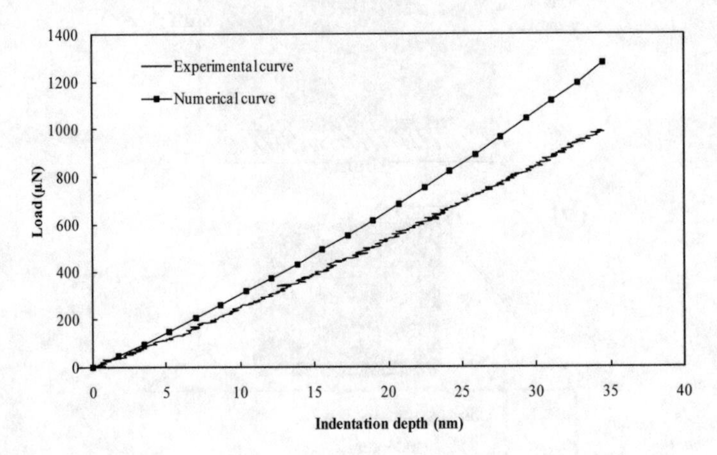

Fig. 5. Comparison between experimental and numerical curves for the $TiN/Zr_{60}Ni_{10}Cu_{20}Al_{10}$ sample

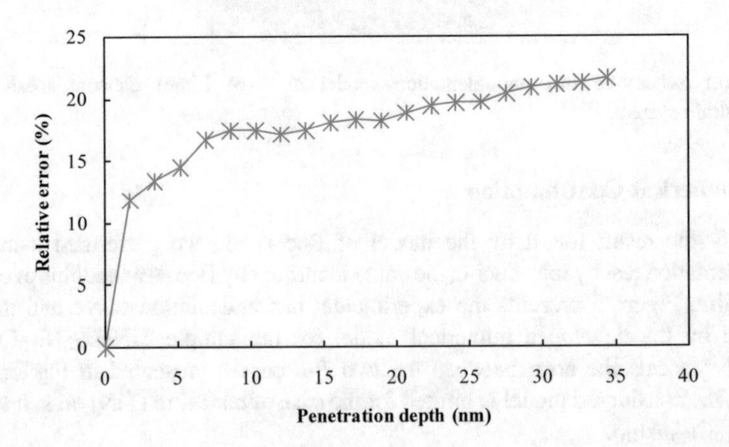

Fig. 6. Relative error between the experimental and numerical nanoindentation curves for the $TiN/Zr_{60}Ni_{10}Cu_{20}Al_{10}$ sample

4 Conclusion

In this paper we studied the mechanical properties of TiN thin film material using an analytical model. Based on the model of Bec et al., the elastic modulus of TiN thin film deposited on $Zr_{60}Ni_{10}Cu_{20}Al_{10}$ substrate is equal to 415.6 GPa. According to numerical confrontation results, we have demonstrated the limit of the considered model for hard film on soft substrate. In this case, the response of the analytical model is quite influenced by the substrate effect.

As perspectives of this research, we propose to develop a new model so as to characterize properly the coating materials for a hard film on soft substrate.

References

Georges, J.-M., Georges, E., Loubet, J.-L.: Improvements in the indentation method with a surface force apparatus. Philos. Mag. A **74**(5), 1061–1072 (1996). https://doi.org/10.1080/01418619608239707

Bull, S.J.: A simple method for the assessment of the contact modulus for coated systems a simple method for the assessment of the contact modulus for coated systems. Philosphical Mag. **95**, 1907–1927 (2015). https://doi.org/10.1080/14786435.2014.909612

Chakroun, N., Tekaya, A., Belhadjsalah, H., Benameur, T.: Measuring elastic properties of the constituent multilayer coatings for different modulation periods. Int. J. Appl. Mech. **10**(4), 1850046 (2018)

Chakroun, N., Tekaya, A., Belhadjsalah, H., Benameur, T.: A new inverse analysis method for identifying the elastic properties of thin films considering thickness and substrate effects simultaneously. Int. J. Appl. Mech. **9**(7), 1750096 (2017)

Cheng, Y.T., Cheng, C.M.: Scaling, dimensional analysis, and indentation measurements. Mater. Sci. Eng. R Rep. **44**(4–5), 91–150 (2004). https://doi.org/10.1016/j.mser.2004.05.001

Doerner, M.F., Nix, W.D.: A method for interpreting the data from depth-sensing indentation instruments. J. Mater. Res. **1**, 601–609 (1986)

Inui, N., Mochiji, K., Moritani, K.: A nondestructive method for probing mechanical properties of a thin film using impacts with nanoclusters. Int. J. Appl. Mech. **8**(3), 1650041 (2016)

Jayaraman, S., Hahn, G.T., Oliver, W.C., Rubin, C.A., Bastias, P.C.: Determination of monotonic stress-strain curve of hard materials from ultra-low-load indentation tests. Int. J. Solids Struct. **35**(5–6), 365–381 (1998)

Jung, Y.G., Lawn, BR., Martyniuk, M., Huang, H., Hu, X.Z.: Evaluation of elastic modulus and hardness of thin films by nanoindentation. J. Mater. Res. **19**(10), 3076–3080 (2004). http://www.scopus.com/inward/record.url?eid=2-s2.0-6344248912&partnerID=40&md5=ebfcd22e4bfd8b74946744cb5e4ab6b0%5Cn, http://journals.cambridge.org/action/displayFulltext?pageCode=100101&type=1&fid=8107730&jid=JMR&volumeId=19&issueId=10&aid=8107728&fromPage=

Kumar, A., Zeng, K.: Alternative methods to extract the hardness and elastic modulus of thin films from nanoindentation load-displacement data. Int. J. Appl. Mech. **2**(1), 41–68 (2010)

Liao, Y., Zhou, Y., Huang, Y., Jiang, L.: Measuring elastic-plastic properties of thin films on elastic-plastic substrates by sharp indentation. Mech. Mater. **41**(3), 308–318 (2009)

Mercier, D., Mandrillon, V., Verdier, M., Brechet, Y.: Mesure de Module d'Young D'un Film Mince À Partir de Mesures Expérimentales de Nanoindentation Réalisées Sur Des Systèmes Multicouches. Matériaux & Techniques **99**, 169–178 (2011)

Oliver, W.C., Pharr, G.M.: An improved technique for determining hardness and elastic modulus using load and displacement sensing indentation experiments. J. Mater. Res. **7**(6), 1564–1583 (1992)

Tekaya, A., Ghulman, H.A., Benameur, T., Labdi, S.: Cyclic nanoindentation and finite element analysis of Ti/TiN and CrN nanocoatings on Zr-based metallic glasses mechanical performance. J. Mater. Eng. Perform. **23**(12), 4259–4270 (2014)

An Inverse Calculation of Local Elastoplastic Parameters from Instrumented Indentation Test

Naoufel Ben Moussa(✉), Ons Marzougui, and Farhat Ghanem

Laboratoire de Mécanique, Matériaux et Procédés LR99ES05,
Ecole Nationale Supérieure d'Ingénieurs de Tunis, Université de Tunis,
5 Avenue Taha Hussein, Montfleury, 1008 Tunis, Tunisia
emailnaoufel@gmail.com, onmar.29@gmail.com,
farhatghanem@gmail.com

Abstract. The aim of this work is to identify the local behavior of a mechanical sample's surface layers from the instrumented indentation test and through optimization techniques. These tools are used to automate the search of the parameters of the mechanical constitutive law used in a finite element (FE) calculation and to reduce the number of calculations necessary to solve the problem of calibration. The experimental database includes mechanical responses of instrumented indentation performed on aluminum Alloy AA2017 and AISI 316L austenitic stainless steel samples. The Hooke and Jeeves iterative optimization method is used due to its efficiency, cost of calculation and precision. The procedure predictive aptitude is validated by determining correctly the behavior law of the aluminum alloy sample. The modification of the local material behavior induced by electrical discharge machining to the AISI 3016L sample is highlighted by identifying the tensile curves of different surface layers using the proposed procedure.

Keywords: Optimization techniques · Finite element method · Inverse analysis · Instrumented indentation · Mechanical behavior of materials

1 Introduction

The control and prediction of the physical mechanisms associated with the different deformation modes are often carried out through monotonic or cyclic tests according to different load paths (tensile, compression, shear). These tests, commonly performed on standardized specimens, are destructive and practically unusable for the characterization of mechanical parts having complex geometries and dimensions ranging from the micron scale to the metric scale. In addition, the prediction of fatigue life of mechanical component requires the determination of the local material behavior of layers affected by processes or treatments which is not possible using classical characterization tests (Zouhayar et al. 2013; Yahyaoui et al. 2015; Sidhom et al. 2014a, b; Ben Moussa et al. 2014a, b). Instrumented indentation, which appeared in the 1980s, is an interesting alternative for characterizing the mechanical and tribological behavior of parts and coatings. The analysis of the load-displacement curve resulting from the test allows,

A. Benamara et al. (Eds.): CoTuMe 2018, LNME, pp. 54–61, 2019.
https://doi.org/10.1007/978-3-030-19781-0_7

using analytical models such as Oliver and Pharr model (Oliver and Pharr 1992), the most conventionally used, to determine the average modulus of elasticity and the average hardness of the material. Nevertheless, it remains difficult to exploit instrumented indentation results to identify the coefficients of models like Chaboche, Ludwig, Hollomon, Voce, etc. commonly used to describe the material behavior law. Several studies have been conducted to identify the behavior of the material from the indentation test. Most of the proposed methods are limited to the case of spherical indenters and are based on the estimation of the indentation representative strain and the ratio of plastic and total energies (Dao et al. 2001). Beghini et al. (Beghini et al. 2006) determined the constitutive law parameters by the inversion of the P(h) curve resulting from a spherical indentation test. Authors assumed that the elastic properties of the material are known and their method led to a good approximation of some metallic materials parameters. Concerning the problem of the solution uniqueness in indentation test, Cheng and Cheng (Cheng and Cheng 2004) proposed a very complete study of the indentation and more particularly the possibility of determining a material's behavior law from an indentation test. Bucaille et al. (Bucaille et al. 2003) proposed a method to determine the tensile curve parameters by using two different conical indenters. In this context Nakamura et al. (Nakamura and Gu 2007) used an inverse analysis based on the Kalman filter technique and two indenters tips in order to obtain as much information about an anisotropic material's response. Although the robustness of their method, the error on the estimation of the hardening coefficient and their hypothesis of its compensation with the yield stress value, presents a limitation of their work. In the current study, a new procedure for identifying an elastoplastic constitutive law's parameters from the instrumented indentation test was established. This procedure is based on a numerical simulation of the instrumented indentation test and Hooke-Jeeves pattern search technique. The procedure ability to predict the local material behavior is validated using an aluminum alloy sample then it was applied to the case of a stainless steel sample machined by electrical discharge.

2 Numerical Study

The determination of the constitutive law coefficients using inverse methods requires performing a large number of indentation simulations before finding the correct values minimizing the differences between the experimental and the simulation results. This requires a significantly reduce of computing time significantly without affecting results quality. One of the alternatives commonly used in numerical simulation based on finite element method, is to take advantage of the problem symmetries to switch from the three-dimensional model to a bi-dimensional configuration. For indentation simulation, Vickers and Berkovich indenter can be assimilated to a conical geometry to make the problem axisymmetric (Fig. 1). The determination of the equivalent conical indenter and more particularly the semi-apex angle was carried out in this work by conducting several simulations with a semi-apex angle variation in 2D models until the results were consistent with those of the 3D models. Since both indenters have similar projected surfaces and generate the same representative strain, an equivalent semi-apex angle of 70.32° is retained.

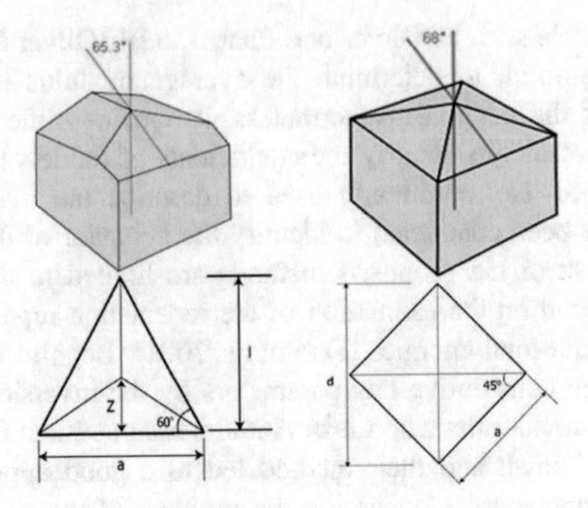

Fig. 1. Berkovich and Vickers indenters

The indenters used are generally made of diamond or sapphire having high mechanical characteristics ($E_i = 1140$ GPa and $\nu_i = 0.07$) compared to the sample material, which justifies the definition of the indenter as a rigid solid. This avoids unnecessary calculations of deformations in the indenter. The specimen behavior law was described in this work by Ludwig's isotropic elastoplastic model linking true stress σ to true strain ε by the relation:

$$\sigma = \sigma_y + k.\varepsilon^n \quad (1)$$

The determination of the elements' mesh size is a crucial step in order to compromise speed and calculation precision. Within this framework we chose a structured quadratic mesh refined of 0.1 µm in the contact zone (under the indenter) and a mesh with quadratic elements dominant in the rest of the geometry (Fig. 2a).

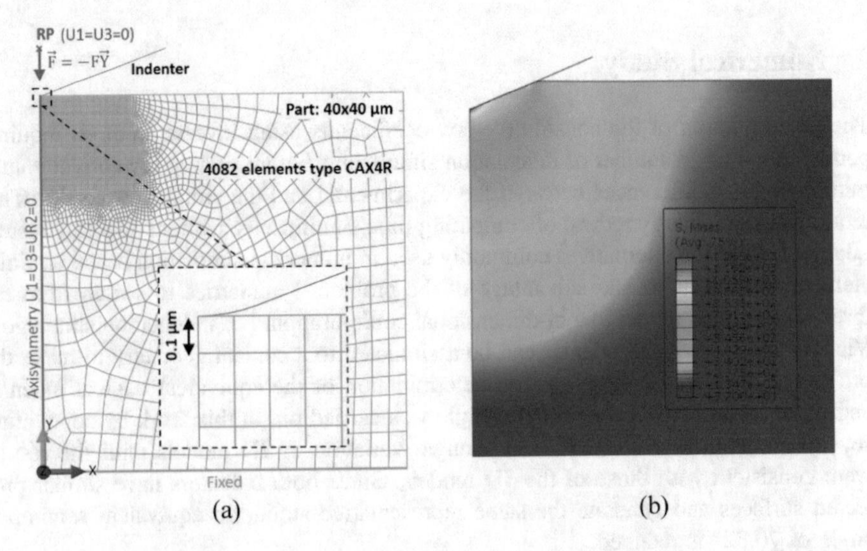

Fig. 2. Instrumented indentation simulation, a: model details, b: Indentation von Mises stress

3 Parameters Identification Procedure

3.1 Inverse Analysis Technique

An inverse analysis procedure is proposed in this work to identify the material's parameters from the load-displacement curve (P-h) obtained during the loading and unloading phases of an instrumented indentation test (Fig. 3). This procedure consists in conducting a first simulation of instrumented indentation with an initial combination of material parameters saved in an "inp" data file. The calculated load displacement curve of the node located at the indenter tip is written in an "odb" file. This curve is read by an optimizer tool and compared to the experimental load displacement curve. Based on the differences between the experimental and calculated curves and according to Hooke and Jeeves pattern search technique (Hooke and Jeeves 1961), the optimizer tool tries a new combination of material parameters by updating the input data file and executes a new instrumented indentation simulation.

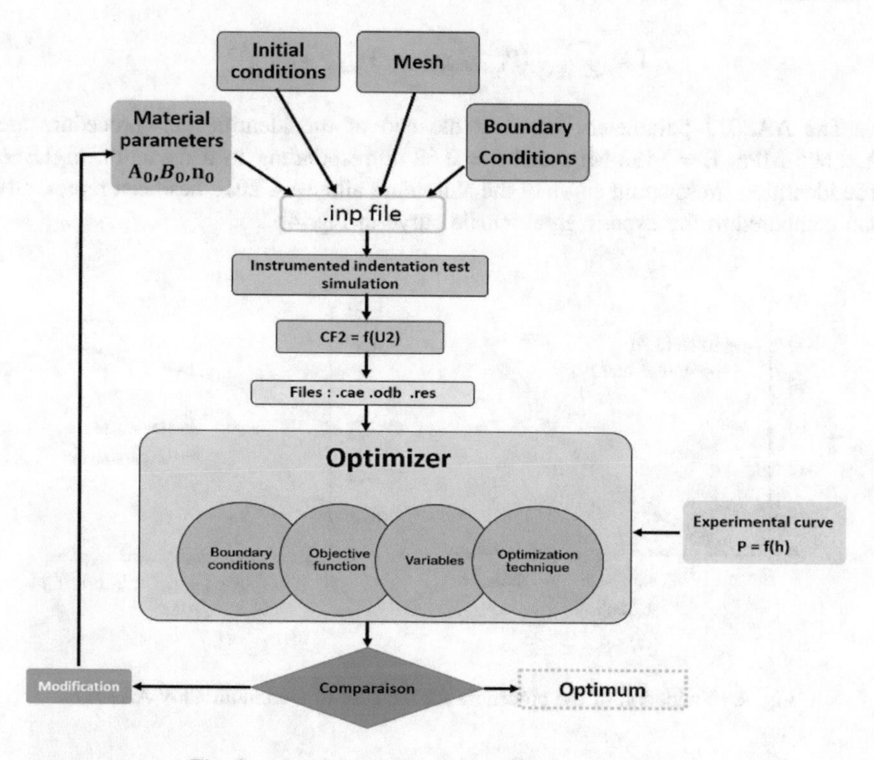

Fig. 3. Material parameters identification procedure

The procedure is repeated automatically until reaching the optimal material coefficients, minimizing the differences between the numerical and experimental results.

3.2 Validation of the Proposed Procedure

The evaluation of the proposed methodology ability to predict correctly the material's behavior from instrumented indentation test, is performed by trying to find out the parameters A, B and n of an aluminum alloy. An AA2017 aluminum prismatic sample with dimensions 20 × 20 mm was mechanically polished with emery papers grad 600, 1000, 2400 and 4000 to remove the layers affected by anterior machining and retrieve the bulk material. Subsequently, an instrumented indentation test with an imposed load of 100 mN using a Berkovich indenter was performed on the sample (Fig. 4a). The numerical simulation of the indentation test was performed with an initial combination of the coefficients (A = 623, B = 1436 and n = 0.66). After a series of simulations performed automatically using the proposed procedure and during which the differences between the calculated and experimental load-displacement curves have been minimized, the values of the behavior model's parameters found to be close to those obtained by tensile test. The optimality criterion chosen is the sum of the squared differences defined as follow:

$$f = \sum_{i=1}^{n} \left(P_{i\_simulation} - P_{i\_experimental}\right)^2 \quad (1)$$

The AA2017 parameters found at the end of the identification procedure are: A = 605 MPa, B = 1452 MPa and n = 0.62 corresponding to a deviation of 1E−6. The identified stress-strain curve of the aluminum alloy AA 2017 has been represented and compared to the experimental tensile curve in Fig. 4b.

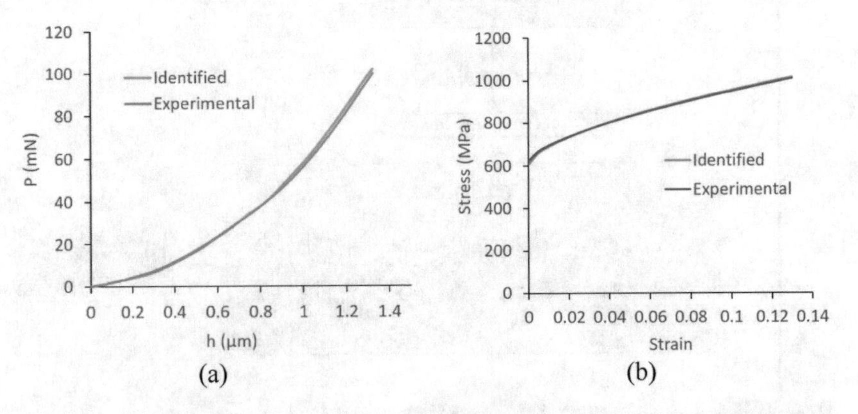

Fig. 4. Validation of the procedure for the case of aluminum alloy AA2017

4 Application

After the validation of the aptitude of the proposed procedure to predict the mechanical behavior law of aluminum sample, the same procedure is used to identify the effect of residual stress and hardening induced by electrical discharge machining (EDM). An AISI 316L stainless steel prismatic sample was machined by EDM sinking using a graphite tool. The intensity, pulse duration and pulse-off time duration of discharge are 5A, 26 μs and 5 μs, respectively. The chemical composition and the mechanical properties of the AISI 316L stainless steel are presented in Table 1.

Table 1. Chemical composition and mechanical properties of the AISI 316L stainless steel.

C	N	Cr	Ni	Mo	Mn	Si	Cu	Co	S	P	Fe
0.018	0.078	16.6	10.2	2.02	1.84	0.38	0.36	0.18	0.029	0.035	Balance

Elastic modulus	Ultimate strength	Yield stress	Elongation	Hardness
E=210.3 GPa	Rm= 578 MPa	$Rp_{0.2}$=302 MPa	A=63%	$HV_{0.05}$=232

In order to characterize the local behavior of the surface layers affected by the EDM process, instrumented indentation tests were carried out at different depths over a thickness of 200 μm. The load-displacement curves P = f (h) for seven depths acquired at a frequency of 10 Hz are presented in Fig. 5a. The differences between the load-displacement curves reveal the modification of the material local behavior resulting from the residual stress and hardening distributions induced by the EDM process. The local material behavior law for each layer is identified by the determination of Ludwig model coefficients using the proposed methodology from the indentation curves and plotted in Fig. 5b. It has been demonstrated that EDM process leads to a hardening resulting in an increase of yield stress with a maximum value at the surface. This effect decreases in deeper layers until reaching the bulk material. The tensile residual stresses induced by EDM process act in the same direction since the material is subject to compressive stress during instrumented indentation.

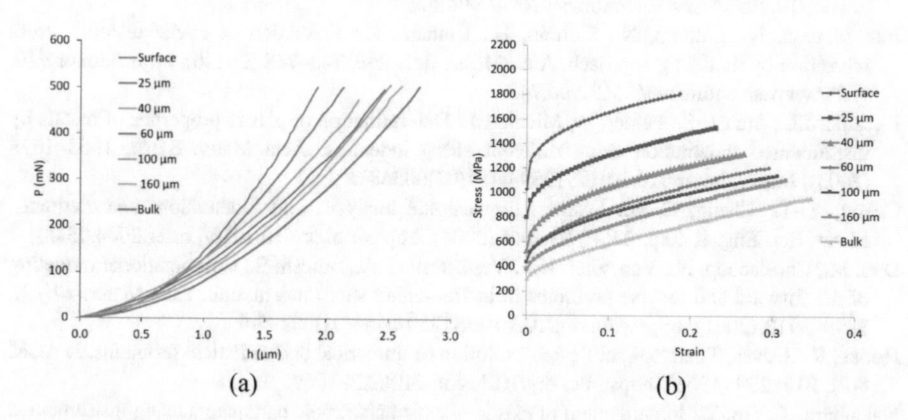

Fig. 5. Application to electro eroded sample, a: experimental indentation curves, b: identified tensile curves

5 Conclusion

In this work, a methodology of identification of local material behavior from instrumented indentation test is proposed. This methodology is based on a numerical simulation of the indentation test coupled to Hooke and Jeeves iterative optimization method. The numerical simulation of the instrumented test was optimized by determining the semi-apex angle of equivalent conical indenter making the problem axisymmetric and by testing several configurations and mesh sizes in order to reduce computation times while improving results precision.

The validation of the methodology accuracy is performed by identifying the tensile curve of aluminum alloy AA2017 from an instrumented indentation test. For the case on AISI 316L sample, the identification of the local material behavior of different surface layers highlighted the effects of hardening and residual stress distribution induced by the EDM process.

The identification of the local material behavior of each affected layers is very useful to enhance the predictive aptitude of fatigue life models by considering the correct behavior resulting from process and surface treatment.

References

Beghini, M., Bertini, L., Fontanari, V.: Evaluation of the stress–strain curve of metallic materials by spherical indentation. Int. J. Solids Struct. **43**(7), 2441–2459 (2006). https://doi.org/10.1016/j.ijsolstr.2005.06.068

Ben Moussa, N., Al-Adel, Z., Sidhom, H., Braham, C.: Numerical assessment of residual stress induced by machining of aluminum alloy. Adv. Mater. Res. **996**, 628–633 (2014a). https://doi.org/10.4028/www.scientific.net/AMR.996.628

Ben Moussa, N., Sidhom, N., Sidhom, H., Braham, C.: Prediction of cyclic residual stress relaxation by modeling approach. Adv. Mater. Res. **996**, 743–748 (2014b). https://doi.org/10.4028/www.scientific.net/AMR.996.743

Bucaille, J.L., Stauss, S., Felder, E., Michler, J.: Determination of plastic properties of metals by instrumented indentation using different sharp indenters. Acta Mater. **51**(6), 1663–1678 (2003). https://doi.org/10.1016/S1359-6454(02)00568-2

Cheng, Y.-T., Cheng, C.-M.: Scaling, dimensional analysis, and indentation measurements. Mater. Sci. Eng. R Rep. **44**(4), 91–149 (2004). https://doi.org/10.1016/j.mser.2004.05.001

Dao, M., Chollacoop, N., Van Vliet, K.J., Venkatesh, T.A., Suresh, S.: Computational modeling of the forward and reverse problems in instrumented sharp indentation. Acta Mater. **49**(19), 3899–3918 (2001). https://doi.org/10.1016/S1359-6454(01)00295-6

Hooke, R., Jeeves, T.A.: "Direct Search" solution of numerical and statistical problems. J. ACM **8**(2), 212–229 (1961). https://doi.org/10.1145/321062.321069

Nakamura, T., Gu, Y.: Identification of elastic–plastic anisotropic parameters using instrumented indentation and inverse analysis. Mech. Mater. **39**(4), 340–356 (2007). https://doi.org/10.1016/j.mechmat.2006.06.004

Oliver, W.C., Pharr, G.M.: An improved technique for determining hardness and elastic modulus using load and displacement sensing indentation experiments, vol. 7 (1992). https://doi.org/10.1557/jmr.1992.1564

Sidhom, H., Ben Moussa, N., Ben Fathallah, B., Sidhom, N., Braham, C.: Effect of surface properties on the fatigue life of manufactured parts: experimental analysis and multi-axial criteria. Adv. Mater. Res. **996**, 715–721 (2014a). https://doi.org/10.4028/www.scientific.net/AMR.996.715

Sidhom, N., Moussa, N.B., Janeb, S., Braham, C., Sidhom, H.: Potential fatigue strength improvement of AA 5083-H111 notched parts by wire brush hammering: experimental analysis and numerical simulation. Mater. Des. **64**, 503–519 (2014b). https://doi.org/10.1016/j.matdes.2014.08.002

Yahyaoui, H., Ben Moussa, N., Braham, C., Ben Fredj, N., Sidhom, H.: Role of machining defects and residual stress on the AISI 304 fatigue crack nucleation. Fatigue Fract. Eng. Mater. Struct. **38**(4), 420–433 (2015). https://doi.org/10.1111/ffe.12243

Zouhayar, A.-A., Naoufel, B.M., Houda, Y., Habib, S.: Surface integrity after orthogonal cutting of aeronautical aluminum alloy 7075-T651. In: Design and Modeling of Mechanical Systems, pp. 485–492. Springer, Heidelberg (2013)

Finite Element Simulation of Single Point Incremental Forming Process of Aluminum Sheet Based on Non-associated Flow Rule

Abir Bouhamed[2(✉)], Hanen Jrad[2], Lotfi Ben Said[2,3], Mondher Wali[1,2], and Fakhreddine Dammak[2]

[1] Department of Mechanical Engineering, College of Engineering, King Khalid University, Abha, Saudi Arabia
mondherwali@yahoo.fr

[2] Laboratory of Electromechanical Systems (LASEM), National Engineering School of Sfax, University of Sfax, Route de Soukra km 4, 3038 Sfax, Tunisia
{abir.bouhamed, Fakhreddine.dammak}@enis.tn, hanen.j@gmail.com, bensaidrmq@gmail.com

[3] Mechanical Engineering Department, College of Engineering, University of Hail, Hail, Saudi Arabia

Abstract. This paper presents a numerical simulation of single point incremental forming (SPIF) process. An elasto-plastic constitutive model with quadratic yield criterion of Hill'48 based on a combination of the non-associated flow rule theory and mixed isotropic-nonlinear kinematic hardening behavior has been implemented in user material subroutine (VUMAT) to describe the behavior of sheet metal during SPIF process. The simulation results included the variation of the thickness along transverse direction and the localization of the major Von Mises stress.

Keywords: Single point incremental forming · Non-associated flow rule · Mixed hardening material model · Elasto-plastic model

1 Introduction

Single Point Incremental Forming (SPIF) is an emerging manufacturing process, which is identified as a potential and economically viable process for sheet metal prototypes and small batch production. This process has attracted an increasing interest in the field of sheet metal forming in the past decades due to its unique advantages, including process flexibility, reduced tooling costs and increased material formability (Lu et al. 2015). During the SPIF process, a flat sheet is incrementally deformed into a desired shape by the action of a tool that follows a defined tool path conforming to the final part geometry (Behera et al. 2017).

Several researchers have an increasing attention on finite element methods which are widely used to develop the manufacturing process and to investigate the effects of process parameters, taking into account of different material behaviors proposed for assessing the formability.

A. Benamara et al. (Eds.): CoTuMe 2018, LNME, pp. 62–68, 2019.
https://doi.org/10.1007/978-3-030-19781-0_8

Han et al. (2013) controlled the springback to design an accurate tool path for ISF using a three-dimensional elasto-plastic finite element model based on the particle swarm optimization neural network which has illustrated a good result in the prediction of springback.

Malhotra et al. (2010) have proposed several material behaviors and element formulations to simulate the SPIF process. It was demonstrated that the material model represents a vital part of sheet metal forming simulations, which are captured varied phenomena occurring during plastic deformation, like anisotropic yielding, nonlinear isotropic hardening and kinematic hardening.

Robert et al. (2012) have investigated a new algorithm based on incremental deformation to simulate elasto-plastic material behavior with anisotropic plasticity criteria, in order to analyze the stress state and thickness distribution of sheet metal studied without taking into account the kinematic hardening in the material behavior.

Ben Said et al. (2016) developed an elasto-plastic constitutive model without kinematic hardening to optimize the SPIF process and to see which tool path strategy makes this process more effective. This formulation, based on the associated flow rule, was coupled with isotropic ductile damage and kinematic hardening to predict the ductile damage in SPIF process presented in Ben Said et al. (2017).

In this paper, finite element model is developed to describe the behavior of blank sheet during single point incremental forming (SPIF) process by taking into account the prediction of both anisotropy and mixed isotropic/kinematic hardening behaviors based on non-associated flow rule and on Hill yield criterion.

2 Elasto-Plastic Constitutive Equations Based on Non-associated Flow Rule

The objective of this section is to present an elasto-plastic model to describe the behavior of the blank sheet during the incremental sheet metal forming process. The non-associated model improves the prediction of both anisotropy yielding and non-linear mixed isotropic/kinematic hardening, even though rather simple quadratic constitutive equations were used.

– Decomposition of the total strain

$$\boldsymbol{\varepsilon} = \boldsymbol{\varepsilon}^{e} + \boldsymbol{\varepsilon}^{p} \tag{1}$$

– Hook's law

$$\boldsymbol{\sigma} = \boldsymbol{D} : \boldsymbol{\varepsilon}^{e} \tag{2}$$

– Back-stress tensor

$$\boldsymbol{X}_k = a_k \boldsymbol{\alpha}_k \tag{3}$$

– Yield function

$$f = \varphi_f(\boldsymbol{\sigma}) - \sigma_P \leq 0 \tag{4}$$

$$\varphi_f(\boldsymbol{\sigma}) = \|\boldsymbol{\sigma}\|_{\boldsymbol{P}} = \sqrt{\boldsymbol{\sigma}^t \boldsymbol{P} \boldsymbol{\sigma}} \tag{5}$$

$$\sigma_P = \sigma_Y + R(r)\,;R(r) = Q\left(1 - e^{-\beta r}\right) \tag{6}$$

– Plastic potential function

$$F = \varphi_F(\boldsymbol{\sigma}) - \sigma_P + \frac{1}{2}\sum_{k=1}^{M} \frac{b_k}{a_k} \boldsymbol{X}_k : \boldsymbol{X}_k \tag{7}$$

$$\varphi_F(\boldsymbol{\sigma}) = \|\boldsymbol{\sigma}\|_{\boldsymbol{Q}} = \sqrt{\boldsymbol{\sigma}^t \boldsymbol{Q} \boldsymbol{\sigma}} \tag{8}$$

– Flow rule

$$\dot{\boldsymbol{\varepsilon}}^p = \dot{\gamma}\frac{\partial F}{\partial \boldsymbol{\sigma}} = \dot{\gamma}\boldsymbol{n};\ \boldsymbol{n} = \frac{1}{\varphi_F}\boldsymbol{Q}\boldsymbol{\sigma} \tag{9}$$

– Kinematic hardening

$$\dot{\boldsymbol{X}}_k = a_k \dot{\boldsymbol{\varepsilon}}^p - b_k \dot{\gamma} \boldsymbol{X}_k \tag{10}$$

– Isotropic hardening

$$\dot{r} = \dot{\gamma} \tag{11}$$

where $\boldsymbol{\varepsilon}^e$ and $\boldsymbol{\varepsilon}^p$ are the elastic and plastic strain tensor respectively, $\boldsymbol{\sigma}$ is the stress tensor, $\boldsymbol{D}$ is the elastic stiffness tensor, $\boldsymbol{\alpha}_k$ and $\boldsymbol{X}_k$ are the kinematic hardening variables, σ_Y is the initial yield stress, r and R are the isotropic hardening variables, a_k, b_k, β and Q are material parameters, $\dot{\gamma}$ is the plastic multiplier and finally, $\boldsymbol{P}$ is a fourth-order tensor presenting the initial plastic anisotropy of the yield function and $\boldsymbol{Q}$ is the fourth-order tensor representing the initial anisotropy of the plastic potential function

$$\boldsymbol{P} = \boldsymbol{P}(F, G, H, N, M, L) = \begin{bmatrix} H+G & -H & -G & 0 & 0 & 0 \\ & H+F & -F & 0 & 0 & 0 \\ & & F+G & 0 & 0 & 0 \\ & & & 2N & 0 & 0 \\ & Sym & & & 2M & 0 \\ & & & & & 2L \end{bmatrix} \tag{12}$$

$$\boldsymbol{Q} = \boldsymbol{Q}\left(F', G', H', N', M', L'\right) \tag{13}$$

The parameters F, G, H, N, M, L, F', G', H', N', M' and L' are the anisotropy of the material and can be obtained by tests of the material in different orientations. r_0, r_{45} and r_{90}. The details of the formulation and numerical implementation of the present model can be found in Bouhamed et al. (2019).

3 Numerical Results

A numerical results set of single point incremental forming (SPIF) process are presented to show the applicability and versatility of the proposed non-associated model. The presented FE model was implemented in a user-defined subroutine (VUMAT) for ABAQUS/Explicit to conduct the simulation of ISF process of the cone. The material considered in this study was AA6022-T43 aluminum alloy sheet having a thickness of 1 mm. Its mechanical properties are listed in Table 1. The punch tool has a diameter of 10 mm and considered as discretized rigid body. The blank sheet is meshed using S3R shell element with five integration points through the thickness. The friction coefficient between the blank and the punch is 0.1 to describe the contact condition (Ben Ayed et al. 2014). The forming strategy during the simulation is depicted in Fig. 1.

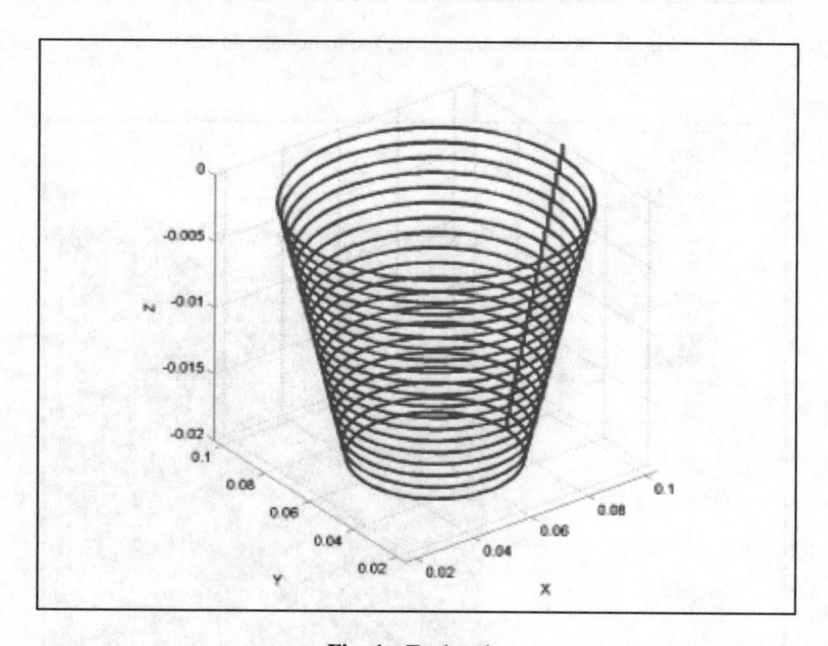

Fig. 1. Tool path.

Numerical results presented in this section are carried out using FE simulations of SPIF process based on the proposed non-associated model incorporated by utilizing yield function and plastic potential as separated functions.

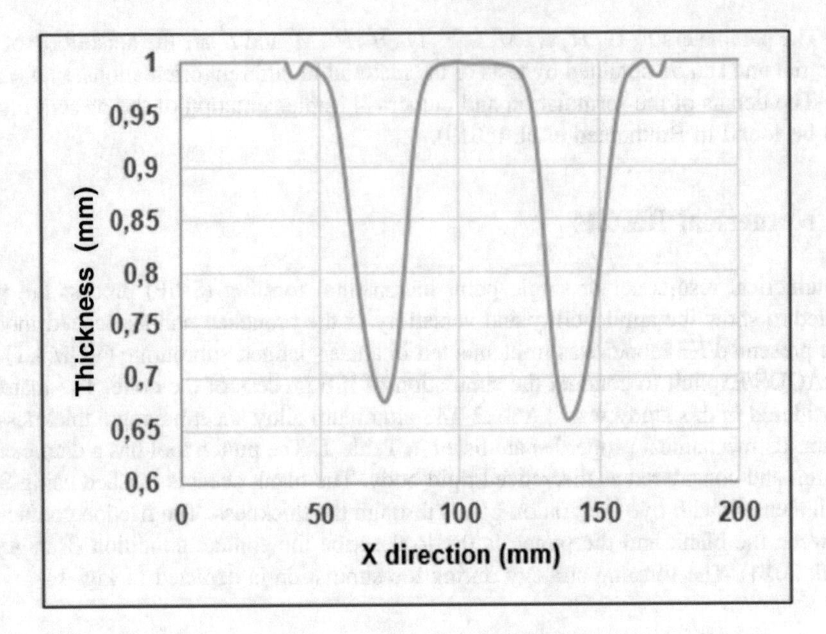

Fig. 2. Thickness strain along the transverse direction.

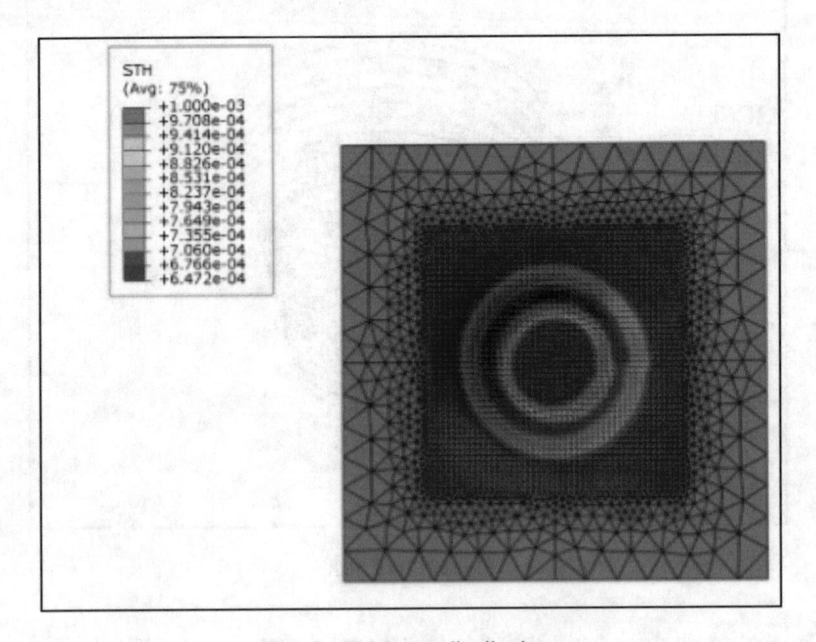

Fig. 3. Thickness distribution.

The evolution of the thickness strain along the transverse direction is illustrated in Fig. 2. From Figs. 2 and 3, it is noted that the thickness on the upper and the lower surfaces of the manufactured part are not influenced, however, the maximum of thinning is located mainly in the vicinity of punch path.

Table 1. Material properties of AA6022-T43 (Brem et al. 2005, Wali et al. 2016)

Young modulus (GPa)	70				
Poisson's ratio	0.33				
Anisotropic coefficients	Yield function coefficients	F	G	H	N
		0.632	0.496	0.504	1.585
	Potential function coefficients	F'	G'	H'	N'
		0.697	0.493	0.507	1.228
Isotropic hardening (MPa)	$\sigma_Y + R(r) = 136 + 110(1 - e^{-7.5r})$				
Kinematic hardening	$a = 1400\,\text{MPa};\ b = 20$				

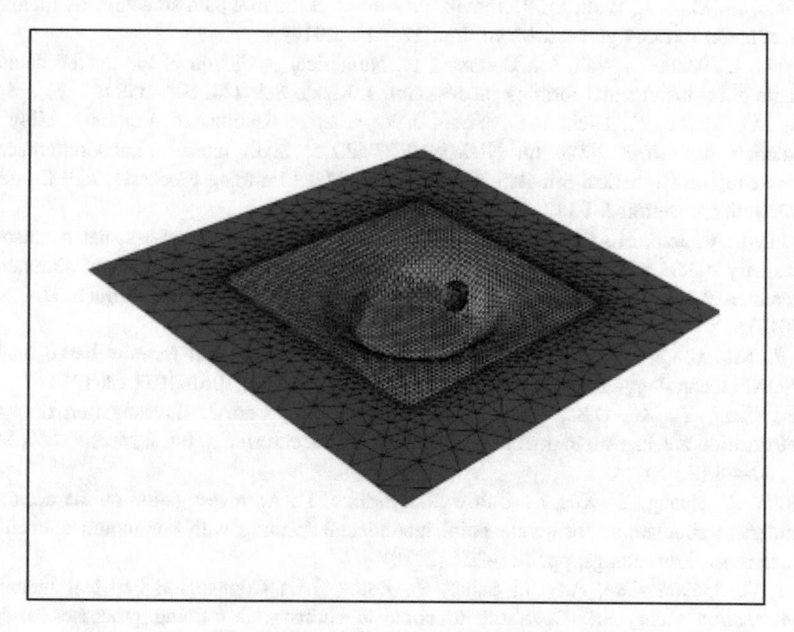

Fig. 4. Von Mises stress distribution.

According to the distribution of the Von Mises stress, Fig. 4 illustrate that the major Von Mises stress located principally along the punch path and minor deformations are located on the borders near to the clamped shape.

4 Conclusion

In this research work, a non-associated flow rule, combined with a mixed non-linear isotropic/kinematic hardening, is implemented on user-defined material subroutine (VUMAT) to simulate an example of single point incremental sheet metal forming. The numerical results illustrate that the major deformations have been located along the contour of the cone. This work mainly emphasizes the advantages of the presented non-associated model to predict more aspects of anisotropy in sheet metal forming.

References

Behera, A.K., de Sousa, R.A., Ingarao, G., Oleksik, V.: Single point incremental forming: an assessment of the progress and technology trends from 2005 to 2015. J. Manuf. Process. **27**, 37–62 (2017)

Ben Ayed, L., Robert, C., Delamézière, A., Nouari, M., Batoz, J.L.: Simplified numerical approach for incremental sheet metal forming process. Eng. Struct. **62–63**, 75–86 (2014)

Ben Said, L., Mars, J., Wali, M., Dammak, F.: Effects of the tool path strategies on incremental sheet metal forming process. Mech. Ind. **17**, 411 (2016)

Ben Said, L., Mars, J., Wali, M., Dammak, F.: Numerical prediction of the ductile damage in single point incremental forming process. Int. J. Mech. Sci. **131**, 546–558 (2017)

Brem, J.C., Barlat, F., Dick, R.E., Yoon, J.W.: Characterizations of aluminum alloy sheet materials numisheet 2005. In: NUMISHEET 2005: Sixth International Conference and Workshop on Numerical Simulation of 3D Sheet Metal Forming Processes, AIP Conference Proceedings, Detroit MI 117–1190 (2005)

Bouhamed, A., Jrad, H., Ben Said, L., Wali, M., Dammak, F.: A non-associated anisotropic plasticity model with mixed isotropic-kinematic hardening for finite element simulation of incremental sheet metal forming process. Int. J. Adv. Manuf. Technol. **100**(1–4), 929–940 (2019)

Han, F., Mo, J., Qi, H.: Springback prediction for incremental sheet forming based on FEM-PSONN technology. Trans. Nonferrous Met. Soc. China **23**, 1061–1071 (2013)

Lu, B., Fang, Y., Xu, D.K., Chen, J., Ai, S., Long, H., Cao, J.: Investigation of material deformation mechanism in double side incremental sheet forming. Int. J. Mach. Tools Manuf. **93**, 37–48 (2015)

Malhotra, R., Huang, Y., Xue, L., Cao, J., Belytschko, T.: An investigation on the accuracy of numerical simulations for single point incremental forming with continuum elements. In: Conference Proceedings, pp. 221–227 (2010)

Robert, C., Delamézière, A., Dal Santo, P., Batoz, J.L.: Comparison between incremental deformation theory and flow rule to simulate sheet-metal forming processes. J. Mater. Process. Technol. **212**, 1123–1131 (2012)

Wali, M., Autay, R., Mars, J., Dammak, F.: A simple integration algorithm for a non-associated anisotropic plasticity model for sheet metal forming. Int. J. Numer. Meth. Eng. **107**, 183–204 (2016)

Dispersive Waves in 2D Second Gradient Continuum Media

Yosra Rahali, Hilal Reda, and Jean-François Ganghoffer(✉)

LEMTA, University of Lorraine, Vandoeuvre-les-Nancy, France
rahali.yosra@gmail.com, hilal_reda@hotmail.com,
jean-francois.Ganghoffer@univ-lorraine.fr

Abstract. We analyze the dispersion of elastic waves in periodic beam networks based on second order gradient models obtained by the homogenization of the initially discrete network, relying on the discrete asymptotic method extended up to the second gradient of the displacement. The lattice beams have a viscoelastic behavior described by Kelvin-Voigt model and the homogenized second gradient viscoelasticity model reflects both the initial lattice topology, anisotropy and microstructural features in terms of its geometrical and micromechanical parameters. The continuum models enriched with the higher-order gradients of the displacement and velocity introduce characteristic lengths parameters which account for microstructural effects at the mesoscopic level. A study of the dispersion relations and damping ratio evolutions for the longitudinal and shear waves has been done for the reentrant lattice. An important increase of the natural frequency due to second order effects is observed.

Keywords: Second-gradient models · Homogenization · Dispersive waves · Periodic lattices

1 Introduction

New classes of cellular solids and lattice materials have over the last decade found a wide range of engineering applications, such as light-weight structures, vibration control devices, systems for energy absorption, relying on the fact that such lattice-like materials enhance the static and dynamic responses in comparison to their solid counterpart. This improvement of the properties depends on the bulk material of the lattice, its relative density, and the internal geometrical structure. Periodic lattices can be considered as prototype models of many systems whose description can be simplified as assemblies of beam elements rigidly connected or joined by hinges. The dynamic behavior of such periodic network has raised the interest of many researchers, especially due to the use of non-destructive techniques for accessing mechanical properties of the investigated material. Due to the prohibitive cost of computing the dynamical response of periodic networks including many elements, it proves more economical to represent the network materials on the macroscale as an equivalent homogeneous material obtained from the homogenization over a suitable unit cell consisting of a rigid joint network of beams.

A. Benamara et al. (Eds.): CoTuMe 2018, LNME, pp. 69–76, 2019.
https://doi.org/10.1007/978-3-030-19781-0_9

The homogenized moduli contain information about the microstructure, although in an average sense. Homogenized models based on classical Cauchy-type elasticity theory are however not able to provide realistic predictions of many effects arising from small scale, amongst of them wave dispersion. Classical theories based upon the sole first order displacement gradient lack indeed internal length parameters, characteristic of the underlying microstructure. This explains the success of gradient-enriched theories in capturing microstructural effect on the macroscopic behavior of materials, by including high-order gradients associated to internal lengths representative of the microstructure.

Generalized continuum theories have been shown to offer an attractive alternative for capturing dynamic behaviors overlooked by classical elasticity, especially dispersion relations, (see Lombardo and Askes 2011) and references therein. Applications of gradient elasticity in dynamics have fostered extensive research (Ostoja-Starzewki 2002; Andrianov et al. 2010; Askes and Aifantis 2011).

Although a wide body of research has been devoted to gradient-enriched theories for both elasticity and phenomena described by internal variables, gradient viscoelasticity theories have deserved much less interest in the literature.

The outline of the present work is as follows: the homogenized viscoelastic behavior of repetitive planar lattices consisting of viscoelastic Kelvin-Voigt type beams is determined in Sect. 2, based on an equivalence between the writing of the principle of virtual work for the lattice and the posited second-gradient continuum. The constitutive relation for general repetitive lattices exhibiting arbitrary anisotropy is also expanded in matrix format in Sect. 2, based on the introduction of stress and hyperstress vectors reflecting the lattice topology and microstructural parameters. The effective constitutive laws are next introduced into the dynamical planar equilibrium equations (Sect. 3). The dispersion relations and damping ratio evolutions versus the wave number are evaluated for the re-entrant lattice in Sect. 4, and the phase velocity for both longitudinal and shear waves in Sect. 5. We conclude in Sect. 5 by a summary of the main results.

2 Homogenized Viscoelastic Second Gradient Behavior of Periodic Beam Lattice

2.1 Discrete Homogenization Method

The discrete homogenization method used for the prediction of the effective mechanical properties of reentrant lattice requires the development of all geometrical variables and kinematic variables as Taylor series expansions versus a small parameter defined as the ratio of unit cell size to a macroscopic length characteristic of the entire lattice. These expansions are thereafter inserted into the equilibrium equation of forces and moments (exerted on the 2D Bernoulli beam extremities) written at the nodes and expressed in weak form. After resolution of the unknown displacements in the localization problem posed over the identified reference unit cell, the stress and hyperstress tensors are constructed versus their conjugated kinematic variables, respectively the first and second deformation gradients, thereby defining the homogenized constitutive law; this allows identifying the effective moduli for the equivalent continuum. We refer the reader to (Reda et al. 2016) for more details related to the different steps of the method.

The constitutive law for a homogeneous anisotropic viscoelastic second order grade continuum, written in index format:

$$\begin{aligned}\{\sigma\} &= \underbrace{[A^e]\{\epsilon\} + [B^e]\{\kappa\}}_{elastic\ part} + \underbrace{[A^v]\{\dot{\epsilon}\} + [B^v]\{\dot{\kappa}\}}_{viscous\ part} \\ \{S\} &= \underbrace{[B^e]\{\epsilon\} + [D^e]\{\kappa\}}_{elastic\ part} + \underbrace{[B^v]\{\dot{\epsilon}\} + [D^v]\{\dot{\kappa}\}}_{viscous\ part}\end{aligned} \tag{1}$$

With σ_{ij}, S_{ijk}, ϵ_{pq}, κ_{pqr}, $\dot{\epsilon}_{pq}$, $\dot{\kappa}_{pqr}$ successively the stress and hyperstress tensors, and their conjugated kinematic quantities, namely the first and second deformation gradients and their time derivatives, the first and second deformation velocity gradients.

The constitutive tensors $A^e_{ijpq}, D^e_{ijkpqr}, B^e_{pqrij}, A^v_{ijpq}, D^v_{ijkpqr}, B^v_{pqrij}$ therein are respectively the first and second order elasticity and viscosity coefficients, the coupling moduli, which all depend on the specific considered lattices.

We represent in Fig. 1, the selected lattice in this study.

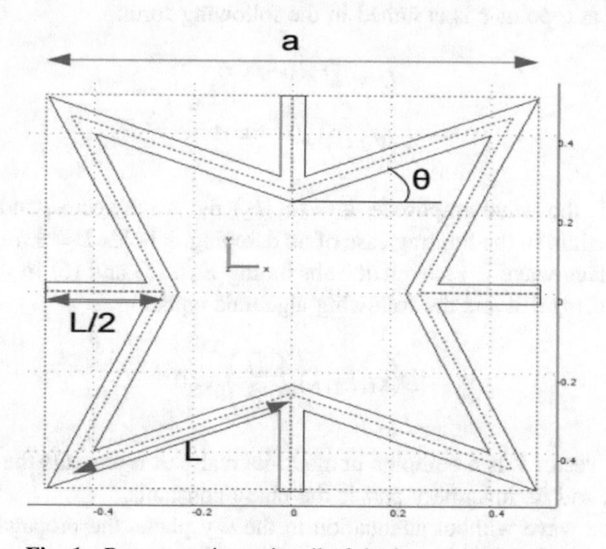

Fig. 1. Representative unit cell of the investigated lattices

For centro-symmetrical lattices the pseudo-tensors $[B^e]$ and $[B^v]$ vanishes. This leads to

$$\begin{aligned}\{\sigma\} &= \underbrace{[A^e]\{\epsilon\}}_{elastic\ part} + \underbrace{[A^v]\{\dot{\epsilon}\}}_{viscous\ part} \\ \{S\} &= \underbrace{[D^e]\{\kappa\}}_{elastic\ part} + \underbrace{[D^v]\{\dot{\kappa}\}}_{viscous\ part}\end{aligned} \tag{2}$$

3 Dynamical Equilibrium and Characteristic Equation

For any homogenized 2D viscoelastic lattice, the equations of motion for a second gradient medium, write in components form as the two following differential equations along the x and y directions of a Cartesian coordinates system,

$$\left(\begin{array}{l}\left(\frac{\partial\sigma_{11}}{x_1}+\frac{\partial\sigma_{12}}{x_2}\right)-\frac{\partial^2 S_{111}}{x_1x_1}-\frac{\partial^2 S_{112}}{x_1x_2}\\ -\frac{\partial^2 S_{121}}{x_2x_1}-\frac{\partial^2 S_{122}}{x_2x_2}\end{array}\right)=\rho^*\ddot{u} \tag{3}$$

$$\left(\begin{array}{l}\left(\frac{\partial\sigma_{21}}{x_1}+\frac{\partial\sigma_{22}}{x_2}\right)-\frac{\partial^2 S_{211}}{x_1x_1}-\frac{\partial^2 S_{212}}{x_1x_2}\\ -\frac{\partial^2 S_{221}}{x_2x_1}-\frac{\partial^2 S_{222}}{x_2x_2}\end{array}\right)=\rho^*\ddot{v} \tag{4}$$

Here, $\ddot{u}$ and $\ddot{v}$ are the horizontal and vertical components of the acceleration vector. The effective density therein is given in general by $\rho^*=\frac{M1}{A_{cell}}$, with M_1 the mass of the set of lattice beams, A_{cell} being the area of the periodic cell.

For a harmonic wave propagating the generalized displacement field with components U, V at a point $\mathbf{r}$ is assumed in the following form:

$$U=\widehat{U}\,e^{(\lambda t-i\mathbf{k}\cdot\mathbf{r})}, \tag{5}$$

$$V=\widehat{V}\,e^{(\lambda t-i\mathbf{k}\cdot\mathbf{r})} \tag{6}$$

where $\widehat{U}$, $\widehat{V}$ is the wave amplitude, $\mathbf{k}=(k_1,k_2)$ the wave vector, and λ a complex frequency function. In the limiting case of no damping, it holds $\lambda=\pm i\omega$, and the usual form of the plane wave is recovered. Substituting Eqs. (5) and (6) in the equation of motion (3) and (4) delivers the following algebraic equation

$$[D(\mathrm{k}_1,\mathrm{k}_2,\lambda)]\left\{\begin{array}{c}\widehat{U}\\ \widehat{V}\end{array}\right\}=0 \tag{7}$$

The wave vector **k** is a complex number: its real part represents the attenuation in the x-y plane, and its imaginary part is the phase constants.

For a plane wave without attenuation in the x-y plane, the propagation constants along the x and y directions are $k_1=i\varepsilon_1=|k|cos(\theta)$ and $k_2=i\varepsilon_2=|k|\sin(\theta)$.

Any triad k_1,k_2,λ obtained by solving the eigenvalues problem in (7) represents plane waves propagating at the frequency λ.

The eigenvalue problem for Eq. (7) yields a characteristic equation developed as:

$$\lambda^4+a\,\lambda^3+b\,\lambda^2+c\,\lambda+d=0 \tag{8}$$

The roots of Eq. (8) may be expressed in the following form:

$$\lambda_s(k) = -\zeta_s(k) \cdot \omega_{ns}(k) \pm i \cdot \omega_{ns}(k)\sqrt{1 - \zeta_s^2} \tag{9}$$

in which s represents the branch type, namely L standing for the longitudinal waves and S for the shear waves. Two pairs of complex conjugates solutions are obtained, corresponding respectively to longitudinal and shear waves.

In Eq. (9), one identifies the natural frequency $\omega_{ns}(k)$, the damped frequency $\omega_{ds}(k)$ and the damping factor ζ_s, viz the following quantities

$$\omega_{ns}(k) = \sqrt{real(\lambda_s)^2 + imag(\lambda_s)^2},$$

$$\omega_{ds}(k) = \omega_{ns}(k)\sqrt{1 - \zeta_s^2}, \ \zeta_s = \frac{real(\lambda_s)}{\omega_{ns}}$$

Relying on these expressions, we can plot the dispersion curve for the dissipated frequency and the damping ratio versus the wave vector **k**.

4 Dispersion Relations and Damping Ratio Evolutions

In Fig. 2, we plot the damping ratio ζ versus the wave number for different viscosity coefficients μ_e. In the following μ_e is given in MPa · sec, w_d and w_n in rad/sec.

The geometrical and mechanical parameters of the four unit cells are given in Table 1.

Table 1. Geometrical and mechanical parameters of the unit cell

Geometrical parameters	Mechanical properties
$L = 10\,\text{mm}$	$E_s = 1400\,\text{MPa}$
$t = 1\,\text{mm}$	$\nu = 0.3$
$\theta = 30°$	$\rho = 1000\,\text{kg/m}^3$

Figure 3 shows the frequency band structure with damping (μ_e = 80) and without damping (μ_e = 0).

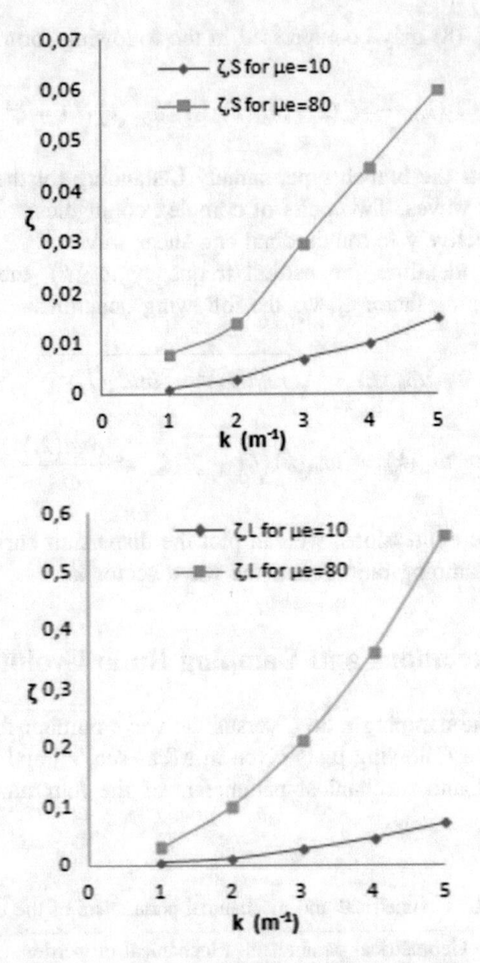

Fig. 2. Damping ratio for two values of the damping coefficient $\mu_e = 10$ and $\mu_e = 80$ for $\theta = \pi/6$

Results show shifts in the frequency band diagrams (the damping frequency decreases) due to the presence of damping; these shifts are more important with an increase of the viscosity coefficient for the longitudinal and shear modes. This behavior has also an impact on the damping ratio diagrams: when the viscosity coefficient increases, the damping ratio increases proportionally. These results are in very good agreement with those found in (Hussein and Frazier 2013; Phani and Hussein 2013).

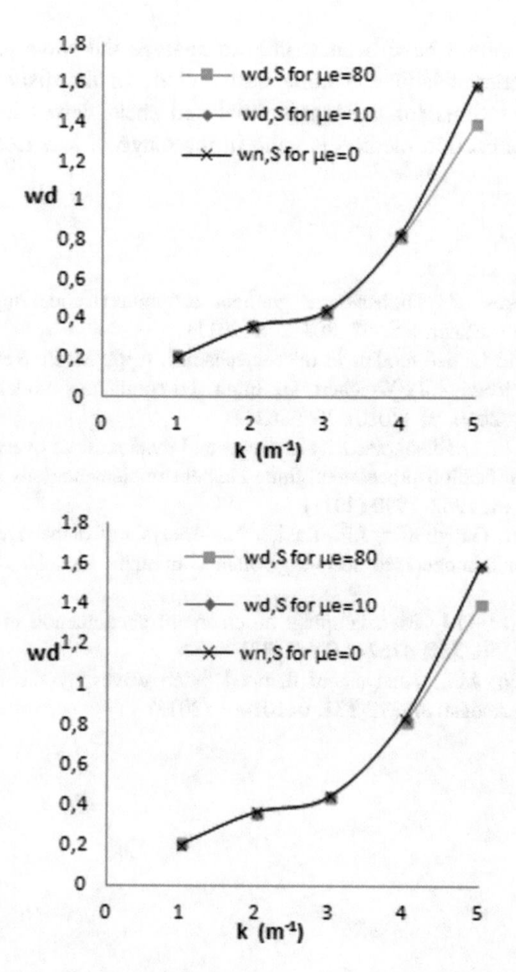

Fig. 3. Dispersion relation for low and high damping situations with a direction of propagation $\theta = \pi/6$

5 Conclusion

This work provides an analysis of the dispersion of elastic waves in periodic beam networks based on second order gradient models obtained by the homogenization of the initially discrete network obtained by the discrete asymptotic method extended up to the second gradient. The lattice beams have a viscoelastic behavior described by Kelvin-Voigt model and the homogenized second gradient viscoelasticity continuum model which has first and second order elasticity coefficients reflecting both the initial lattice topology, anisotropy and microstructural features in terms of geometrical and micromechanical parameters. The dynamical equations of motion for the equivalent second order continuum have been written to analyze the wave propagation characteristics of the reentrant lattice. A comparative study of the dispersion relations and damping ratio evolutions for the longitudinal and shear waves has been done. The developed homogenization method is valid in the range of low frequencies.

References

Lombardo, M., Askes, H.: Higher-order gradient continuum modelling of periodic lattice materials. Comput. Mater. Sci. **52**, 204–208 (2011)

Ostoja-Starzewki, M.: Lattice models in micromechanics. Appl. Mech. Rev. **55**, 35–60 (2002)

Andrianov, I., Awrejcewicz, J., Weichert, D.: Improved continuous models for discrete media. Math. Prob. Eng. **2010**, 35 (2010). ID 986242

Askes, H., Aifantis, E.: Gradient elasticity in statics and dynamics: an overview of formulations, length scale identification procedures, finite element implementations and new results. Int. J. Solids Struct. **48**, 1962–1990 (2011)

Reda, H., Rahali, Y., Ganghoffer, J.F., Lakiss, H.: Analysis of dispersive waves in repetitive lattices based on homogenized second-gradient continuum models. Compos. Struct. **152**, 712–728 (2016)

Hussein, M.I., Frazier, M.J.: Metadamping: an emergent phenomenon in dissipative metamaterials. J. Sound Vib. **332**, 4767–4774 (2013)

Phani, A.S., Hussein, M.I.: Analysis of damped bloch waves by the rayleigh perturbation method. J. Vibr. Acoust. ASME **135**, 041014–1 (2013)

Piezoelastic Behavior of Adaptive Composite Plate with Integrated Sensors and Actuators

Hanen Mallek[2(✉)], Hana Mellouli[2], Hanen Jrad[2], Mondher Wali[1,2], and Fakhreddine Dammak[2]

[1] Department of Mechanical Engineering, College of Engineering, King Khalid University, Abha, Saudi Arabia
mondherwali@yahoo.fr

[2] Laboratory of Electromechanical Systems (LASEM), National Engineering School of Sfax, University of Sfax, Route de Soukra km 4, 3038 Sfax, Tunisia
{hanen.mallek, hana.mellouli, Fakhreddine.dammak}@enis.tn, hanen.j@gmail.com

Abstract. This paper is concerned with a piezoelectric shell element to analyze smart structures. The finite element formulation is based on discrete double directors shell elements. The implementation is applicable to the analysis of laminated shells with integrated piezoelectric layers. The third-order shear deformation theory is used in the present method to remove the shear correction factor and improve the accuracy of transverse shear stresses. The element has four nodes with eight nodal degrees of freedoms: three displacements, four rotations and one electric potential, which is assumed to be a linear function through the thickness of each active sub-layer. The piezoelastic behavior of smart composite plate is examined. The obtained results are compared to existing solutions available in literature. An excellent agreement among the results confirms the high accuracy of the current piezoelastic model.

Keywords: Piezoelastic behavior · Double directors' shell element · Sensors/actuators · Piezoelectric material

1 Introduction

Smart structures are considered as new design philosophy and engineering approach that integrates the actions of distributed actuators and sensors into the structural system (Marinković et al. 2006; Rama et al. 2018; Jrad et al. 2018; Mallek et al. 2019a, b). In recent years, the study of these structures has attracted many researchers because of their potential for use in advanced aerospace as well as hydro space, nuclear and automotive structural applications due to their excellent electromechanical properties, easy fabrication, design flexibility, and efficiency to convert electrical energy into mechanical energy (Foda et al. 2010; Chesne and Pezerat 2011; Zhang et al. 2011; Rama 2017; Gabbert 2002). In fact, their intrinsic electromechanical coupling effect

A. Benamara et al. (Eds.): CoTuMe 2018, LNME, pp. 77–84, 2019.
https://doi.org/10.1007/978-3-030-19781-0_10

produces mechanical deformations under the application of electrical loads (i.e. the direct effect) and electrical fields under the application of mechanical loads (i.e. the converse effect).

Modeling a linear analysis of intelligent structures using First order Shear Deformation Theory (FSDT) is widespread in the literature. Neto et al. (2012) proposed a 3-node finite shell element to predict piezoelastic static and dynamic response of smart laminated structures. Recently, another 3-node shell element is developed by Marinković and Rama (2017) in order to predict the static and dynamic response of piezoelectric laminated composite shells. The enhancements in the form of strains smoothing technique and discrete shear gaps were applied in this element formulation. Lammering and Yang (2009) also presented a 4-node degenerate shell elements based on the FSDT. This element is implemented according to the two-field formulation with linearly distributed electric potential.

Nevertheless, FSDT theory does not allow a good analysis through the thickness because it considers constant transverse shear strains across the thickness. This theory requires the introduction of transverse shear correction factors which can be restrictive and may cause inaccuracies. This limitation can be overcome using High order Shear Deformation Theory HSDT (Valvano and Carrera 2017) or modified FSDT theory (Mellouli et al. 2019a; Trabelsi et al. 2018). Higher-order theories take into account the variations of in-plane displacements, transverse shear deformation and transverse normal strain depending in thick structures, through the thickness, without the need for any shear correction coefficients. Investigations using the HSDT model have proved their excellent performance in several studies including the modeling of thick piezo-laminated structures and sandwich structures (Sudhakar and Kamal 2003; Correia et al. 2002; Wu et al. 2002).

Some investigations concerning the linear and nonlinear analysis of shell structures in areas like static, free vibration, and forced vibration of FGM and FG-CNT structures, using discrete double directors shell elements, have been reported in the literature (Mallek et al. 2018; Zghal et al. 2017; Frikha et al. 2016; Wali et al. 2015; Mellouli et al. 2019b). Inspired from these investigations, the piezoelastic response of laminated structures with integrated smart layers is elaborated in this paper, using 3D- piezoelectric shell model based on a discrete double directors shell elements.

2 Theoretical Formulations

The piezoelectric double directors finite shell element is developed in this section, based on the kinematics of high order shear deformation theory. The initial C_0 and the deformed C_t configurations of the shell is assumed to be smooth, continuous and differentiable. Variables associated to C_0 (respectively C_t) are denoted by upper-case letters (respectively lower-case letters). Vectors and tensors are expressed using bold letters.

2.1 Kinematic Assumptions

Considering the hypothesis of a double director shell model, the position vector of any material point (q) in the deformed configuration C_t defined in terms of curvilinear coordinates $\boldsymbol{\xi} = (\xi, \eta, \varsigma = z)$ is given as:

$$\boldsymbol{x}_q(\xi,\eta,z) = \boldsymbol{x}_p(\xi,\eta) + f_1(z)\boldsymbol{d}_1(\xi,\eta) + f_2(z)\boldsymbol{d}_2(\xi,\eta), z \in [-h/2\,,\,h/2] \tag{1}$$

where p represents the material point located on the midsurface surface of the shell, $\boldsymbol{d}_1$ and $\boldsymbol{d}_2$ are the unit shell director vectors and h is the thickness.

The general functions $f_1(z)$ and $f_2(z)$ which reflects a double director's theory can be expressed in function of the thickness variable as follows:

$$f_1(z) = z - 4z^3/3h^2 \;\; , \;\; f_2(z) = 4z^3/3h^2 \tag{2}$$

The vectors of membrane, bending and shear strains are given by:

$$\boldsymbol{e} = \left\{ \begin{array}{c} e_{11} \\ e_{22} \\ 2e_{12} \end{array} \right\}, \boldsymbol{\chi}^k = \left\{ \begin{array}{c} \chi_{11}^k \\ \chi_{22}^k \\ 2\,\chi_{12}^k \end{array} \right\} \;, \boldsymbol{\gamma}^k = \left\{ \begin{array}{c} \gamma_1^k \\ \gamma_2^k \end{array} \right\} , k = 1,2 \tag{3}$$

These virtual components are computed in the initial configuration C_0 as:

$$\left\{ \begin{array}{l} \delta e_{\alpha\beta} = 1/2(\boldsymbol{A}_\alpha.\delta\boldsymbol{x}_{,\beta} + \boldsymbol{A}_\beta.\delta\boldsymbol{x}_{,\alpha}) \\ \delta\gamma^k = \boldsymbol{A}_\alpha.\delta\boldsymbol{d}_k + \delta\boldsymbol{x}_{,\alpha}.\boldsymbol{d}_k; \quad \alpha,\beta = 1,2; \quad k = 1,2 \\ \delta\chi_{\alpha\beta}^k = 1/2(\boldsymbol{A}_\alpha.\delta\boldsymbol{d}_{k,\beta} + \boldsymbol{A}_\beta.\delta\boldsymbol{d}_{k,\alpha} + \delta\boldsymbol{x}_{,\alpha}.\boldsymbol{d}_{k,\beta} + \delta\boldsymbol{x}_{,\beta}.\boldsymbol{d}_{k,\alpha}) \end{array} \right. \tag{4}$$

The electrical field $\boldsymbol{E}$ is evaluated based on the gradient of the electric potential φ. Its expression is given by:

$$\boldsymbol{E} = \; - \, \varphi_{,\alpha}, \; \alpha = 1..3 \tag{5}$$

2.2 Weak Form and Finite Element Approximation

In order to obtain the numerical solution using the finite element method, the weak form of equilibrium equations is formulated as:

$$G = \int_A \left(\boldsymbol{N}.\delta\boldsymbol{e} + \sum_{k=1}^{2} (\boldsymbol{M}_k.\delta\boldsymbol{\chi}^k) + \boldsymbol{T}_1.\delta\gamma^1 + \tilde{\boldsymbol{q}}.\delta\boldsymbol{E} \right) dA - G_{ext} = 0 \tag{6}$$

where $\boldsymbol{N}$, $\boldsymbol{M}_k$ and $\boldsymbol{T}_1$ represent the membrane, bending and shear stresses resultants respectively. $\tilde{\boldsymbol{q}}$ is the electric displacement and $\boldsymbol{G}_{ext}$ is the external virtual work. These vectors can be written in the form:

$$N = \int_{-h/2}^{h/2} \begin{bmatrix} \sigma_{11} \\ \sigma_{22} \\ \sigma_{12} \end{bmatrix} dz,\ M_k = \int_{-h/2}^{h/2} f_k(z) \begin{bmatrix} \sigma_{11} \\ \sigma_{22} \\ \sigma_{12} \end{bmatrix} dz$$

$$T_1 = \int_{-h/2}^{h/2} f_1'(z) \begin{bmatrix} \sigma_{13} \\ \sigma_{23} \end{bmatrix} dz,\ \widetilde{q} = \int_{-h/2}^{h/2} q\, dz\ ,\ k = 1, 2 \tag{7}$$

The generalized resultant of stress and strain vectors are expressed as

$$R = \begin{bmatrix} N & M_1 & M_2 & T_1 & \widetilde{q} \end{bmatrix}^T_{14\times1},\ \Sigma = \begin{bmatrix} e & \chi^1 & \chi^2 & \gamma^1 & -E \end{bmatrix}^T_{14\times1} \tag{8}$$

The linear constitutive equations of piezoelasticity expressing the coupling between the elastic and electric fields relevant to present problem can be defined as:

$$\begin{cases} \sigma = C\,\varepsilon - p^T E \\ q = p\,\varepsilon + k\,E \end{cases} \tag{9}$$

Using Eqs. (6) and (9), the stress resultant R is related to the strain field.

$$R = H_T \Sigma,\ H_T = \begin{bmatrix} H_{11} & H_{12} & H_{13} & 0 & H_{15} \\ & H_{22} & H_{23} & 0 & H_{25} \\ & & H_{33} & 0 & H_{35} \\ & & & H_{44} & H_{45} \\ Sym & & & & H_{55} \end{bmatrix} \tag{10}$$

with H_T is the linear coupling elastic and electric matrix expressed as:

$$\begin{cases} (H_{11}, H_{12}, H_{13}, H_{22}, H_{23}, H_{33}) = \int_{-h/2}^{h/2} \left(1, f_1, f_2, f_1^2, f_1 f_2, f_2^2\right) C dz \\ H_{44} = \int_{-h/2}^{h/2} \left(f_1'\right)^2 C_\tau\, dz \\ (H_{15}, H_{25}, H_{35}) = \int_{-h/2}^{h/2} (1, f_1, f_2)\, p_1^T\, dz \\ H_{45} = \int_{-h/2}^{h/2} f_1'\, p_2^T dz \\ H_{55} = \int_{-h/2}^{h/2} k\, dz \end{cases} \tag{11}$$

where C and C_τ are in plane and out-of-plane linear elastic sub-matrices. p_1^T, p_2^T and k represent the in plane and out-of-plane piezoelectric coupling sub-matrices and dielectric permittivity matrix, respectively.

In the finite element approximation, the geometry, the displacements and the electric potential are approximated by means of the isoparametric concept. The double director vectors δd_1 and δd_2 are approximated with the same functions as in (Mallek et al. 2018). Therefore, the discrete form of Eq. (7) leads to the discretized static linear piezoelastic equilibrium equation for the structure.

3 Numerical Results

In this section, the static behavior of adaptive composite with surface bonded actuators/sensor is performed to demonstrate the accuracy and the performance of the proposed piezoelectric shell element. A simply supported cross-ply square made of S-glass/Epoxy, with the internal sequence of layers [$45^{\circ}/-45^{\circ}/45^{\circ}$], acts as actuators and sensors. Two piezoelectric layers, made of PXE-52, bonded to the top and bottom surfaces. The side dimension is a = 0.1 m, the thickness of each S-glass/Epoxy layer is 0.0004 m and of active layer is 0.0002 m. The material properties of S-glass/Epoxy used for the calculation are $Y_1 = 55\,\text{GPa}$, $Y_2 = 16\,\text{GPa}$, $G_{12} = 7.6\,\text{GPa}$, $\nu_{12} = 0.28$. The material and piezoelectric properties of PXE-52 are: $Y_1 = Y_2 = 62.5\,\text{GPa}$, $G_{12} = 24\,\text{GPa}$, $\nu_{12} = 0.3$, $e_{11} = e_{22} = -280.10^{-12}$ m/V and $p_{33} = -3.45^{-8}$ F/m. An 8×8 finite element mesh is applied.

Case 1: Sensing:

The plate is studied as a sensor case (see Fig. 1). Deformation caused by external mechanical loads results in electric charges due to the direct piezoelectric effect. A uniformly distributed load $L_{mech} = \mu p_o$ (μ represents the load level) is initially subjected to the plate, which leads to a linear distribution of the bending moment along the length, where $p_0 = 10\,\text{kN/m}^2$. The present predictions and solutions provided by Moita et al. (2002) are shown in Table 1. The results are in good agreement with the alternative solutions.

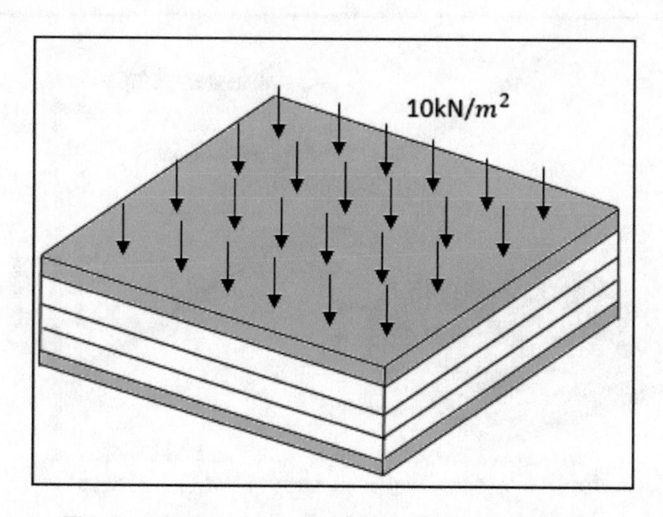

Fig. 1. Adaptive composite plate under mechanical load.

Table 1. Central deflection W_c [mm] for different mechanical loads.

Load level μ	Mechanical load $L_{mech} = \mu\, p_0$	
	(a)	(b)
0.5	0.1085	0.1096
1.0	0.2170	0.2193
1.5	0.3255	0.3290
2.0	0.4340	0.4386
2.5	0.5425	0.5483
3.0	0.6510	0.6580

(a) Moita et al. (2002); (b) Present Model

Case 2: Actuation:

Starting from $\varphi_0 = \pm 135.35\,\text{V}$, an increase in the potential voltage up to $\varphi = \pm 454.05\,\text{V}$ is applied across the thickness, the induced internal stresses result in a bending moment which causes deflection of the plate. The center deflection W_c of simply supported composite plate is depicted in Fig. 2 for different load levels, defined by $\mu = L_{electr}\, \varphi_0$. The obtained results are in good agreement with the alternative linear solution obtained by Moita et al. (2002), using Kirchhoff classical piezo-laminated 3-node plate/shell element.

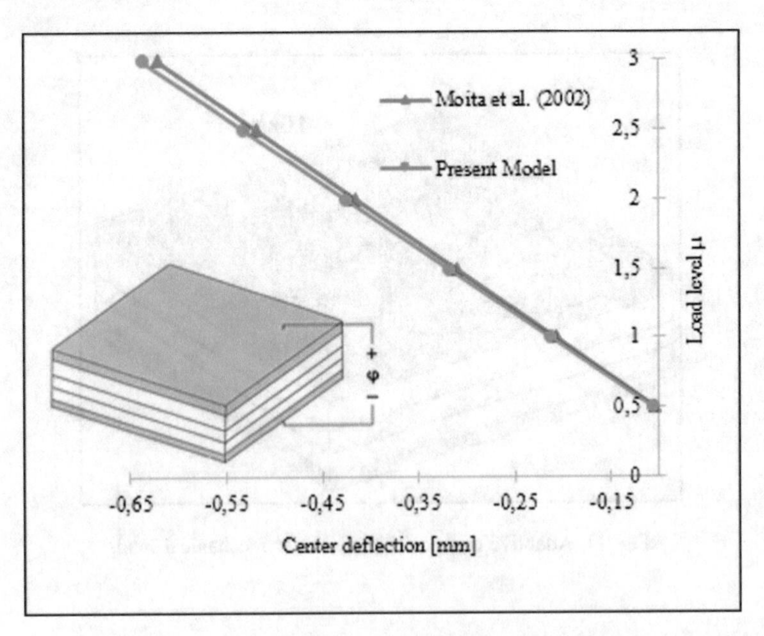

Fig. 2. Center deflection of the adaptive composite plate under different electric loads.

4 Conclusion

This paper presents a developed finite element model based on a discrete double-directors shell elements that can be gainfully used for modeling and simulation of the behavior of smart structure with integrated sensors/actuators. The electric potential is assumed to be a linear function through the thickness of each active sub-layer. The static behavior is performed in terms of deflection with varying the load level. The validation of the present model is based on comparing the obtained results to the literature once for a simply supported laminated plate subjected to mechanical and electric load. A good agreement is obtained between the present results and the reference results.

References

Chesne, S., Pezerat, C.: Distributed piezoelectric sensors for boundary force measurements in Euler-Bernoulli beams. Smart Mater. Struct. **20**(7), 075009 (2011)

Correia, I.F.P., Soares, C.M.M., Soares, C.A.M., Herskovits, J.: Active control of axisymmetric shells with piezoelectric layers: a mixed laminated theory with a high order displacement field. Comput. Struct. **80**, 2256–2275 (2002)

Foda, M.A., Almajed, A.A., ElMadany, M.M.: Vibration suppression of composite laminated beams using distributed piezoelectric patches. Smart Mater. Struct. **19**(11), 115018 (2010)

Frikha, A., Wali, M., Hajlaoui, A., Dammak, F.: Dynamic response of functionally graded material shells with a discrete double directors shell element. Compos. Struct. **154**, 385–395 (2016)

Gabbert, U., Koppe, H., Seeger, F., Berger, H.: Modeling of smart composite shell structures. J. Theoret. Appl. Mech. **40**(3), 575–593 (2002)

Jrad, H., Mallek, H., Wali, M., Dammak, F.: Finite element formulation for active functionally graded thin-walled structures. Comptes Rendus Mec. **346**(12), 1159–1178 (2018)

Lammering, R., Yang, F.: A four-node finite element for piezoelectric shell structures in convective coordinates. Mech. Adv. Mater. Struct. **16**, 198–209 (2009)

Mallek, H., Jrad, H., Algahtani, A., Wali, M., Dammak, F.: Geometrically non-linear analysis of FG-CNTRC shell structures with surface-bonded piezoelectric layers. Comput. Methods Appl. Mech. Eng. **347**, 679–699 (2019a)

Mallek, H., Jrad, H., Wali, M., Dammak, F.: Geometrically nonlinear finite element simulation of smart laminated shells using a modified first-order shear deformation theory. J. Intell. Mater. Syst. Struct. (2018). https://doi.org/10.1177/1045389x18818386

Mallek, H., Jrad, H., Wali, M., Dammak, F.: Piezoelastic response of smart functionally graded structure with integrated piezoelectric layers using discrete double directors shell element. Compos. Struct. **210**, 354–366 (2019b)

Marinković, D., Koppe, H., Gabbert, U.: Numerically efficient finite element formulation for modeling active composite laminates. Mech. Adv. Mater. Struct. **13**, 379–392 (2006)

Marinković, D., Rama, G.: Co-rotational shell element for numerical analysis of laminated piezoelectric composite structures. Compos. Part B Eng. **125**, 144–156 (2017)

Mellouli, H., Jrad, H., Wali, M., Dammak, F.: Meshless implementation of arbitrary 3D-shell structures based on a modified first order shear deformation theory. Comput. Math Appl. **77**, 34–49 (2019a)

Mellouli, H., Jrad, H., Wali, M., Dammak, F.: Meshfree implementation of the double director shell model for FGM shell structures analysis. Eng. Anal. Bound. Elem. **99**, 111–121 (2019b)

Moita, J.M.S., Soares, C.M.M., Soares, C.A.M.: Geometrically non-linear analysis of composite structures with integrated piezoelectric sensors and actuators. Compos. Struct. **57**(1–4), 253–261 (2002)

Neto, M.A., Leal, R.P., Yu, W.: A triangular finite element with drilling degrees of freedom for static and dynamic analysis of smart laminated structures. Compos. Struct. **108–109**, 61–74 (2012)

Rama, G.: A 3-node piezoelectric shell element for linear and geometrically nonlinear dynamic analysis of smart structures. Facta Univ. Ser. Mech. Eng. **15**(1), 31–44 (2017)

Rama, G., Marinković, D., Zehn, M.: Efficient three-node finite shell element for linear and geometrically nonlinear analyses of piezoelectric laminated structures. J. Intell. Mater. Syst. Struct. **29**(3), 345–357 (2018)

Sudhakar, A.K., Kamal, M.: Finite element modeling of smart plates/shells using higher order shear deformation theory. Compos. Struct. **62**, 41–50 (2003)

Trabelsi, S., Frikha, A., Zghal, S., Dammak, F.: Thermal post-buckling analysis of functionally graded material structures using a modified FSDT. Int. J. Mech. Sci. (2018). https://doi.org/10.1016/j.ijmecsci.2018.05.033

Valvano, S., Carrera, E.: Multilayered plate elements with node-dependent kinematics for the analysis of composite and sandwich structures. Facta Univ. Ser. Mech. Eng. **15**(1), 1–30 (2017)

Wali, M., Hentati, T., Jaraya, A., Dammak, F.: Free vibration analysis of FGM shell structures with a discrete double directors shell element. Compos. Struct. **125**, 295–303 (2015)

Wu, X.H., Chen, C., Shen, Y.P., Tian, X.G.: A high order theory for functionally graded piezoelectric shells. Int. J. Solids Struct. **39**, 5325–5344 (2002)

Zghal, S., Frikha, A., Dammak, F.: Static analysis of functionally graded carbon nanotube-reinforced plate and shell structures. Compos. Struct. **176**, 1107–1123 (2017)

Zhang, W.M., Tabata, O., Tsuchiya, T., Meng, G.: Noise-induced chaos in the electrostatically actuated MEMS resonators. Phys. Lett. A **375**(32), 2903–2910 (2011)

Design Methodology and Manufacturing Process

Cycle Time and Hole Quality in Drilling Canned Cycle

Monia Ben Meftah[1(✉)], Hassen Khlifi[1], Bassem Gassara[1], Maher Baili[2], Gilles Dessein[2], and Wassila Bouzid[1]

[1] Unité de Génie de Production Mécanique et Matériaux, ENIS, Route Soukra Km 3, 5, B.P. 1173, 3038 Sfax, Tunisia
{monia.benmeftah,bassem.gassara}@enis.tn, hassenreal@hotmail.fr, wassilabouzid@yahoo.fr
[2] LGP-ENIT, Université de Toulouse, 47 Avenue d'Azereix, BP 1629, 65016 Tarbes Cedex, France
{maher.baili,gilles.dessein}@enit.fr

Abstract. The aim of this work is to present a feed rate modeling in drilling cycle taking into account the kinematic behavior of machine tool and an experimental study is carried out to validate it. Based on this model, a cycle time for the chip-breaking cycle and the deep hole cycle has been calculated. In particular, a precise estimation of cutting time has been predicted. Then, the effects of the value of the programmed feed rate and the incremental distance size on the kinematic behavior of the tool has been investigated. Moreover, a comparative study between this two-cycle type has been shown in term of kinematic profile and quality of the drilled hole (diameter error and cylindricity).

Keywords: High speed drilling · Deep hole canned cycle · Chip-breaking canned cycle · Feed rate modeling · Cycle time · Diameter error · Cylindricity

1 Introduction

Drilling is the most current operation of machining in manufacturing industries. In CNC machine, this operation is programmed using a single instruction called a canned cycle. It is a convenient way of performing a series of operations. The most frequently canned cycles used in drilling are the following: a spot drilling canned cycle (G81), a peck drilling canned cycle (G83) used for deep hole drilling, and the chip-breaking cycle (G73) used for drilling a material that has the tendency to produce a stringy chip. Islam et al. (2016) have investigated the effects of these three canned cycles and cutting parameters on the drilled hole quality. Xavier et al. (2016) have studied the influence of drilling cycle on the roughness of the drilled holes. Aized and Amjad (2013) have presented the influence of the incremental distance on the imposed quality. This study has proved that the quality of hole increases by decreasing the value of incremental distance.

The deep hole drilling is a delicate operation because the cutting is carried out at the bottom of the hole and the chips are evacuated with difficulty. This phenomenon

A. Benamara et al. (Eds.): CoTuMe 2018, LNME, pp. 87–94, 2019.
https://doi.org/10.1007/978-3-030-19781-0_11

generates a decrease in the quality of the hole (Siddiquee et al. 2014). Generally, in this case, the manufactures use a peck drilling cycle (G83). This cycle consists in retracting the drill regularly to evacuate the chips. The movements of retraction tend to increase the cycle time. Then, a precise estimate of the cutting time becomes necessary to calculate the machining cost (Othmani et al. 2011).

In high speed machining, particularly in drilling, the machine does not always reach the programmed feed rate. The actual feed rate calculation depends on kinematical parameters of the machine and tool path geometry (Pessoles et al. 2010) and (Gassara et al. 2013). This implies an underestimation of machine time.

In this paper, a feed rate modeling for the chip-breaking cycle (G73) and deep hole cycle (G83) is presented. Based on this model, a drilling cycle time is calculated taking into account the variation of the feed rate. A comparative study between these two cycles is investigated based on kinematic behavior during the tool path and quality of the drilled hole (diameter error and cylindricity).

2 Feed Rate Modelling

In this part, the feed rate evolution for tool path containing linear interpolation (G00 or G01) is modeled by taking into account the specific parameters of different cycle type, such as the chip-breaking cycle (G73) and deep hole cycle (G83) and kinematic parameters of the machine.

2.1 Tool Path and Specific Parameters in Drilling Cycle

In deep hole drilling cycle (G83) the tool retracts all the way out of the hole with each increment distance (Fig. 1), while, in chip breaking cycle (G73) the tool allows only dwells at the bottom for each incremental distance (Fig. 1). The tool path and the specific parameters in drilling with G83 cycle and G73 cycle are shown in Fig. 1. The number n of the incremental distance is calculated as follows:

$$\left\{\begin{array}{lll} n = Int\left(\frac{L_u - P}{Q}\right) + 2 \text{ and } Q_f = L_u - P - (n-2)Q & \text{if } \left(\frac{L_u - P}{Q}\right) \text{ decimal} \\ n = \left(\frac{L_u - P}{Q}\right) + 1 \quad \text{and } Q_f = 0 & \text{if not} \end{array}\right\} \quad (1)$$

Where P is the size of the first incremental distance (mm), Q is the size of the following incremental distance, and Q_f is the last incremental distance (mm).

2.2 Feed Rate Modeling for Linear Interpolation

The feed rate modeling is developed considering the value of jerk as a constant. In this case, the kinematic profile during linear interpolation is composed by seven phases (Fig. 2), where the equations of feed rate and the periods (T_1, T_2, T_3, T_4, T_5, T_6, T_7) of each phase are determined based on the model developed by (Pessoles et al. 2010).

In drilling cycle, the blocks follow on from each other and the input and output feed rates of the blocks are rarely null. Moreover, according to the length of the interpolation

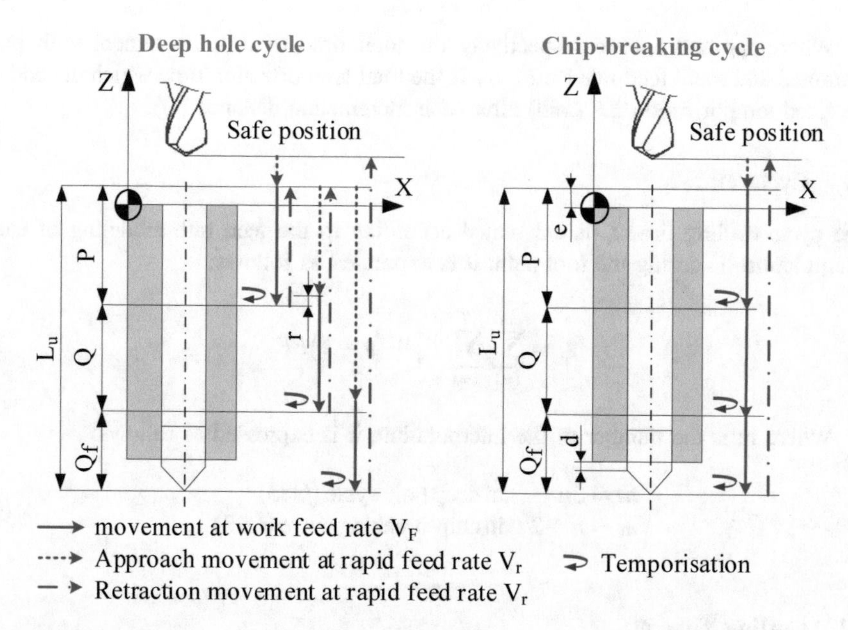

Fig. 1. The drilling cycle (deep hole cycle (G83) and chip-breaking cycle (G73)).

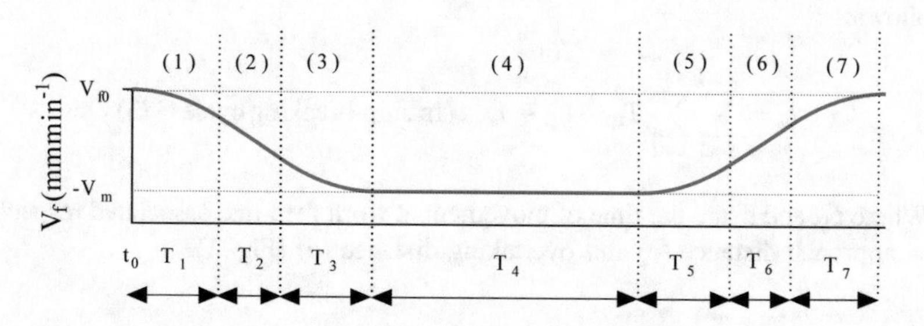

Fig. 2. The kinematic profile for linear interpolation.

and the programmed feed rate, maximum acceleration or the programmed feed rate are not always reached. Therefore, some of the seven phases do not exist.

3 Modeling of Drilling Cycle Time

The classification of the tool path leads to express the cycle time as a sum of three parts:

$$t_c = t_{V_F} + t_{Vr} + t_{EF} \tag{2}$$

Where t_{VF} and t_{Vr} are respectively the total time of drill movement with programmed and rapid feed rate (min). t_{EF} is the total temporization time which depend of the fixed temporization EF (min) after each incremental distance (Q).

3.1 Cycle Time t_c

The cycle drilling time t_c is calculated according to the feed rate modeling of each interpolation 'i' during the tool path, it is expressed as follows:

$$t_c = \sum_{i=1}^{m} \sum_{k=1}^{7} T_{k_i} + (n-1)EF \tag{3}$$

Where m is the number of the interpolation, it is expressed as follows:

$$\begin{aligned} m &= 3n \quad \text{in deep hole cycle (G83)} \\ m &= n+2 \quad \text{in chip-breaking cycle (G73)} \end{aligned} \tag{4}$$

3.2 Cutting Time t_u

The cutting time t_u, during which the tool is in contact with the work piece, is expressed as follows:

$$t_u = \sum_{i=2}^{n+1} \sum_{k=1}^{7} T_{k_i} - t_{ap} - t_d \quad \text{(In chip-breaking cycle G73)} \tag{5}$$

Where t_{ap} and t_d are the time of movement at work feed rate associated respectively to the approach distance (e) and overtaking distance (d) (Fig. 1).

$$t_u = \sum_{i=0}^{n-1} \sum_{k=1}^{7} T_{k(2+3i)} - t_{ap} - t_d \quad \text{(In deep hole cycle G83)} \tag{6}$$

Where t_{ap} and t_d are the time of movement at work feed rate associated respectively to the approach distance (e and r) and overtaking distance (d) (Fig. 1).

4 Results and Discussions

4.1 Experimental Work

The study of the quality of drilled hole was performed on drilling of AISI 4140 alloy steel with a carbide drill (Guhring). A blind holes Ø16 × 56 mm were drilled with a pre-drill diameter of 10 mm. Holes were drilled on a high-speed machine Huron Kx10. The precision measurement data for circularity and diameter error were obtained by a

ASLI SOV-3020 3D Vision Measurement System. Eight points were probed to determine the diameter in four sections at 2 mm height increments.

4.2 Drilling Cycle Time

The Figs. 3 and 4 show the experimental and theoretical feed rate evolution respectively for deep hole drilling cycle (G83) and chip-breaking drilling cycle (G73). They present a good correlation between experimental and theoretical results. Based on these results, it is noted that the feed rate does not always reach the maximum value during drilling operation and this justifies the difference between the cycle time calculated by CAM software (the trajectory length/programmed feed rate) and the estimated cycle time by the developed model, as shown in Fig. 5. It is the same case of the cutting time (Fig. 6). This difference depends on the programmed feed rate value (Fig. 7). For example, the percentage of feed rate for a programmed feed rate V_F = 798.9 mm/min is equal to 95.63% in G83 drilling cycle and 89.88% in G73 drilling cycle, and for V_F = 3195.5 mm/min, it has decreased to 69.56% in G83 drilling cycle and 40.25% in G73 drilling cycle. This difference depends also on the length of incremental distance, where the percentage of feed rate decreases with the short incremental distance (Fig. 8).

It is noted that in each incremental distance the cutting movement in chip-breaking canned cycle (G73) begins with the acceleration phase of the feed rate (Fig. 4), while in deep hole canned cycle (G83) it begins with programmed feed rate value (Fig. 3). Therefore, the cutting time t_u in G83 is less than that in G73 cycle (Fig. 6).

As a result, it is interesting to note that the chip-breaking canned cycle (G73) leads to the least cycle time but, the deep hole canned cycle (G83) may be satisfying more the tool change time and cost of machining.

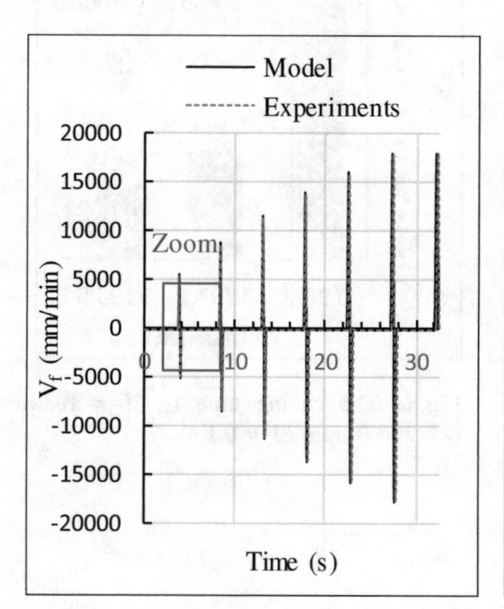

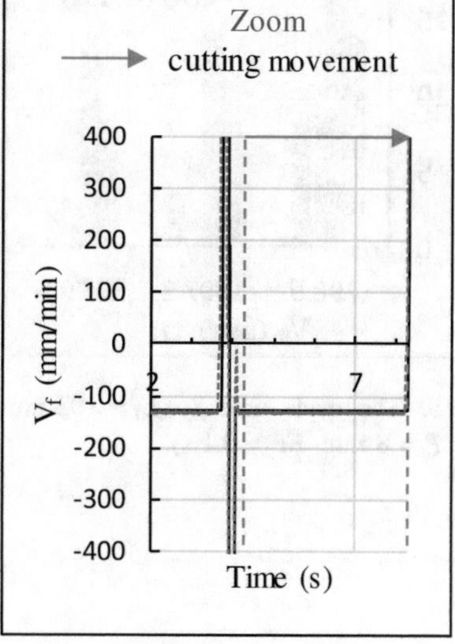

Fig. 3. The variation of feed rate in deep hole canned cycle (G83) (D = 16 mm, V_c = 55.82 m/min, f = 0.118 mm/rev, pre-drill diameter = 10 mm).

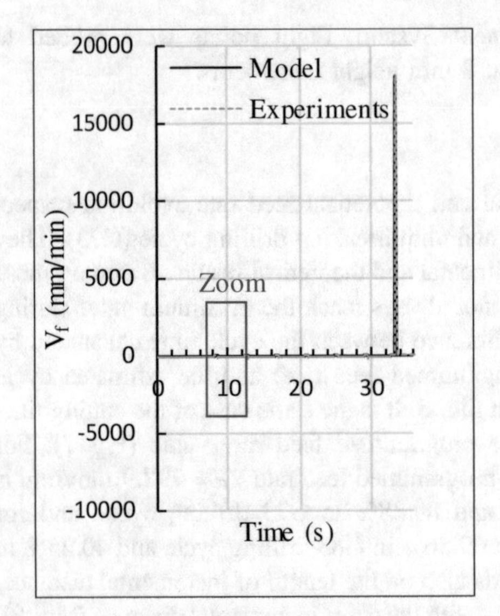

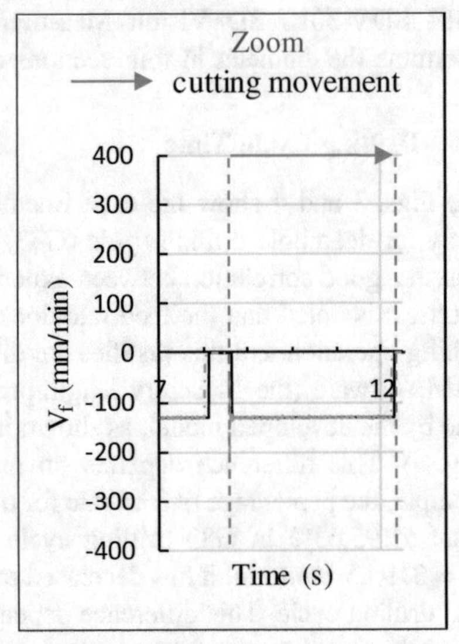

Fig. 4. The variation of feed rate in chip-breaking canned cycle (G73) (D = 16 mm, V_c = 55.82 m/min, f = 0.118 mm/rev, pre-drill diameter = 10 mm).

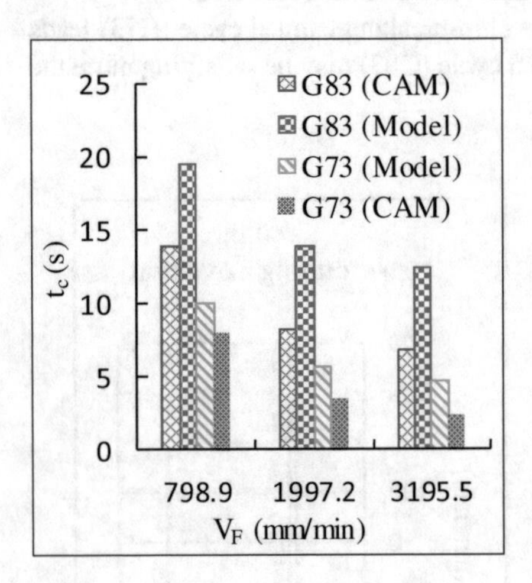

Fig. 5. The cycle time t_c (L_u = 102 mm, $Q = P$ = 8 mm, EF = 0.1 s).

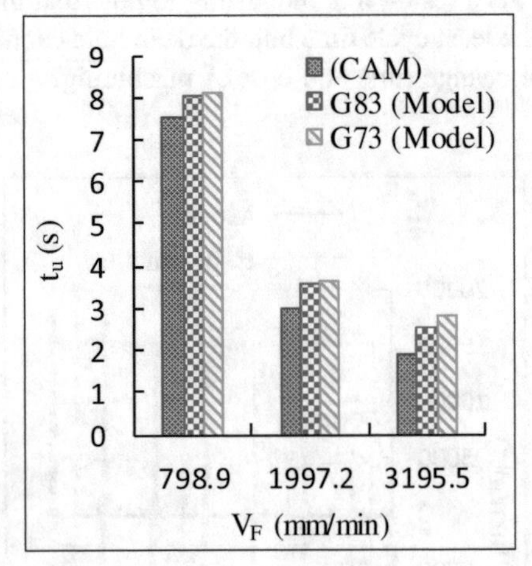

Fig. 6. The cutting time t_u. (L_u = 102 mm, $Q = P$ = 8 mm, EF = 0.1 s).

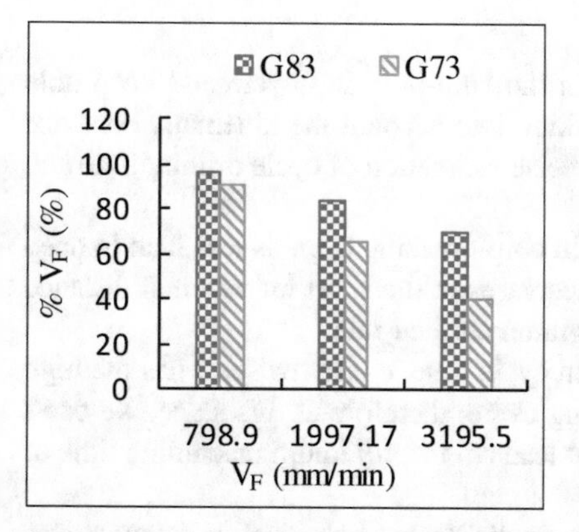

Fig. 7. The percentage of the programmed feed rate (L_u = 102 mm, $Q = P$ = 8 mm)

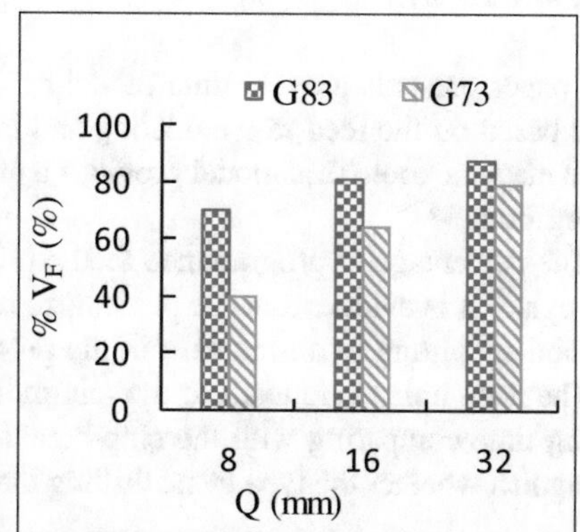

Fig. 8. The percentage of the programmed feed rate (L_u = 102 mm, D = 16 mm, V_c = 400 m/min, f = 0.402 mm/rev)

4.3 Drilling Quality

The Fig. 9 shows the effect of the canned cycle on the hole drilled quality, it demonstrates that the chip-breaking canned cycle produced the best results in terms of diameter error than the deep hole canned cycle along hole axis. In G73 cycle, the maximum diameter error reached are 167.91 µm while in G83 cycle are 286.710 µm.

The determination of the cylindricity of drilled hole shows that the G73 drilling cycle leads also to the best cylindricity than G83 drilling cycle, the values of cylindricity calculated in G73 and G83 cycle are respectively 283.54 µm and 380.37 µm.

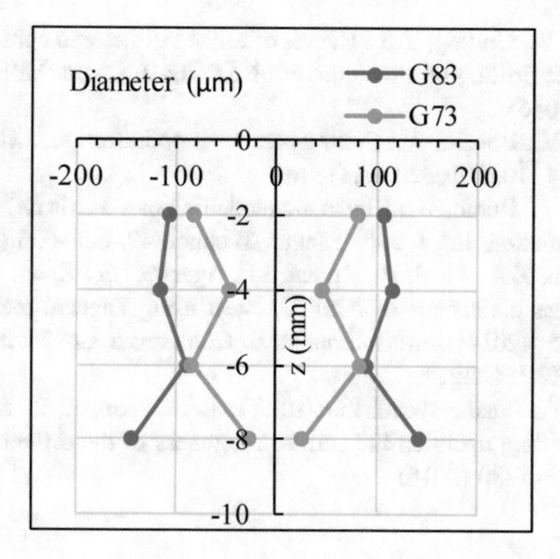

Fig. 9. Change of diameter along hole axis (D = 16 mm, V_c = 55.82 m/min, f = 0.118 mm/rev, pre-drill diameter = 10 mm).

5 Conclusion

This paper presents a cycle time modeling in the chip-breaking cycle and deep hole cycle based on the feed rate modeling and taking into account the kinematic behavior of the machine tool. This model provides a precise estimation of cycle drilling time and cutting time.

The percentage of programmed feed rate in chip-breaking cycle is less than in deep hole cycle. It is evident that the percentage decreases in the short incremental distance size and according to an increase in the programmed feed rate.

The deep hole cycle leads to the minimum cycle time, contrariwise, it has the high cutting time comparing with the chip-breaking cycle, therefore at this stage, we can't distinguish whether the type cycle drilling that leads to the minimum machining time or cost.

The study of the effects of canned cycles on the drilled hole quality demonstrates that the chip-breaking canned cycle (G73) produces a smaller diameter error and cylindricity than the deep hole canned cycle (G83).

Acknowledgements. The work is carried out thanks to the support and funding allocated to the Unit of Mechanical and Materials Production Engineering (UGPM2/UR17ES43) by the Tunisian Ministry of Higher Education and Scientific Research.

References

Aized, T., Amjad, M.: Quality improvement of deep-hole drilling process of AISI D2. Int. J. Adv. Manuf. Technol. **69**, 2493–2503 (2013)

Gassara, B., Baili, M., Dessein, G., Hbaieb, M., Saï, W.B.: Feed rate modeling in circular–circular interpolation discontinuity for high-speed milling. Int. J. Adv. Manuf. Technol. **65**, 1619–1634 (2013)

Islam, M.N., Boswell, B., Ginting, Y.R.: Effects of canned cycles and cutting parameters on hole quality in cryogenic drilling of aluminum 6061-6T. Int. J. Chem. Mol. Nucl. Mater. Metall. Eng. **10**(7), 857 (2016)

Othmani, R., Hbaieb, M., Bouzid, W.: Cutting parameter optimization in NC milling. Int. J. Adv. Manuf. Technol. **54**, 1023–1032 (2011)

Pessoles, X., Landon, Y., Rubio, W.: Kinematic modelling of a 3-axis NC machine tool in linear and circular interpolation. Int. J. Adv. Manuf. Technol. **47**, 639–656 (2010)

Siddiquee, A.N., Khan, Z.A., Goel, P., Kumar, M., Agarwal, G., Khan, N.Z.: Optimization of deep drilling process parameters of AISI 321 steel using Taguchi method. Procedia Mater. Sci. **6**, 1217–1225 (2014). 3rd International Conference on Materials Processing and Characterisation (ICMPC 2014)

Xavier, L.F., Suresh, P., Balaragavendheran, R., Yeshwanth Kumar, P., Deepak, S.: Studies on the influence of drilling cycle on the surface roughness of the drilled holes. ARPN J. Eng. Appl. Sci. **11**, 1277–1280 (2016)

DMST Investigation of the Effect of Cambered Blade Curvature on Small H-Darrieus Rotor Performance

Khaled Souaissa[1(✉)], Moncef Ghiss[2], Mouldi Chrigui[3], Hatem Bentaher[1], and Aref Maalej[1]

[1] Electro-Mechanical Systems Laboratory (LASEM), National Engineering School of Sfax, University of Sfax, Sfax, Tunisia
khaled.souaissa@gmail.com, bentaherh@yahoo.com, aref.maalej@gmail.com

[2] Mechanical Laboratory of Sousse (LMS), National Engineering School of Sousse, University of Sousse, Sousse, Tunisia
moncef.ghiss@eniso.u-sousse.tn

[3] Research Unit in Mechanical Modeling Materials and Energy (M2EM), National Engineering School of Gabès, University of Gabès, Gabès, Tunisia
mouldi.cherigui@gmx.de

Abstract. A Double Multiple Stream Tube (DMST) model has been carried out in order to investigate the effect of the leading-trailing blade edge curvature on the performance of a small H-Darrieus rotor vertical axis wind turbine (VAWT). Four shaped blades (linear, concave, convex and semi-convex) with cambered airfoils profile (NACA4312) are considered in this study. The reliability of the DMST model has been demonstrated through good agreement between the calculated and measured efficiency of an H-Darrieus. The VAWT performance is primarily investigated according to tip speed ratios, aspect ratio and solidities. The numerical results showed that leading-trailing edge (LTE) curvature has substantially enhanced the turbine's efficiency at large range of TSR. As compared to straight LTE, up to 31.58% of the performance increases in concave blade shape with 61.85% of mass reduction, up to 23.16% of improvement is found in semi-concave blade shape with a mass reduction of 23.16% and up to 15.8% of efficiency improvement with a mass reduction of 7.45%.

Keywords: H-Darrieus rotor · DMST · Blade pitch angle · Power coefficient · Convex · Concave · Shaped blade

1 State-of-the-Art of Wind Turbine Rotor Design

When it comes to wind energy prediction, various simulation tools have been developed to predict the performance of wind turbine rotor (Ghasemian et al. 2017; Simão and Madsen 2014). They can be ranged from very simple analytical models like the Blade Element Momentum (BEM) model, which are still used in the early stages of wind turbine design process. In addition, they require low computational cost. Which can be integrated in iterative rotor design process (Lekou 2013). However, the methods

A. Benamara et al. (Eds.): CoTuMe 2018, LNME, pp. 95–102, 2019.
https://doi.org/10.1007/978-3-030-19781-0_12

which are considered very advanced models, are three-dimensional Computational Fluid Dynamics (3D-CFD), which give us deep insight in aerodynamics behaviors during the design process. Furthermore, they require high computational cost (Cao and Zhu 2018). It should be stated here, despite the fast progress of computer power, 3D-CFD is still time-consuming; these methods are more suitable to be used in the evaluating stage of the rotor design rather than in the iterative design process. That is why the use of based BEM method tools developed to predict rotor aerodynamics performance comes in handy.

2 QBlade Simulation Tool

QBlade is as an open source framework, is being developed since 2010 at the chair of fluid mechanics of the TU Berlin (Marten and Wendler 2013). QBlade is used to design and predict either Horizontal axis wind turbines (HAWTs) or Vertical Axis Wind Turbines' (VAWTs) efficiency. Indeed, QBlade is both based on the BEM method for the prediction of HAWTs and a Double Multiple Stream tubes (DMS) model for the prediction of VAWTs' performance (Paraschivoiu 1988; Beri and Yao 2011). QBlade embedded the XFOIL code based on the viscous-inviscid coupled panel method to evaluate the airfoil lift and drag coefficients polar which can also be extrapolated beyond the static stall point along 360° range of angles of attack (Marten *et al.* 2015).

3 DMST Model Validation

To validate the DMST model used in this study, a comparison of computational results with those of wind tunnel experiment (Danao et al. 2013) is performed with various TSR s. The VAWT baseline was a 3–bladed Darrieus rotor with NACA0022 airfoils. The blade chord c was 0.04 m with a rotor radius R set to 0.35 m. The chord-based Reynolds number is about 19,169. As we can see from Fig. 2, a comparison between DMST model and the two 2D-CFD model and experiment results, respectively. Its clear that the DMST model is able to reproduce the aerodynamic behavior of the Darrieus rotor. Indeed, the variation of the (power, TSR) curves of both DMST and experiment results are broadly the same. Nevertheless, the power is overpredicted by both DMST model and 2D-CFD model which is basically due to the tip vortex shedding, the wake-blades interaction and spanwise velocity which are not considered in these models (Fig. 1).

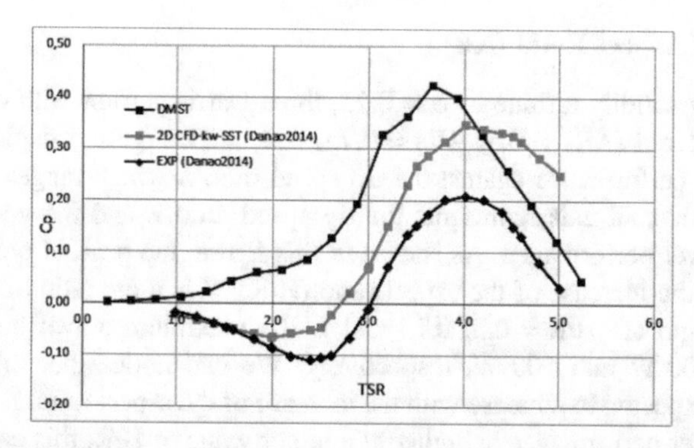

Fig. 1. Comparison between DMST, 2D CFD and experimental results

4 Results and Discussion

4.1 Effect of Pitch Angle

The performance of the VAWT at different pitch angles ranging from −3° to 3° is carried out at different Tip Speed Ratio (TSR) from 1 to 5 and for three different aspect ratios ($AR_1 = 0.5$, $AR_2 = 0.75$, $AR_3 = 1$). The numerical results are illustrated at Fig. 2a–c. From Fig. 2a, at low aspect ratio (AR1 = 0.5) the Darrieus rotor experiences a negative power output whatever the pitch angle is for the entire range of TSRs. As can be seen from Fig. 2b, c, for positive pitch angles ranging from 0° to 3°, the peak power efficiency is expected at low tip speed ratio. However, for negative pitch angle ranging from −3° to 0°, the peak power efficiency is expected at high tip speed ratio. Furthermore, the negative pitch angles allowed a larger operating range for the turbine than the positive pitch angle. Figure 2c, it is noted that the optimum TSR is shifted toward high value since the pitch angle is growing.

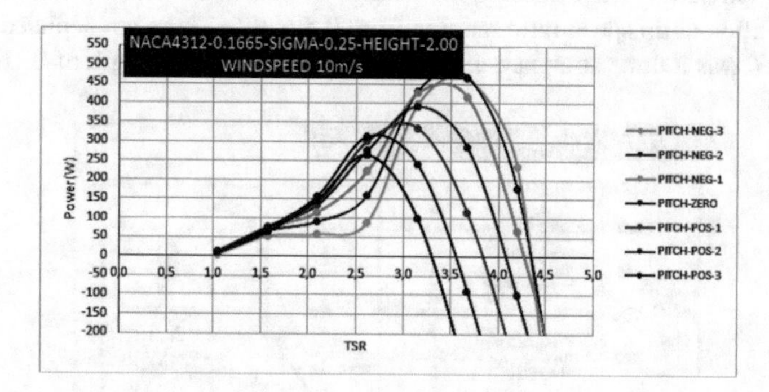

Fig. 2. (Power vs TSR) curves for low solidity ($\sigma = 0.25$) turbine with different pitch angle

4.2 Effect of Aspect Ratio (AR)

For a fixed low solidity turbine of $\sigma = 0.25$, three Darrieus rotor with different aspect ratio are considered ($AR_1 = 0.5$, $AR_2 = 0.75$, $AR_3 = 1$). Figure 5 depicts the fluctuations of power performance against the tip speed ratio λ which ranges from $\lambda = 1$ to $\lambda = 5$. The horizontal axis represents the tip speed ratio λ and the vertical axis represents the power performance. As shown in this figure, the peak of power coefficient increases with the increase of the aspect ratio (H/R). When the ratio of the radius and blade span length are H/R = 0.5, 0.75 and 1, the maximum power output are about P = −20 W, 200 W and 500 W, respectively. We can notice that the optimum tip speed ratio is expected to increase with the increase of the aspect ratio. It means that the maximum power performance is higher at a larger value of H/R. this can be attributed to the generated small blade tip vortex since the aspect ratio is increasing, so that the power performance of the rotor blade is improved (Fig. 3).

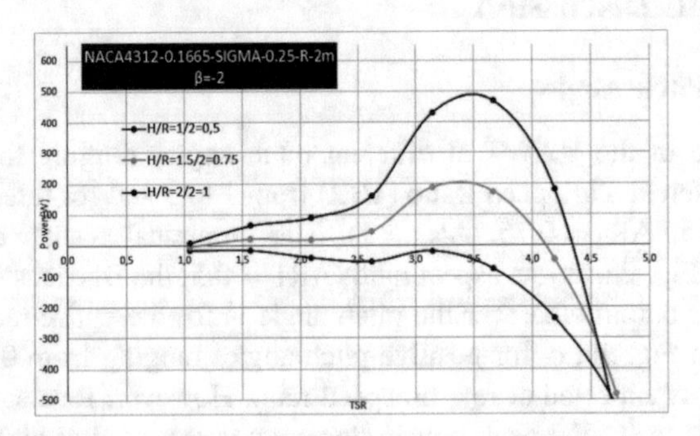

Fig. 3. Effect of aspect ratio on low solidity turbine performance for pitch angle of $\beta = -2^{\circ}$

4.3 Effect of Freestream Velocity

The power efficiency was evaluated for a range of freestream velocity from 2.5 m/s to 15 m/s at different tip speed ratio ranging from 0.5 to 4.5. These are depicted in Fig. 4. The VAWT was a three straight cambered blade with a chord length of 0.1665 m and

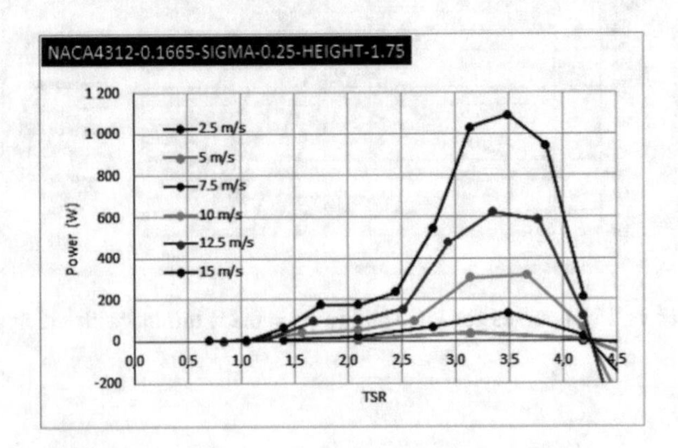

Fig. 4. Effect of wind speed on VAWT's performance

rotor height of 1.75 m. As can be seen from this figure, the produced power amount is closely related to freestream velocity. For a fixed low solidity solidity turbine ($\sigma = 0.25$), since the freestream is going to rise, the power peak will increases without altering its optimum tip speed ratio. Figure 4 highlights that at low TSR < 1.15 the freestream velocity changes have no effect on the turbine performance.

4.4 Effect of Solidity

Figure 5 shows that for two freestream velocity of 10 m/s and 15 m/s, the low solidity turbine ($\sigma_1 = 0.25$) offers the largest range of operation and the best power performance. The power peak is expected at high tip speed ratio (TSR $\approx$ 3.4). As the turbine solidity increases, the turbine experiences a much lower power performance compared to the lower solidity turbines. the power peak occurs at a lower TSR. It can be seen from these figures that there is a large lower range, in which the turbine solidity variation has no significant effect on the generated turbine power.

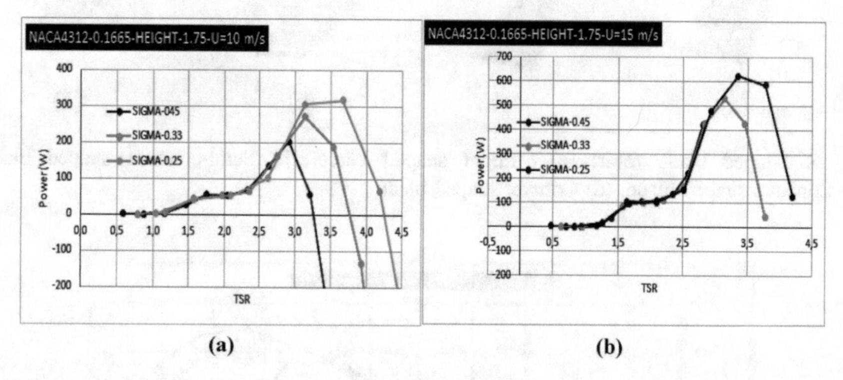

Fig. 5. Effect of solidity on VAWT's performance with different wind speed: (a) wind speed 10 m/s. (b) wind speed 15 m/s

4.5 Effect of Variable Chord Length

Four geometrical models (Fig. 7) were considered to investigate the effect of blade shape on the Darrieus rotor turbine. All four turbines' configurations share the same low solidity of ($\sigma = 0.25$) and the blade length of H = 2 m. the rotor blades are pitched with a negative angle value $\beta = -2°$. The variation of chord length of each blade is reported in Table 1. It should be mentioned here that the constant equation parameters are chosen in order to compare them to the previous analysis configuration.

The DMST turbine performance results for the different configurations are depicted in Fig. 6. First of all, it's quite clearly that the turbine' performance is closely related to the shape of the blades. In addition, the above conclusion about the relation between low solidity turbine and the high operating range of the rotor is still exists. It is clear from this figure that the performance of the curvature shaped blade (concave or convex) experiences a better performance than the straight blade. The peak power of the

Table 1. Chord length equation

Configuration	Chord equation	Airfoil profile	Height	Radius	Blades number
Straight blade	$C(z) = 0.1665$	NACA4312	2	2	3
Concave blade	$C(z) = 0.1z^2 + 0.067$	NACA4312	2	2	3
Convex blade	$C(z) = -0.1z^2 + 0.1665$	NACA4312	2	2	3
Semi concave blade	$C(z) = 0.1\sqrt{z} + 0.067$	NACA4312	2	2	3

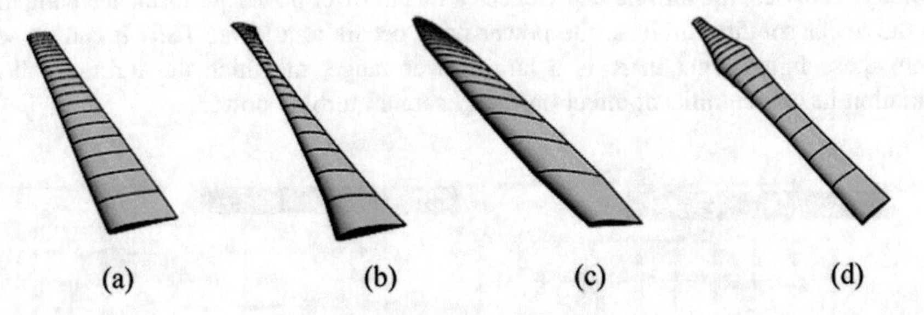

Fig. 6. Shaped blade rotor: (a) Straight shaped blade, (b) Semi-concave shaped blade, (c) Concave shaped blade, (d) Convex shaped blade

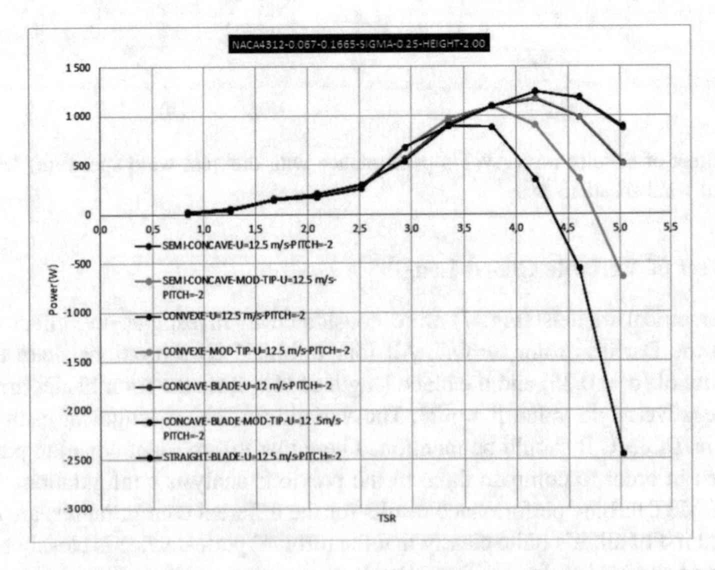

Fig. 7. Effect of blade shape on low solidity turbine performance ($\sigma = 0.25$)

curvature shaped blade and the optimum tip speed ratio are much higher than those of the straight blade. Furthermore, a comparison between convex and concave shaped blade reveals that the later gives more power efficiency than the convex one at high tip speed ratio. Additionally, it is clear that the convex shaped blade performs better than the semi convex shaped blade.

The numerical results showed that leading-trailing edge (LTE) curvature has substantially enhanced the turbine's efficiency at large range of TSR. As compared to straight LTE, up to 31.58% of the performance increases in concave blade shape with 61.85% of mass reduction, up to 23.16% of improvement is found in semi-concave blade shape with a mass reduction of 23.16% and up to 15.8% of efficiency improvement is found in convex shaped blade with a mass reduction of 7.45% (Table 2).

Table 2. Results comparisons between rotor with curved LTE and straight blade

Blade shape	Opt TSR	Power out (W)	Volume	Mass gain	Power gain
Straight	3,6	950	4579,672		
Convex	3,77	1100	3017,444	7,45%	15.8%
Semi-concave	4,1	1170	1747,231	61,85%	23,16%
Concave	4,3	1250	1747,231	61,85%	31,58%

5 Conclusion

This work focuses on the curvature effect of leading-trailing edge (LTE) of straight bladed H-Darrieus rotor turbine with regard to cambered blades profile. A preliminary work has been conducted in order to choose the appropriates features of the VAWT such as pitch angle, aspect ratio and turbine solidity. The DMST results show that the LTE curvature has a double advantage on the VAWT. First, the blade curvature has a significant effect on the turbine's performance. Second, the blade curvature reduces the blade mass and then the turbine's cost.

6 Future Work

While we have proved the effect of Leading-Trailing Edge curvature on the efficiency of the VAWT and on the blade mass reduction by handling a simple quadratic equation, it seems to be more appropriate to carry out an optimization scheme of LTE curves function (chord length function) which maximizes the power output of the VAWT.

References

Beri, H., Yao, Y.: Double multiple streamtube model and numerical analysis of vertical axis wind turbine. Energy Power Eng. **03**(03), 262–270 (2011). https://doi.org/10.4236/epe.2011.33033

Cao, J., Zhu, W.: Applied sciences development of a CFD-based wind turbine rotor optimization tool in considering wake effects. Appl. Sci. **8**, 1056 (2018)

Danao, L.A., Edwards, J., Howell, R.: A numerical study on the effects of unsteady wind on vertical axis wind turbine performance, pp. 1–9 (2013). http://doi.org/10.1115/IMECE2013-62493

Ghasemian, M., Ashrafi, Z.N., Sedaghat, A.: A review on computational fluid dynamic simulation techniques for Darrieus Vertical Axis Wind Turbines. Energy Convers. Manage. **149**(July), 87–100 (2017). https://doi.org/10.1016/j.enconman.2017.07.016

Lekou, D.J.: Advances in Wind Turbine Blade Design and Materials, 2013 edn. Woodhead Publishing (2013). https://doi.org/10.1533/9780857097286.2.325

Marten, D., et al.: Integration of a WT Blade Design Tool in XFoil/XFLR5, January 2010

Marten, D., Wendler, J.: QBLADE: an open source tool for design and simulation of horizontal and Vertical Axis Wind Turbines. Int. J. Emerg. Technol. Adv. Eng. **3**(3), 264–269 (2013)

Simão, C., Madsen, A.: Comparison of aerodynamic models for Vertical Axis Wind Turbines (2014). https://doi.org/10.1088/1742-6596/524/1/012125

A New CAD-CAM Approach Using Interacting Features for Incremental Forming Process

Sofien Akrichi(✉), Amira Abbassi, and Noureddine Ben Yahia

University of Tunis, URSSMDT, ENSIT,
5 Avenue Taha Hussein, Montfleury, 1008 Tunis, Tunisie
ak.sofien@gmail.com, abassiamira@gmail.com,
nourdine.benyahia@yahoo.com

Abstract. Single Point Incremental Forming (SPIF) is a process of progressive deformation for sheet metal, which is obtained by plastic deformation of thin sheets and low volume production applications. This paper focuses on the development of interaction of features manufacturing using three-axis CNC machine. The integration of the CAD/CAM system is a powerful, predictive and accurate tool, providing for a reduction in high production costs. In this article, a new support system has been developed for G-Code Generation that is not commercialized in CAD/CAM software. It allows to calculate and display each point of tool displacement according to the deformation time in order to manage the CNC machine program.

Keywords: SPIF · API-CATIA · Automatic tool path · CAD/CAM system

1 Introduction

As far as the current development of thin sheet work is concerned, the metal deformation process is widely used to produce complex system components used in automotive and paramedical products. The new current processes are in continuous competition with old global shaping processes such as stamping or deep drawing. In this context, incremental forming techniques make it possible to deform the sheet by means of a spherical head tool with numerically controlled machine tools from three to five axes (Daleffe et al. 2013), (Fritzen et al. 2013). Indeed, SPIF is often used as a negative incremental forming, which can produce a wide range of sheet metal parts without the need for costly tools, as well as complicated die.

The integration of knowledge-based system, and the aid systems for CAD/CAM software is used in several studies. The Application Programming Interface of CAD programs are capable of optimizing certain parts of the work (Paniti et al. 2010). These aid systems are not generally marketable and form a number of research topics. They are particularly useful in situations where the amount of information available is prohibitive for the intuition of an unaided human decision-maker and where accuracy and optimization are important.

The literature offers several recent researches in the field of stamping and some articles on the incremental forming. The work of Paniti et al. (2010), which presents a CAD-based approach for a new sheet-forming technology that consists of a Dieless

A. Benamara et al. (Eds.): CoTuMe 2018, LNME, pp. 103–111, 2019.
https://doi.org/10.1007/978-3-030-19781-0_13

Incremental Sheet Forming (DSF). This is achieved by developing a program that allows the creation of a 3D model using a CAD Application Programming Interface (API) under CATIA and an offline slave tool path calculator based on the Output of a commercial CAM program. The author will show how to apply parameterizations in the same model using the CAD API to overcome the problems of a commercial post-processor and how to calculate slave toolpaths for different formatting strategies (Tisza 2012) (Ambrogio et al. 2012).

Jie et al. (2004) used the normal vectors of triangle facets with slicing of a STL model to make Z-level tool paths. Hu et al. (2012) developed an integrated CAD/CAM system for incremental forming and 5-axis laser cutting based on machining features. Tekkaya et al. (2007) developed a correction module based on a CATIA/Unigraphics CAM module, determining an offset depending on the work-piece geometry, while Skjoedt et al. (2007) created a program to convert profile-milling code into a helical Tool Path.

This paper is a research conducted on improvement of the newly formed part process using SPIF, with API Optimization developed under CAD-CAM software. The main objective is to treat the automation of SPIF processes, by integrating a CAD/CAM system. The system modeling is done using API-CATIA. The automation of the SPIF process is determined taking into account the geometrical parameters of the part, the machine parameter and the diameter of the tool, as well as the type of tool path to be used. Five variables are used as SPIF parameters; tool diameter, incremental step size, spindle speed, feed rate, and forming angle.

The contribution to the state-of-art will be given by the application of the new SPIF CAD-features, which aims to improve the quality of the SPIF process.

2 SPIF Process

Generally, the production of parts by SPIF is led by a set of steps to follow. first, the presentation of desired shape (CAD model). Based on a CAD model, the generation of deformation process planning of the sheet is performed by a CAM system in order to obtain the tool path on geometric surfaces. After toolpath correction, transfers the CN code in the three-axis CNC machine for production of the desired part. In Fig. 1 we present all methodology from CAD modeling to produce part with CNC machine.

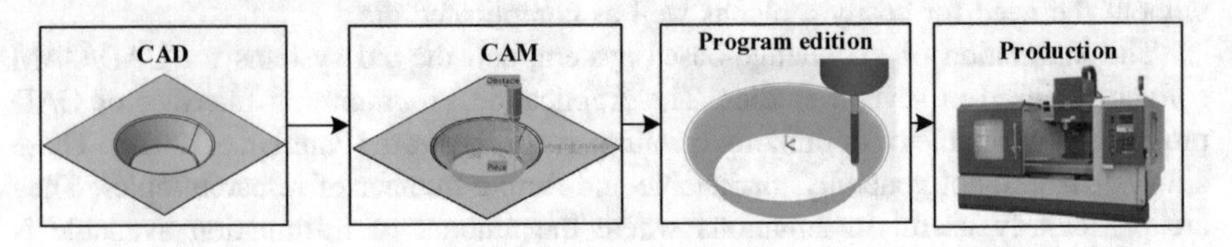

Fig. 1. SPIF methodology

3 API-CATIA Implementation

The newly created model implements several tools making the analysis, the preparation and the programming of the manufacture possible. These tools can be integrated within a complete CAD/CAM system or in the form of independent modules specialized in certain tasks, such as generation of manufacturing trajectories.

Decision support systems are powerful tools integrating scientific methods to support complex and targeted decisions with techniques developed in the information sciences in many fields (Haddad et al. 2015). Aid Systems offer a theoretically correct and attractive way of managing uncertainty and preferences in decision problems. They are based on carefully studied empirical principles underlying the discipline of decision analysis and have been applied successfully in many practical systems. Besides, the role of an aid system is to automate tool path strategy generation from robust models and significantly reduce user interactions and tool path preparation time.

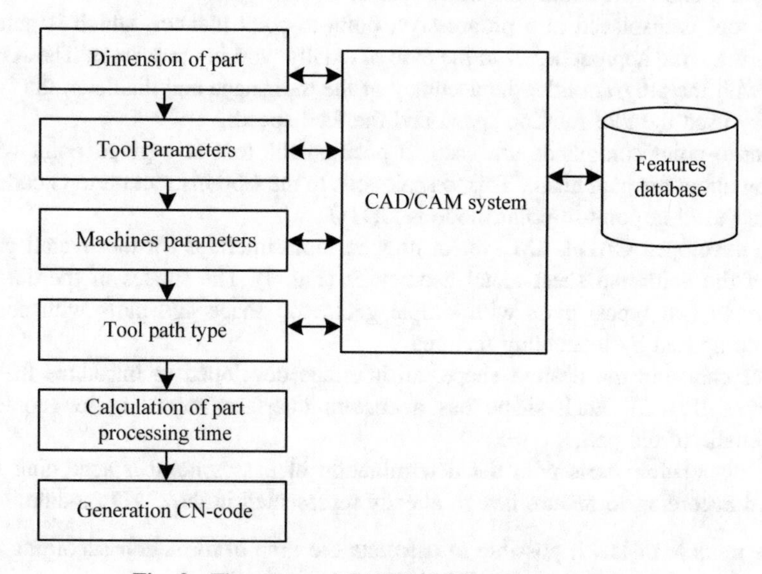

Fig. 2. The proposed CAD/CAM System architecture

In the product development stages, generally direct or inverted engineering is used. In our case, a model of the 3D solid is first conceived in a CAD platform (CATIA) following geometric parameters (diameters, thickness, cones depth, radius), tool parameters (type, diameter) parameters machine (speed rate, feed rate, vertical increment) and tool path type. The CAD model is based on design by Feature. Based on the surface model, the tool path information is generated accordingly. The overall overview of the aid system is shown in a diagram representing the relationships between its components, which are illustrated in Fig. 2.

4 Development of the Proposed CAD System

The proposed CAD system is designed to improve SPIF processes by optimizing parts preparation time and providing faster responses to changes in product design. The system is structured in the form of two modules, namely:

- 3D design module is intended for the automatic modeling of the final part under CATIA V5. The 3D part is visualized under the "Generative Sheet Metal Design" workshop. this module is based in "Features approach". Features approach has been proposed to increase the capabilities of geometric modelers. It reduces the time spent in the development of a product during the design and the manufacturing phase. A feature is defined as a final part surface.
- Calculation module allows to manage all the calculations necessary to generate the NC program and the manufacturing time.

Figure 3 shows flowchart for CAD system with axisymmetric part.

The tool is displaced in a progressive, point-to-point manner, which is generally used during rapid approaches or in the case of axially working processes. The APT file begins with the program title, the accuracy of the tool shape and diameter, the type of trajectory used the tool rotation speed and the feed speed.

Point-to-point commands are used to position the tool at a given point without worrying about the path taken. This is analogous to the G00 instruction in G code. The command used in point-to-point mode is GOTO.

The developed CAD/CAM system uses an input interface for the overall presentation of the deformed sheet metal geometries (Fig. 4). The shapes of the deformed parts are in two types: parts with simple geometric shape and parts with complex shapes composed by interacting features.

After choosing the desired shape, an interface developed to introduce the SPIF parameters (Fig. 5). Each shape has a custom interface according to geometrical characteristic of the part.

The calculation starts with the determination of passes number according to the (Δz) and according to feature design already represented in the CAD module.

- This module makes it possible to calculate the time of formation according to the evolution of the deformation of each point for toolpath.
- This module also allows defining the couple tool/part for each feature and dimensions.
- The outputs of this module are the text file (.txt).

Figure 6 shows a portion of an CN program APT file that must be processed before sending it to the Numeric Control machine.

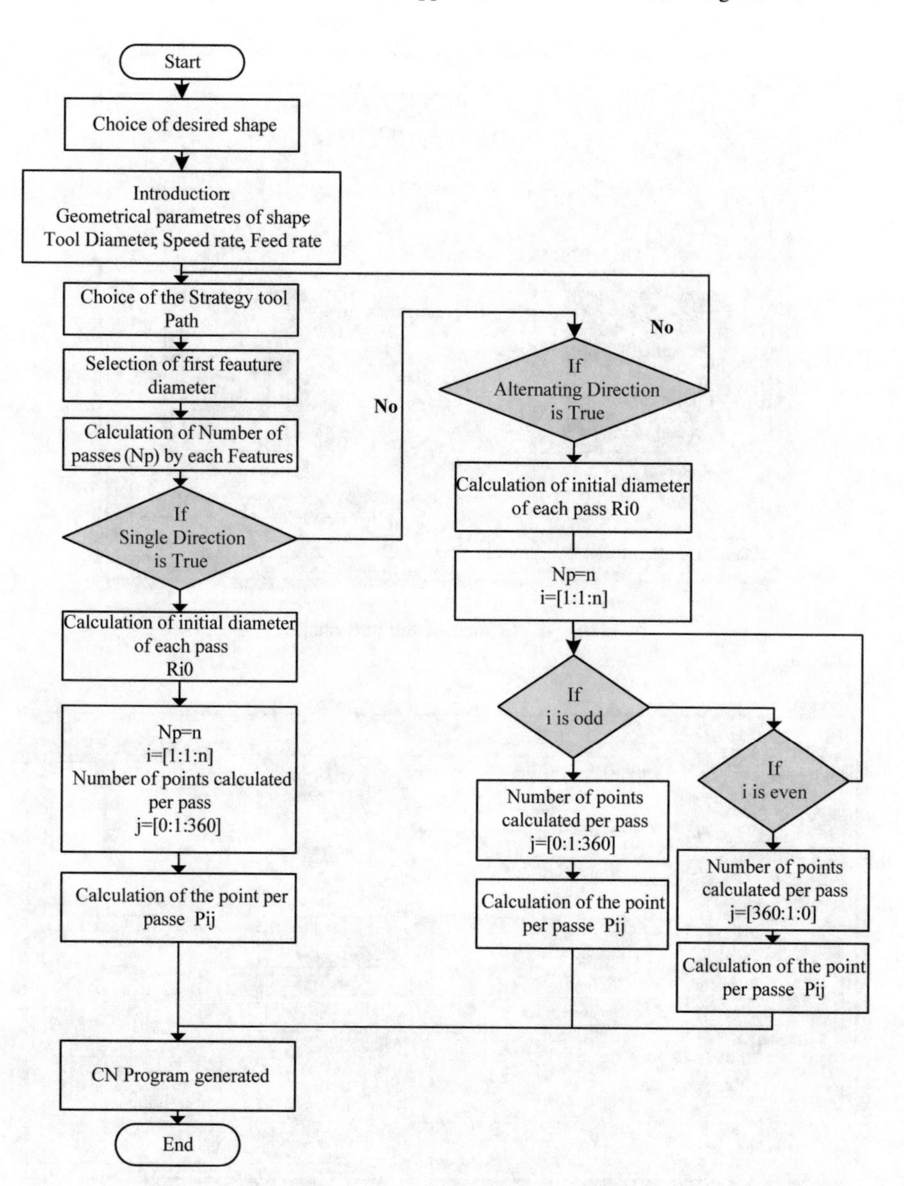

Fig. 3. Flowchart of System created for axisymmetric part

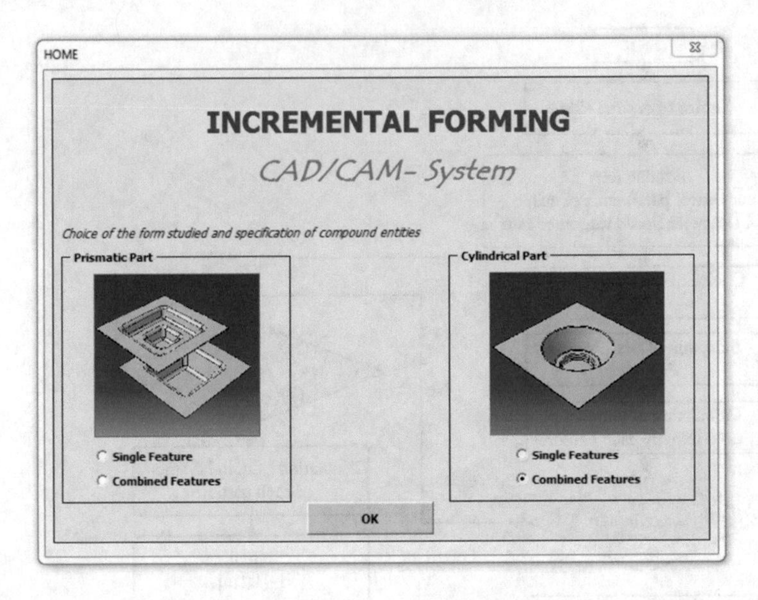

Fig. 4. Choice of the part shape

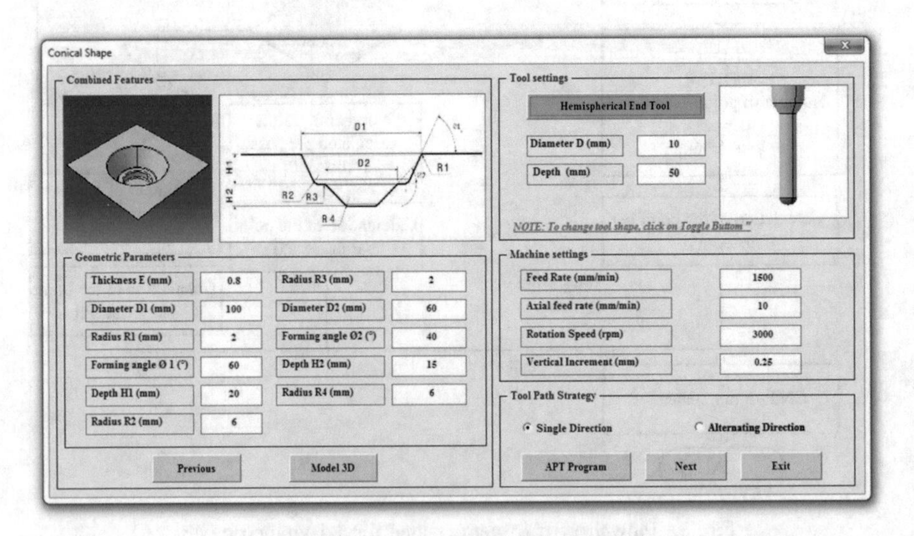

Fig. 5. Geometrical parameters dialog box

```
$$ ____________________________________________
$$      Manufacturing program N°1
$$ ____________________________________________

$$ Manufacturing program.1
MACHIN/ MILL , 1
NOPOST
UNITS/ MM

TLAXIS/ 0, 0, 1
$$ Loading and setting tool 1
TPRINT/Hemispherical Punch D 10
$$
SPINDL,    200,MILL
TOOLPATH_TYPE/3AXIS
TLCOMP/1,ADJUST,1
SPINDL/    1000.0000,RPM,CLW
$$ Starting point
GOTO  /  0,     0,      0
INTOL /      0.03000
OUTTOL/      0.00000
AUTOPS
FEDRAT/ 1000.0000,MMPM

GOTO  /  -16.97344,   61.81081,    0.8
GOTO  /  -19.86841,   60.93911,    0.8
GOTO  /  -22.75925,   59.91483,    0.8
GOTO  /  -26.08668,   58.55057,    0.8
```

Fig. 6. Start of APT program

4.1 Case Study

The SPIF is a complex and nonlinear process due to the multiple variables govern-in the process. SPIF is a process of progressive deformation of the sheet, which consists in obtaining parts of fairly complex shapes. In this work, one complex shape has been tested by the CAD/CAM system. This part is formed of two interacting features. With reference to the geometrical profile, a double truncated cone with circular generatrix was chosen as specimen shape. The first truncated cone is of initial base diameter is equal D1 = 100 mm with a depth of 20 mm, the second truncated cone of diameter D = 60 mm with a depth of 15 mm (Fig. 7a). Experiments were performed on three axis CNC vertical milling machine (SPINER VC 650), equipped with a Siemens SINUMERIK numerical control that assured hemispherical end tool in diameter 10 mm (Fig. 7b).

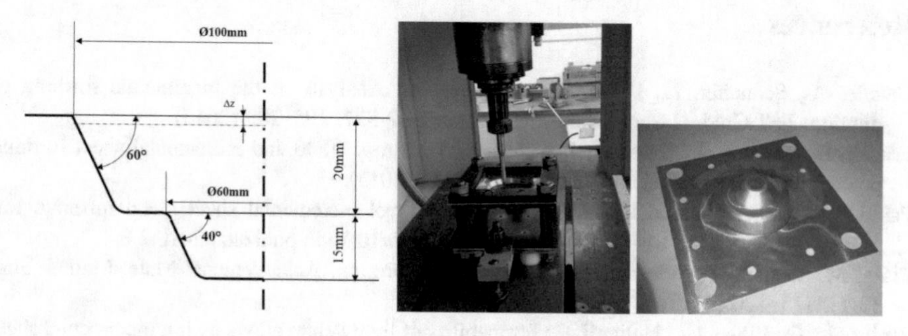

Fig. 7. (a) Dimension of tested part, (b) Manufacturing of part using three-axis CNC machine (c) Shape obtained

The selected forming tool path is in a single direction trajectory with a constant vertical increment of 0.25 mm. During this test, the feed rate is fixed in 1500 mm/min, axial feed rate in 10 mm/min and speed rate in 3000 tr/min. The results obtained are illustrated in Fig. 7c.

5 Conclusion

In this research, we studied the influence of the various parameters of the SPIF incremental forming process throughout the numerical chain from CAD design to manufacturing by CAM system. An original study was conducted on the Incremental Forming Feature and was developed in the CAD part that facilitated the numerical control program for the test machine.

The developed system can automatically manage the following tasks:

- Modeling of the part to be deformed by SPIF process,
- Total time execution of SPIF process operations,
- Toolpath points generation for post-processor.

Acknowledgements. The authors are grateful for the support provided by the CNC laboratory of the Superior Institute of Technological Studies of KEF to use testing machine and validate results of the CAD-CAM approach.

Nomenclatures

n	Number of passes
H [mm]	Part Depth (mm)
j [degree]	Angle of measurement by each pass
P_{ij}:	Point P for the position i and the angle j
x_{ij}: [mm]	Coordinate of the point P_{ij} along the axis (ox)
y_{ij} [mm]	Coordinate of the point P_{ij} along the axis (oy)
z_{ij} [mm]	Coordinate of the point P_{ij} along the axis (oz)
R_{n0} [mm]	Calculated radius of each pass

References

Daleffe, A., Schaeffer, L., Fritzen, D., Castelan, J.: Analysis of the incremental forming of titanium F67 Grade 2 Sheet. Key Eng. Mater. **554–557**, 195–203 (2013)

Fritzen, D., Daleffe, A., Castelan, J., Schaeffer, L.: Brass 70/30 and incremental sheet forming process. Key Eng. Mater. **554–557**, 1419–1431 (2013)

Paniti, I.: CAD API based tool path control for novel incremental sheet metal forming. Int. J. Eng. Inf. Sci. **5**(2), 81–90 (2010). https://doi.org/10.1556/pollack.5.2010.2.8

Tisza, M.: General overview of sheet incremental forming. Achievements Mate. Manuf. Eng. **55**(1), 113–120 (2012)

Ambrogio, G., Filice, L., Gagliardi, F.: Formability of lightweight alloys by hot incremental sheet forming. Mater. Des. **34**, 501–508 (2012). https://doi.org/10.1016/j.matdes.2011.08.024

Jie, L., Jianhua, M., Shuhuai, H.: Sheet metal dieless forming and its tool path generation based on STL files. Int. J. Adv. Manuf. Technol. **23**(2004), 696–699 (2004)
Zhu, H., Li, N., Bai, J.: A CAD/CAM system for integrated process of CNC incremental forming and 5-axis laser cutting. In: Advanced Materials Research, vol. 421, pp. 325–328. Trans Tech Publications (2012). https://doi.org/10.4028/www.scientific.net/AMR.421.325
Tekkaya, A.E., Shankar, R., Sebastiani, G., Homberg, W., Kleiner, M.: Surface reconstruction for incremental forming. Prod. Eng. Res. Devel. **1**(2007), 71–78 (2007)
Skjoedt, M., Hancock, M.H., Bay, N.: Creating helical tool paths for single point incremental forming. Key Eng. Mater. **344**(2007), 583–590 (2007)
Haddad, E., Khalifa, R.B., Yahia, N.B., Zgha, A.: Intelligent generation of a STEP-NC program for machining prismatic workpiece. In: Lecture Notes in Mechanical Engineering, pp. 103–114. Springer International Publishing (2015). https://doi.org/10.1007/978-3-319-17527-0_11

A Novel Approach for Robust Design of Sewing Machine

Najlawi Bilel[1(✉)], Nejlaoui Mohamed[1], Affi Zouhaier[1], and Romdhane Lotfi[2]

[1] LGM, ENIM, University of Monastir, Monastir, Tunisia
najlawibilelali@gmail.com, nejlaouimohamed@gmail.com, zouhaier.affi@enim.rnu.tn

[2] Department of Mechanical Engineering, American University of Sharjah, Sharjah, UAE
lotfi.romdhane@gmail.com

Abstract. In recent years, there has been increasing interest on optimizing the robust design of sewing machines in order to improve their performances. This study presents a novel approach to the multi-objective robust design optimization of sewing machine. A combined multi-objective colonial competitive algorithm (MOCCA) and the Polynomial Chaos expansion (PCE) method is developed and used for the robust multi-objective optimization of the sewing machine. This robust optimization considers simultaneously the motor current, the current fluctuation and their standard deviations. The obtained results showed that the robust design reduces significantly the sensitivity of the sewing machine performances to the design parameters (DPs) uncertainties compared to the deterministic one.

Keywords: Uncertainty · Current fluctuation · Sewing machine

1 Introduction

It is a common practice in the sewing machines design to consider the nominal values only as input variables for design optimization. Najlaoui et al. (2017) developed an optimization problem in order to minimize the consumed energy of the needle bar and thread take up lever (NBTTL) mechanism. Najlawi et al. (2016) developed a multi-objective optimization strategy to minimize the tracking error and the transmission angle index of the NBTTL mechanism. Najlawi et al. (2015) presented an optimization problem based on the imperialist competitive algorithm for optimizing the needle jerk in a sewing mechanism. To estimate the effect of the design parameters uncertainty on the performances of a mechanical systems, several methods have been proposed. In particular, the Polynomial Chaos expansion (PCE) is a popular tool because of its relative accuracy (Rajabi et al. 2015).

The rest of the paper is organized as follows: In Sect. 2, a description of different parts of the needle bar and thread take up lever (NBTTL) mechanism is presented. Then, a dynamic model of the motor driven NBTTL mechanism is developed. In Sect. 3, a multi-objective optimization problem for the robust design of the motor

A. Benamara et al. (Eds.): CoTuMe 2018, LNME, pp. 112–119, 2019.
https://doi.org/10.1007/978-3-030-19781-0_14

driven NBTTL mechanism is formulated. The obtained results are discussed in Sect. 4 and some concluding remarks are shown in Sect. 5.

2 Modeling of the Motor Driven NBTTL System

2.1 The NBTTL Mechanism

The sewing machine is presented in Fig. 1. The NBTTL mechanism, used in sewing machines, consists of a slider crank mechanism and a four-bar mechanism driven by the same crank (Fig. 2).

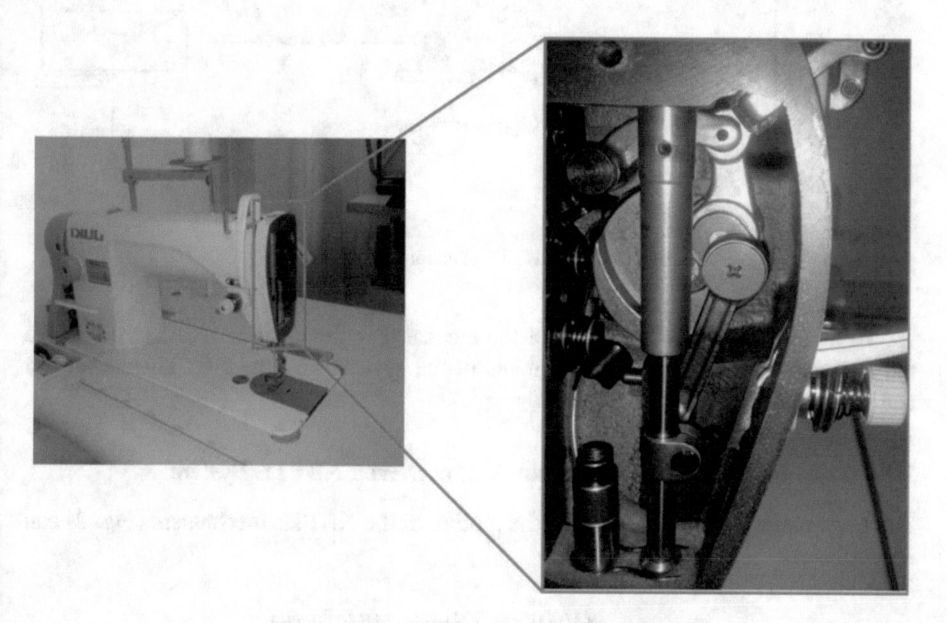

Fig. 1. The NBTTL of a sewing machine

The design of the NBTTL is one of the most important studies in textile industry (Najlawi et al. 2018a). The thread take-up lever mechanism is the four-bar linkage OABC. During the formation of a loop, the take-up lever eye D pulls the upper thread vertically. OEF represents the slider-crank mechanism in which point F denotes the needle. The role of the thread take-up lever in the stitch formation process is to ensure appropriate thread feeding (Nejlaoui et al. 2017). The function of the needle, which is fixed to the needle bar, is to penetrate the fabric. The rotation of the input link (OA) is transmitted to the needle bar through the coupler link EF. The displacement of the needle bar is represented by the distance ‘S’.

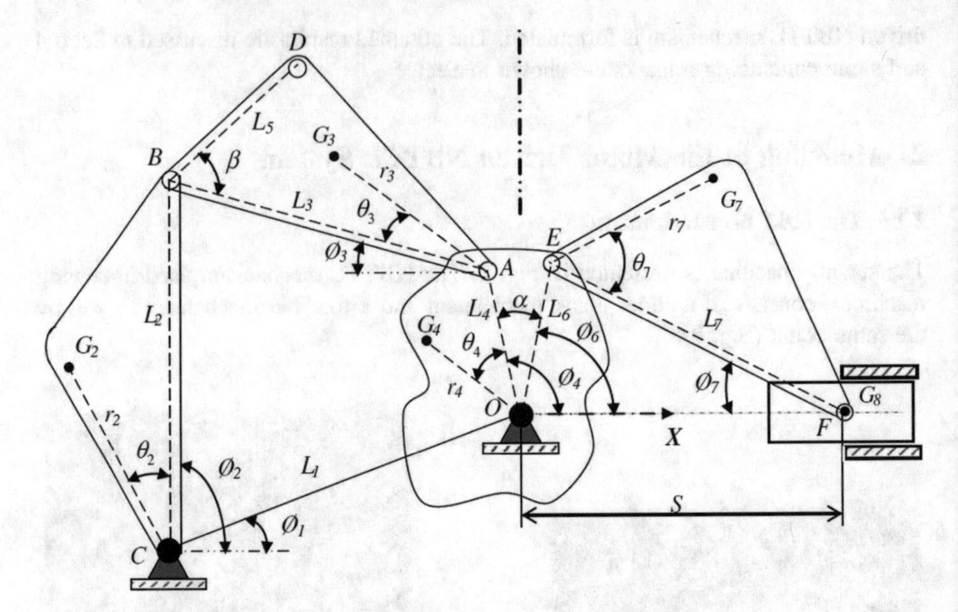

Fig. 2. The NBTTL mechanism parameters

To investigate the performance of the mechatronic device it is necessary to derive the dynamic model of every components of the system, i.e., the NBTTL mechanism and the DC motor.

2.2 The Mechatronic Model of the Motor Driven NBTTL System

The Lagrange's equation describing the motion of the NBTTL mechanism (Fig. 2) can be written as:

$$T_{me} = \frac{d}{dt}\left(\frac{\partial K}{\partial \dot{\phi}_4}\right) - \frac{\partial K}{\partial \phi_4} + \frac{\partial U}{\partial \phi_4} + \frac{\partial D}{\partial \dot{\phi}_4} \tag{1}$$

T_{me}, K, P and D denote, respectively, the driving torque applied to the crank, the kinetic energy, the potential energy and the dissipative energy of the system.

Replacing all these quantities by their expressions yields the following expression:

$$T_{me} = A\ddot{\phi}_4 + \frac{1}{2}\frac{dA}{d\phi_4}\dot{\phi}_4^2 + B \tag{2}$$

Where the expressions of A, and B are given in the appendix.

We present also in Fig. 3, the different characteristics of the used electrical motor.

T_b, R, L, n, K_m, K_g, J, T_L and μ are respectively the motor torque the armature resistance, the inductance, the ratio of the geared speed-reducer, the motor torque constant, the motor voltage constant, the moment of inertia of the rotor, the constant mechanical load torque and the viscous damping at the bearings friction.

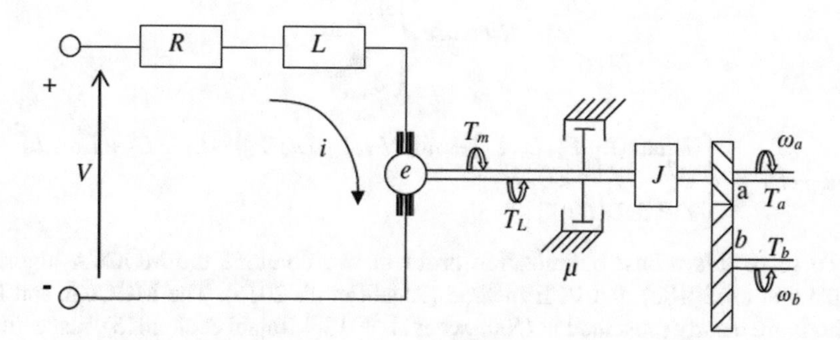

Fig. 3. The DC motor system

The dynamic model of DC motor, as shown in (Najlaoui et al. 2017), is given by:

$$T_b = nK_m i(t) - nT_L - n^2 \mu \dot{\phi}_4 - n^2 J \ddot{\phi}_4 \tag{3}$$

For a mechatronic system, the torque given by the DC motor (Eq. (3)) should be equal to the torque needed by the mechanical system (Eq. (2)).

$$nK_m i(t) - nT_L - n^2 \mu \dot{\phi}_4 - n^2 J \ddot{\phi}_4 = A\ddot{\phi}_4 + \frac{1}{2}\frac{dA}{d\phi_4}\dot{\phi}_4^2 + B \tag{4}$$

At the steady state, the velocity of the crank is assumed to be constant. Thus:

$$i(t) = \frac{1}{nK_m}\left(\frac{1}{2}\frac{dA}{d\phi_4}\dot{\phi}_4^2 + B + nT_L + n^2 \mu \dot{\phi}_4\right) \tag{5}$$

From Eq. (5) we can write:

$$\frac{di(t)}{dt} = \frac{1}{nK_m}\left\{\frac{1}{2}\dot{\phi}_4^2 \frac{d}{dt}\left(\frac{dA}{d\phi_4}\right) + \frac{dB}{dt}\right\} \tag{6}$$

See (Najlaoui et al. 2017) for more details.

3 Robust Design of the Motor Driven NBTTL System

A robust design of the system minimizes, simultaneously, $i_{\max}$, $\left|\frac{di}{dt}\right|_{\max}$ and also their sensitivity to the DP uncertainty. The variability of these objective functions can be quantified by their corresponding standard deviations. The robust design problem of the motor driven NBTTL system can be formulated as:

$$Minimize \begin{cases} \overline{i_{\max}} \\ \overline{\left|\frac{di}{dt}\right|_{\max}} \\ \sigma_{\left|\frac{di}{dt}\right|_{\max}} \\ \sigma_{i_{\max}} \end{cases} \quad (7)$$

$$Subject\,to: \begin{cases} 2[\max(L_1, L_2, L_3, L_4) + \min(L_1, L_2, L_3, L_4)] < L_1 + L_2 + L_3 + L_4 \\ \phi_4^j - \phi_4^{j+1} < 0 \\ DP \in D(DP) \end{cases} \quad (8)$$

To solve this robust optimization problem, we combine the MOCCA algorithm (Najlawi et al. 2018b) and PCE method (Rajabi et al. 2015). The MOCCA and PCE methods are clearly presented in (Najlawi et al. 2018b; Rajabi et al. 2015), respectively. The combined MOCCA-PCE algorithm is presented in (Fig. 4). Thus, the PCE method is integrated into the MOCCA algorithm as given in Fig. 4.

4 Results and Discussion

Using the MOCCA–PCE algorithm, we obtain the robust optimal solutions presented in the Pareto front (Fig. 5).

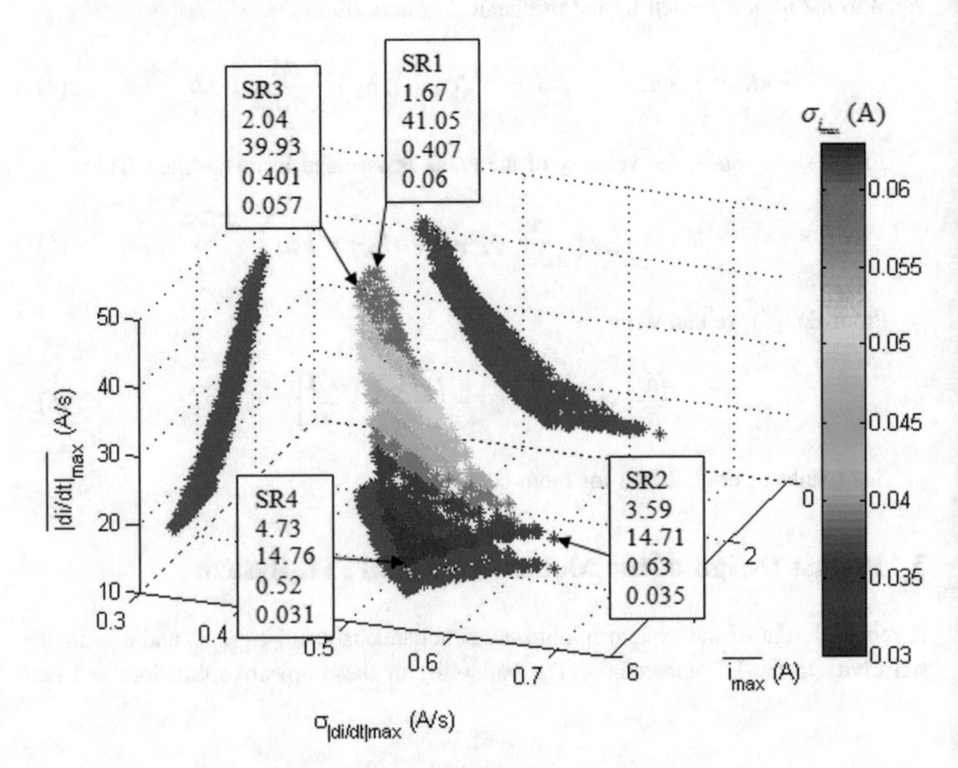

Fig. 4. Pareto front of robust optimal solutions

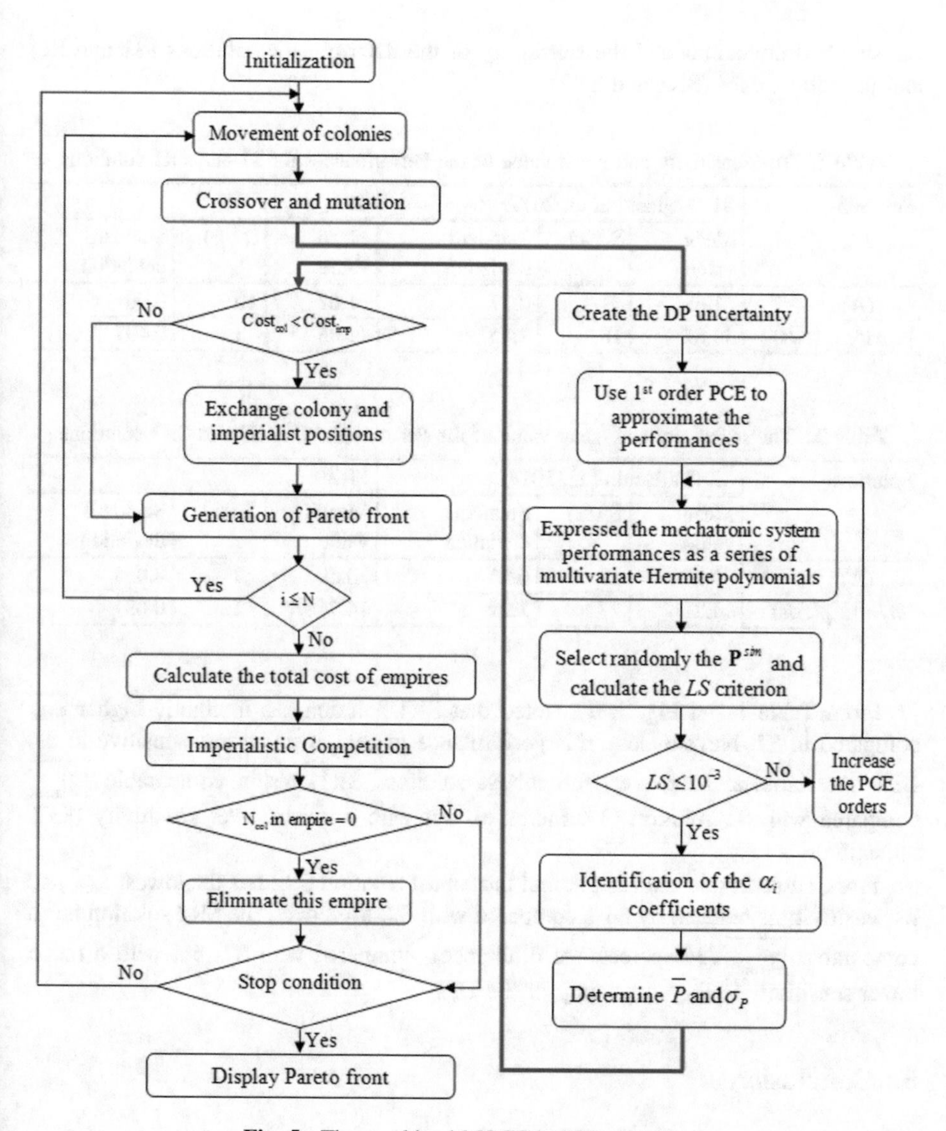

Fig. 5. The combined MOCCA–PCE algorithm.

All the obtained solutions represent a set of robust motor driven NBTTL mechanism design vector and characterized by four objective function values. The solutions are more robust if the corresponding performance standard deviation is reduced. For more clarity, the values of the fourth objective function $\sigma_{i_{\max}}$ were represented by different colors in Fig. 5.

In this section, we try to analyze the advantage of the obtained robust optimal solutions. These robust solutions were compared with the deterministic ones given by (Najlaoui et al. 2017). For more details, we present in Tables 1 and 2 the mean values, the standard deviations and the sensitivity of the deterministic solutions (S1 and S2) and the robust ones (SR1 and SR2).

Table 1. The sensitivity and mean value of the Performances for S1 and SR1 solutions

Solutions	S1 (Najlaoui et al. 2017)			SR1		
	Mean value	S (%)	Standard deviation	Mean value	S (%)	Standard deviation
i_{max} (A)	1.63	31	0.17	1.67	10	0.06
$\|di/dt\|_{max}$ (A/s)	39.87	18	2.45	41.05	3	0.407

Table 2. The sensitivity and mean value of the Performances for S2 and SR2 solutions

Solutions	S2 (Najlaoui et al. 2017)			SR2		
	Mean value	S (%)	Standard deviation	Mean value	S (%)	Standard deviation
i_{max} (A)	3.67	19	0.23	3.59	3	0.035
$\|di/dt\|_{max}$ (A/s)	14.21	27	1.29	14.71	12	0.63

From Table 1 and Fig. 5, it's noted that SR1 solution has a slightly higher $\overline{i_{max}}$ compared to S1. Nevertheless, this performance of SR1 is much less sensitive to the DPs uncertainties. Moreover, the robust solution, SR1, has a comparable $\overline{\left|\frac{di}{dt}\right|_{max}}$ compared with S1 solution (3% increase), but with a much lower sensitivity (83% reduction).

From Table 2, one can remark that the robust solution SR2 has the lowest $\overline{i_{max}}$ and its sensitivity is reduced by 84% compared with S2. Moreover, the SR2 solution has a comparable $\overline{\left|\frac{di}{dt}\right|_{max}}$ (3% percentage difference), compared with S2, but with a much lower sensitivity to the uncertainty of the DPs.

5 Conclusion

This paper introduced a multi-objective robust design optimization of the motor driven NBTTL system, used in sewing machines, under uncertain design parameters. An algorithm, which combines the multi-objective colonial competitive algorithm and the polynomial chaos expansion method, was introduced for the multi-objective robust design optimization of the system where the standard deviation of the objective functions is considered as objective functions.

The obtained results showed that the robust design reduces significantly the sensitivity of the motor driven NBTTL system performances to the DPs uncertainties compared to the determinist results.

Appendix

The expressions of A and B are given, respectively, by:

$$\begin{aligned}A = {} & c_1 + c_2\gamma_2^2 + c_3\gamma_3^2 + c_4\gamma_3\cos(\phi_4 + \phi_3 + \theta_3) + c_5\gamma_5^2 + c_6\gamma_5\cos(\phi_4 + \phi_5 - \theta_5 - \alpha) \\ & + c_7\sin^2(\phi_4 - \alpha) + c_8\gamma_5^2\sin^2(\phi_5) + c_9\gamma_5\sin(\phi_4 - \alpha)\sin(\phi_5)\end{aligned} \tag{A.1}$$

$$\begin{aligned}B = {} & m_2 g r_2\gamma_2\cos(\phi_2 + \theta_2) + m_3 g[L_4\cos\phi_4 + r_3\gamma_3\cos(\phi_3 + \theta_3)] \\ & + m_4 g r_4\cos(\phi_4 + \theta_4) + m_5 g[b\cos(\phi_4 - \alpha) - r_5\gamma_5\cos(\theta_5 - \phi_5)]\end{aligned} \tag{A.2}$$

Where the expressions of c_i $(i = 1, .., 9)$, γ_j and ϕ_j $(j = 2, .., 5)$ are given in (Najlaoui et al. 2017).

References

Najlawi, B., Nejlaoui, M., Affi, Z., Romdhane, L.: Optimal design of the needle bar and thread take up lever mechanism using a multi-objective imperialist competitive algorithm. In: 10th International Symposium on Mechatronics and its Applications (ISMA), 2015, Sharjah (2015). https://doi.org/10.1109/isma.2015.7373463

Najlawi, B., Nejlaoui, M., Affi, Z., Romdhane, L.: Multi-objective robust design optimization of a sewing mechanism under uncertainties. J Intell. Manuf. (2016). https://doi.org/10.1007/s10845-016-1284-0

Najlaoui, B., Nejlaoui, M., Affi, Z., Romdhane, L.: Mechatronic design optimization of the mechanism in a sewing machine. Proc. IMechE, Part C: J. Mech. Eng. Sci. (2017). https://doi.org/10.1177/0954406216687786

Najlawi, B., Nejlaoui, M.: Analytical modeling of needle temperature in an industrial sewing machine. Heat Transf. Res. 1–14 (2018a). https://doi.org/10.1615/heattransres.2018018051

Najlawi, B., Nejlaoui, M., Affi, Z., Romdhane, L.: An efficient evolutionary algorithm for engineering design problems. Soft Comput. 1–17 (2018b). https://doi.org/10.1007/s00500-018-3273-z

Nejlaoui, M., Najlawi, B., Affi, Z., Romdhane, L.: Multi-objective robust design optimization of the mechanism in a sewing machine. Mech. Ind. (2017). https://doi.org/10.1051/meca/2017004

Rajabi, M.M., Behzad, A.A., Craig, T.S.: Polynomial chaos expansions for uncertainty propagation and moment independent sensitivity analysis of seawater intrusion simulations. J. Hydrol. **520**, 101–122 (2015)

Meshfree Analysis of 3-D Double Directors Shell Theory

H. Mellouli[2(✉)], H. Mallek[2], H. Jrad[2], M. Wali[1,2], and F. Dammak[2]

[1] Department of Mechanical Engineering,
College of Engineering, King Khalid University, Abha, Saudi Arabia
mondherwali@yahoo.fr

[2] Laboratory of Electromechanical Systems (LASEM),
National Engineering School of Sfax, University of Sfax,
Route de Soukra km 4, 3038 Sfax, Tunisia
{hana.mellouli,hanen.mallek,
Fakhreddine.dammak}@enis.tn, hanen.j@gmail.com

Abstract. A meshless implementation of arbitrary 3D-model based on a double directors shell element is developed in this work. The meshless technique is based on radial point interpolation method (RPIM) used for the construction of the shape functions for arbitrarily distributed nodes of the shell geometry. The high order shear deformation theory is adopted in this work in order to remove the shear correction coefficient. The convergence of the proposed model is compared to other well-known formulations found in the literature in order to outline the accuracy and performance of the present model.

Keywords: Meshfree method · 3D-model · Double directors shell element · RPIM

1 Introduction

Meshless methods have gained popularity for finding approximate solutions of boundary-value problems due to the ease node placement and accuracy of computed results. Meshless methods were introduced in order to eliminate part of the difficulties associated with reliance on mesh to construct the approximation such as problems with moving discontinuities like crack propagation, mesh alignment sensitivity and problems with large deformations. There are many meshless methods such as the smooth particle hydrodynamics (SPH) (Monaghan (1988)), The reproducing kernel particle (RKPM) (Chen el al. (1996)), and the element free galerkin based on the global weak form (Krysl and Belytschko (1996)).

The main approximation used in meshfree method is the radial basis functions approximation (RBF) used within the interpolator meshless method: the radial point interpolation method (RPIM) (Wang and Liu (2002)) to satisfy the Kronecker delta property.

The two commonly used theories for plates and shells are the Kirchhoff-Love theory (Ivannikov et al. (2014)) and the shear deformation theories (the first-order shear deformation theory (FSDT) (Costa et al. (2013)) and higher-order shear deformation

A. Benamara et al. (Eds.): CoTuMe 2018, LNME, pp. 120–127, 2019.
https://doi.org/10.1007/978-3-030-19781-0_15

theories (HSDT) (Ferreira et al. (2005)). The inefficiency of the Kirchhoff-Love hypothesis appears with neglecting the effects of transverse shear and normal strains of the structure and using the FSDT, shear correction factors should be included to adjust the transverse shear stiffness. The high-order shear deformation theory was established to get better results concerning the shear deformation with parabolic shear strain distribution through the thickness and to avoid the use of the transverse shear correction coefficients. Within the high order shear deformation theory, the double directors shell element is developed with Wali et al. (2014); Frikha et al. (2016) and Mallek et al. (2019a) with finite element in where the vanishing of transverse shear strains on top and bottom faces is considered in a discrete form.

The purpose of this work is to investigate the accuracy of the meshless method in the case of 3D shell using the double directors shell element. With this approach, the quadratic distribution of the shear strain is satisfied and the RPIM is considered as approximation functions. Static analysis of isotropic pinched hemispherical shell with 18° hole with meshless method is examined in this work in order to check the accuracy and performance of the present model.

2 Kinematics of Double Directors Shell Model

In this section, the basic formulations of the free double directors shell elements are presented. For convenience of presentation, the Cartesian coordinate system $(\mathbf{E}_i)$, i = 1, 2, 3, is adopted to describe the shell geometry in the 3D space. To distinguish the initial configuration C_0 from the deformed C_t, capital letters (respectively lowercase letters) are used for quantities relative to the configuration C_0 (respectively C_t). Vectors will be denoted by bold letters.

2.1 Displacement Field and Strains of the Shell Model

According to the double directors shell element, all material points of the shell are defined using parameterizations in terms of curvilinear coordinates $\xi = \left(\xi^1, \xi^2, \xi^3 = z\right)$.

The triple $\left(\boldsymbol{X}_p, \boldsymbol{d}_1, \boldsymbol{d}_2\right)$ defines the position of an arbitrary point 'q' of the shell, $\boldsymbol{X}_p$ gives the position of a point 'p' on the shell midsurface and $\boldsymbol{d}_1$, $\boldsymbol{d}_2$ are the directors unit vectors. The position vector of the point q in the deformed configuration is given by (Mellouli et al. (2019a)):

$$\boldsymbol{x}_q\left(\xi^1, \xi^2, z\right) = \boldsymbol{x}_p\left(\xi^1, \xi^2\right) + f_1(z)\boldsymbol{d}_1\left(\xi^1, \xi^2\right) + f_2(z)\boldsymbol{d}_2\left(\xi^1, \xi^2\right) \tag{1}$$

Where the expressions of $f_1(z)$ and $f_2(z)$ are defined, using the double directors shell model and the quadratic distribution of the shear stress (Wali et al. (2014)), as:

$$f_1(z) = z - 4z^3/3h^2, \quad f_2(z) = 4z^3/3h^2 \tag{2}$$

The virtual membrane, bending and shear strains in the reference state C_0 are expressed as: $(k = 1, 2, \alpha, \beta = 1, 2)$:

$$\begin{cases} \delta e_{\alpha\beta} = \frac{1}{2}\left(\boldsymbol{A}_\alpha.\delta\boldsymbol{x}_{,\beta} + \boldsymbol{A}_\beta.\delta\boldsymbol{x}_{,\alpha}\right) \\ \delta\chi^k_{\alpha\beta} = \frac{1}{2}\left(\boldsymbol{A}_\alpha.\delta\boldsymbol{d}^0_{\mathrm{k},\beta} + \boldsymbol{A}_\beta.\delta\boldsymbol{d}^0_{\mathrm{k},\alpha} + \delta\boldsymbol{x}_{,\alpha}.\boldsymbol{d}^0_{\mathrm{k},\beta} + \delta\boldsymbol{x}_{,\beta}.\boldsymbol{d}^0_{\mathrm{k},\alpha}\right) \\ \delta\gamma^k_\alpha = \boldsymbol{A}_\alpha.\delta\boldsymbol{d}^0_k + \delta\boldsymbol{x}_{,\alpha}.\boldsymbol{d}^0_k, k = 1, 2, \boldsymbol{d}^0_k = \boldsymbol{D}, \alpha, \beta = 1, 2 \end{cases} \tag{3}$$

In matrix notations, these components can be written as:

$$\boldsymbol{e} = \begin{Bmatrix} e_{11} \\ e_{22} \\ 2e_{12} \end{Bmatrix}, \quad \boldsymbol{\chi}^k = \begin{Bmatrix} \chi^k_{11} \\ \chi^k_{22} \\ 2\,\chi^k_{12} \end{Bmatrix}, \boldsymbol{\gamma}^k = \begin{Bmatrix} \gamma^k_1 \\ \gamma^k_2 \end{Bmatrix}, \quad k = 1, 2 \tag{4}$$

2.2 The Weak Form

The numerical solution with the meshfree method is based on the weak form of equilibrium equations. Its form is given as:

$$G = \int_A \left(\boldsymbol{N}.\delta\boldsymbol{e} + \sum_{k=1}^{2} \left(\boldsymbol{M}_k.\delta\boldsymbol{\chi}^k + \boldsymbol{T}_k.\delta\boldsymbol{\gamma}^k\right) \right) dA - G_{ext} = 0 \tag{5}$$

where G_{ext} is the external virtual work and $\boldsymbol{N}$, $\boldsymbol{M}^k$ and $\boldsymbol{T}^k$ represent respectively the membrane, bending and shear stress resultants $(k = 1, 2)$, defined as:

$$\boldsymbol{N} = \int_{-h/2}^{h/2} \begin{bmatrix} \sigma_{11} \\ \sigma_{22} \\ \sigma_{12} \end{bmatrix} dz, \boldsymbol{M}_k = \int_{-h/2}^{h/2} f_k(z) \begin{bmatrix} \sigma_{11} \\ \sigma_{22} \\ \sigma_{12} \end{bmatrix} dz, \boldsymbol{T}_k = \int_{-h/2}^{h/2} f'_k(z) \begin{bmatrix} \sigma_{13} \\ \sigma_{23} \end{bmatrix} dz \tag{6}$$

The generalized resultants of stress $\boldsymbol{R}$ and strain $\boldsymbol{\Sigma}$ are defined as:

$$\boldsymbol{R} = \{\boldsymbol{N} \quad \boldsymbol{M}_1 \quad \boldsymbol{M}_2 \quad \boldsymbol{T}_1 \quad \boldsymbol{T}_2\}^T_{13\times1}, \quad \boldsymbol{\Sigma} = \{\boldsymbol{e} \quad \boldsymbol{\chi}^1 \quad \boldsymbol{\chi}^2 \quad \boldsymbol{\gamma}^1 \quad \boldsymbol{\gamma}^2\}^T_{13\times1} \tag{7}$$

These two components are related by the following equation (Mallek et al. (2018) and Mallek et al. (2019b)):

$$\boldsymbol{R} = \boldsymbol{H}_T\boldsymbol{\Sigma}, \quad \boldsymbol{H}_T = \begin{bmatrix} H_{11} & H_{12} & H_{13} & H_{14} & H_{15} \\ & H_{22} & H_{23} & H_{24} & H_{25} \\ & & H_{33} & H_{34} & H_{35} \\ & & & H_{44} & H_{45} \\ sym & & & & H_{55} \end{bmatrix} \tag{8}$$

where $\mathbf{H}_T$ is the material tangent modulus.

$$(\boldsymbol{H}_{11}, \boldsymbol{H}_{12}, \boldsymbol{H}_{13}, \boldsymbol{H}_{22}, \boldsymbol{H}_{23}, \boldsymbol{H}_{33}) = \int_{-h/2}^{h/2} (1, f_1, f_2, f_1^2, f_1 f_2, f_2^2)\boldsymbol{H} dz \tag{9}$$

$$(\boldsymbol{H}_{14}, \boldsymbol{H}_{24}, \boldsymbol{H}_{34}, \boldsymbol{H}_{44}, \boldsymbol{H}_{54}) = \int_{-h/2}^{h/2} \left(0, 0, 0, (f_1')^2, f_1' f_2'\right)\boldsymbol{H}_\tau dz \tag{10}$$

$$(\boldsymbol{H}_{15}, \boldsymbol{H}_{25}, \boldsymbol{H}_{35}, \boldsymbol{H}_{45}, \boldsymbol{H}_{55}) = \int_{-h/2}^{h/2} \left(0, 0, 0, f_1' f_2', (f_2')^2\right)\boldsymbol{H}_\tau dz \tag{11}$$

where $\boldsymbol{H}$ and $\boldsymbol{H}_\tau$ represent respectively the in plane and out-of-plane linear elastic sub-matrices.

2.3 Meshfree Approximation of High Order Shear Deformation Theory Considering the RPIM

The radial point interpolation method (RPIM) based on radial basis function approximation (RBF) is presented in this section. The RPIM shape functions combine a radial basis function $R^{\mathrm{I}}(X)$ with a polynomial basis function $P^{\mathrm{J}}(X)$. Thus, the approximation of the displacement vector is defined at a point of interest $X = (x, y)$, located in the support domain, as (Mellouli et al. (2019b)):

$$U(x) = \sum_{\mathrm{I}=1}^{N} R^{\mathrm{I}}(X)a_{\mathrm{I}} + \sum_{\mathrm{J}=1}^{M} P^{\mathrm{J}}(X)b_{\mathrm{J}} = \boldsymbol{R}^T(X)\boldsymbol{a} + \boldsymbol{P}^T(X)\boldsymbol{b} \tag{12}$$

where a_{I} and b_{J} represent respectively the non-constants coefficients of $R^{\mathrm{I}}(X)$ and $P^{\mathrm{J}}(X)$. N is the nodal number in the support domain and M denotes the number of monomial terms with M < N. In matrix form, the displacement vector can be rewritten as:

$$\boldsymbol{U} = \boldsymbol{R}\,\boldsymbol{a} + \boldsymbol{P}\,\boldsymbol{b} \tag{13}$$

where $\boldsymbol{R}$ is the radial moment matrix, $\boldsymbol{P}$ represent the polynomial moment matrix, $\boldsymbol{a}$ is the vector of coefficients for RBFs and $\boldsymbol{b}$ illustrates the vector of coefficients for polynomial matrix.

An RBF may have many forms depending on the two shape parameters c and q. Considering a set of nodes $X_1, X_2, \ldots, X_{\mathrm{N}} \in \mathbb{R}^n$. The radial basis functions centered at X_{J} are defined at a point X as:

$$R^{\mathrm{J}}(X) = R(\|X_{\mathrm{J}} - X\|, c, q), \ \mathrm{J} = 1, \ldots, N \tag{14}$$

where $\|X_{\mathrm{J}} - X\|$ is the Euclidean norm. The Multiquadric radial basis function approximation is defined as (Liu et al. (2005)):

$$R^{J}(X) = (\|X_J - X\|^2 + c^2)^q \quad (15)$$

The shape parameter c characterizes the average nodal spacing for all nodes in the local support domain and the shape parameter q is used in this work to be equal to $q = 1.03$ (Liu X et al. 2005) in order to obtain accurate fitting results.

3 Numerical Results and Discussions

A pinched hemispherical shell with 18° hole at the top is analyzed in this section in order to outline the performance and the efficiency of the proposed model. A four concentrated radial loads (two inward and two outward forces apart) are applied to a quadrant of the shell since the symmetry of the problem as seen in Fig. 1. Material and geometrical properties are given as: elastic modulus $E = 6.825 \times 10^7 Pa$, Poisson's ratio $v = 0.3$, radius $R = 10$ mm, thickness $h = 0.04$ mm, and the radial load $P = 1N$. The Multiquadric radial basis function approximation is used in this test, where the shape parameter c is equal to 0.66.

To perform the numerical integrations in the mfree global weak form method, the global background cells are formed by total nodes compared to the finite element method. The performance of the proposed meshfree method using a double directors shell element is evaluated considering numerical integration with 3 × 3 Gaussian quadrature on the background elements for the studied problem. The polynomial basis adopted for this problem is quadratic.

An analytical solution obtained by Steele (1987) and equal to 0.0931 mm, is used to validate the present results compared with those obtained by the double directors finite shell element of Wali et al. (2014). In order to highlight the effect of the number of elements per side, comparison between deflections measured in points where the radial loads are applied, is represented in Table 1. Table 2 summarizes the variation of deflections for several ratio R/h obtained by the present method with comparison of those obtained by the SHO4 finite element of Wali et al. (2014) using 12 × 12 number of elements per side. These last results are normalized with the analytical result of Steele (1987) and represented in Fig. 2.

It can be seen from Table 1 that the present results closely match with the reference results with high performance from 10 elements per side. As seen in Table 2 and Fig. 2, the radius-to-thickness ratio R/h affects the deflection's convergence and it gives good result from a value of 250.

Table 1. Results of the hemispherical shell with 18° hole.

Number of Elements per Side	Deflections (10^{-2} mm)		
	Wali et al. (2014)	The present method	The present method %
8	9,429	8,047	86.43
10	9,374	9,247	99.32
12	9,347	9,33	100.21
16	9,330	9,336	100.27

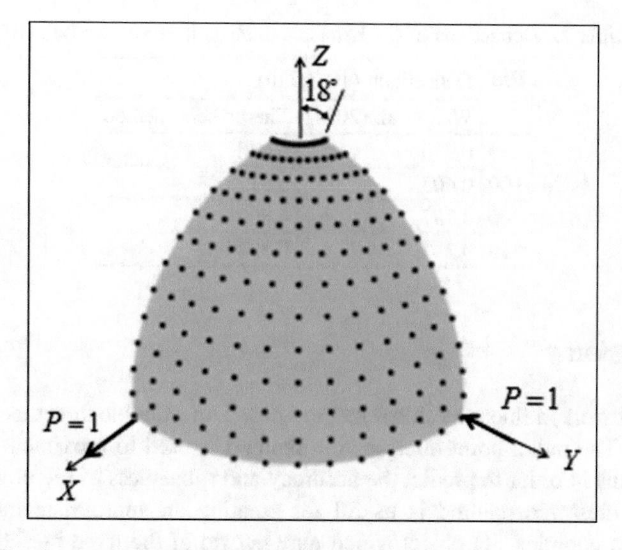

Fig. 1. Geometry of the pinched hemispherical shell with 18° hole.

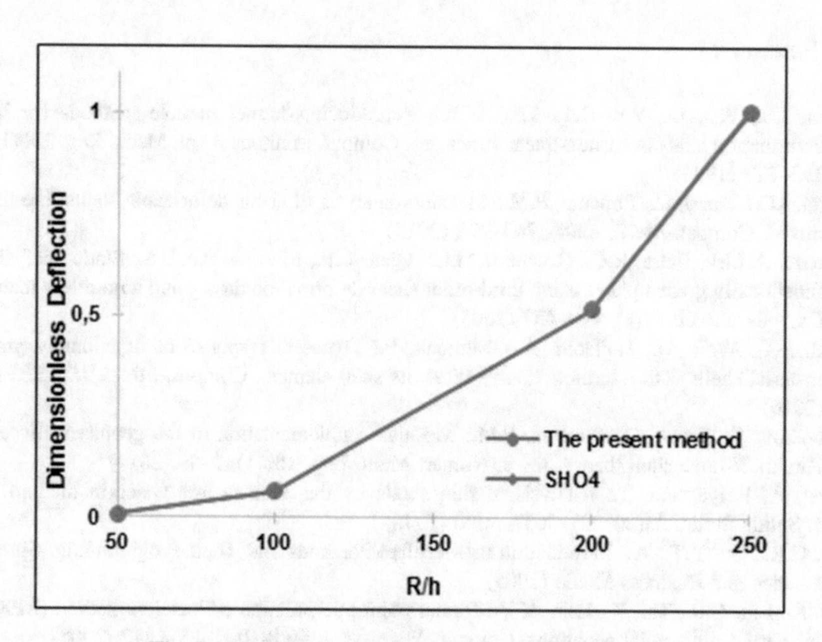

Fig. 2. Dimensionless deflections versus the ratio R/h.

Table 2. Deflections of the hemispherical shell versus the ratio R/h.

R/h	Deflections $\left(10^{-2}\,\mathrm{mm}\right)$	
	Wali et al. (2014)	The present method
50	0,076	0,078
100	0,607	0,611
200	4,797	4,803
250	9,347	9,334

4 Conclusion

In the present work, a linear meshless method based on a double directors shell element is presented. The radial point interpolation method is used to proximate the meshless shape functions in order to predict the accuracy and robustness of the proposed model. The present meshfree method is useful for creating an appropriate mesh from the geometry of a complex 3D object which may get rid of the fixed mesh used in finite element. A good agreement is obtained between the present and the references results.

References

Chen, J.S., Pan, C., Wu, C.T., Liu, W.K.: Reproducing kernel particle methods for large deformation analysis of non-linear structures. Comput. Methods Appl. Mech. Eng. **139**(1–4), 195–227 (1996)

Costa, J.C., Tiago, C., Pimenta, P.M.: Meshless analysis of shear deformable shells: the linear model. Comput. Mech. **52**(4), 763–778 (2013)

Ferreira, A.J.M., Batra, R.C., Roque, C.M.C., Qian, L.F., Martins, P.A.L.S.: Static analysis of functionally graded plates using third-order shear deformation theory and a meshless method. Compos. Struct. **69**(4), 449–457 (2005)

Frikha, A., Wali, M., Hajlaoui, A., Dammak, F.: Dynamic response of functionally graded material shells with a discrete double directors shell element. Compos. Struct. **154**, 385–395 (2016)

Ivannikov, V., Tiago, C., Pimenta, P.M.: Meshless implementation of the geometrically exact Kirchhoff-Love shell theory. Int. J. Numer. Meth. Eng. **100**(1), 1–39 (2014)

Krysl, P., Belytschko, T.: Analysis of thin shells by the element-free Galerkin method. Int. J. Solids Struct. **33**(20–22), 3057–3080 (1996)

Liu, G.R., Gu, Y.T.: An Introduction to Meshfree Methods and Their Programming. Springer Science and Business Media (2005)

Liu, X., Liu, G.R., Tai, K., Lam, K.Y.: Radial point interpolation collocation method (RPICM) for partial differential equations. Comput. Math Appl. **50**(8–9), 1425–1442 (2005)

Mallek, H., Jrad, H., Wali, M., Dammak, F.: Geometrically nonlinear finite element simulation of smart laminated shells using a modified first-order shear deformation theory. J. Intell. Mater. Syst. Struct. (2018). https://doi.org/10.1177/1045389x18818386

Mallek, H., Jrad, H., Wali, M., Dammak, F.: Piezoelastic response of smart functionally graded structure with integrated piezoelectric layers using discrete double directors shell element. Compos. Struct. **210**, 354–366 (2019a)

Mallek, H., Jrad, H., Algahtani, A., Wali, M., Dammak, F.: Geometrically non-linear analysis of FG-CNTRC shell structures with surface-bonded piezoelectric layers. Comput. Methods Appl. Mech. Eng. **347**, 679–699 (2019b)

Mellouli, H., Jrad, H., Wali, M., Dammak, F.: Meshfree implementation of the double director shell model for FGM shell structures analysis. Eng. Anal. Boundary Elem. **99**, 111–121 (2019a)

Mellouli, H., Jrad, H., Wali, M., Dammak, F.: Meshless implementation of arbitrary 3D-shell structures based on a modified first order shear deformation theory. Comput. Math Appl. **77**, 34–49 (2019b)

Monaghan, J.J.: An introduction to SPH. Comput. Phys. Commun. **48**(1), 89–96 (1988)

Steele, C.R.: Private Communication (1987)

Wali, M., Hajlaoui, A., Dammak, F.: Discrete double directors shell element for the functionally graded material shell structures analysis. Comput. Methods Appl. Mech. Eng. **278**, 388–403 (2014)

Wang, J.G., Liu, G.R.: A point interpolation meshless method based on radial basis functions. Int. J. Numer. Meth. Eng. **54**(11), 1623–1648 (2002)

Application of Artificial Intelligence to Predict Circularity and Cylindricity Tolerances of Holes Drilled on Marble

Amira Abbassi(✉), Sofien Akrichi, and Noureddine Ben Yahia

RUSSMTD, University of Tunis, 5 Av Taha Husein, BP 56 Bab Mnara, 1008 Tunis, Tunisia
abassiamira@gmail.com, ak.sofien@gmail.com, Noureddine.benyahia@ensit.rnu.tn

Abstract. High quality marble processing is increasingly needed to ensure surface integrity and meet tight geometric and dimensional tolerances encountered in structural, sculpture and decorative industry. The paper aims at determining optimal drilling parameters for white marble in order to minimize the quality characteristic, namely, the circularity and the cylindricity of holes.

The cutting parameters have an influence on the quality of the machined holes. In order to predict the surface integrity of the parts, a calculation method based on Artificial Neural Networks (ANN) has been developed. An architecture comprising six inputs the rotation speed (N), the feed speed (F), the drill bit diameter (BD), the drill bit height (BH), the number of pecking cycles (P), and the drilling depth (BH) and two outputs (circularity and cylindricity) was used. The choice of cutting parameters has an influence on the convergence of the algorithm. The trained ANNs are monitored as regards the mean square error (MSE).

Keywords: Calacatta-Carrara white marble · Drilling process · Artificial Neural Network (ANN) · Cylindricity · Circularity

1 Introduction

Marble is a rock resulting from the metamorphism of sedimentary carbonate rocks that causes a variable recrystallization of the original mineral carbonate grains. Generally the marble rock is composed of a mosaic of carbonate crystals. This material is widely used in many areas for a long time until today with its different colors and natural patterns. Marble has been commonly used in carving statues, building construct buildings and monuments since ancient times. It is a material used in tiles, countertops and interior flooring (El-Gammal 2011). Most rocks can be classified as brittle materials in which breakage occurs with the formation of small micro-cracks that coalesce to form large cracks. Gunaydin et al. (2004) investigated correlation between the marble sawability and fragility using regression analysis. The study concluded that the sawability of the carbonate rocks can be determined using the fragility of the rock, which is half the product of compressive tensile strengths. Abdullah et al. (2016) conducted a study to examine the influence of cutting parameters on process performance in terms of Kerf

A. Benamara et al. (Eds.): CoTuMe 2018, LNME, pp. 128–134, 2019.
https://doi.org/10.1007/978-3-030-19781-0_16

surface roughness, surface roughness and cone rate for both types of Carrara white marble pieces and Indian green. Wang and Clausen (2002) conducted a detailed study on Carrara marble. The operation of cutting diamond wire depends essentially on the physical, chemical, mechanical and mineralogical-petrographic properties of the rocks, and several researchers who have developed research in this area can be cited (Qiuming et al. 2016); (Elena et al. 2017); (Yilmaz 2011). In addition, the use of Intelligence Artificial (IA) systems in machining research has been developed. Several researchers have studied the influence of machining parameters on the surface condition during different processes (Shanmuga et al. 2009). Gunaydin et al. (2004) used an ANN model as to the sawability prediction of carbonate rocks with large diameter circular saws. The network inputs are the shear strength parameters. Findings are compared with simple and multiple regression models. Drilling is another common process operation. Karataş et al. (2009) opted for an ANN with the algorithm for retro error gradient propagation in learning. They used the results of experimental measurements as training and test data to determine the geometrical defects according to the cutting conditions. The results of the mathematic modeling were approvable. The NN with LM algorithm represents an easy and quick method to explore a nonlinear model. H.-L. Lin developed A neural network (NN) with the Levenberg–Marquardt back-propagation algorithm was adopted to develop the nonlinear relationship between factors and the response (Lin 2012). In this context, we propose a methodological approach for the optimization of an RNA configuration, adapted to predict the circularity and the cylindricity of the drilled holes. The network is powered from the experimental results of several milling trial partners of the one CALACATA marble plot using a diamond tool. The studied neural network is a multi-layer, feed- forward network with backpropagation, comprising an input layer, a hidden layer and an output layer. Henceforth, the quality characteristics of cylindricity and circularity will be referred to as HC and RE, respectively. The ANN is implemented using the ToolboxMatlabTM.

2 Experimental Study

There are many parameters which affect geometric tolerances. The structural parameters of the machine tool are constant for each experiment in this experimental study. The six parameters of the process the Rotation speed (N), the feed speed (F), the drill bit diameter (BD), the drill bit height (BH), the number of pecking cycles (P), and the drilling depth (DD), are summarized in the Table 1. 16 different cutting conditions have been considered.

The drilling process was performed on Calacatta-Carrara white marble using 5-axes CNC vertical machining center (OMAG) with a high precision. The experimental study was carried out in the Tunisian society 'MARBLE TUNIS-CARTHAGE'. The test part used was 300L × 300W × 30H and was Calacatta-Carrara white marble (see Fig. 1).

In the experiments, cylindricity and circularity of the holes are determined using a Coordinate Measuring Machine (CMM) (five-axis CMM; Brown & Sharpe Global Status 9128 5PDEA CMM). The results of the experiment are shown in Table 3. 16 different Machining settings have been considered.

Table 1. The Machining settings

Rotation speed (rpm) (N)	Feed speed (mm/min) (F)	Bit Diam. (mm) (BD)	Bit Height, (mm) (BH)	Pecking cycles (P)	Drilling Depth. (mm) (DD)
4800	350	20	60	3	30
4200	200	15	45	0	10

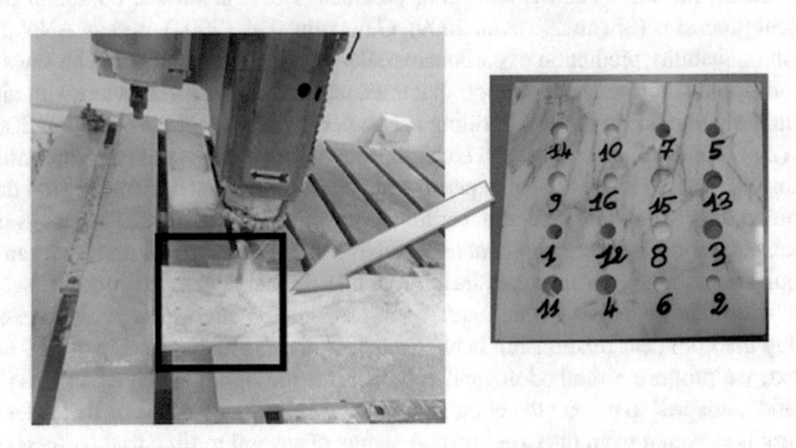

Fig. 1. Experimental set up: 5 axes CNC machining center

From the experimental results, it is observed that the cylindricity and the circularity are sensitive to the pecking cycles (P), the drilling depth (DD) and the rotation speed (N), respectively.

3 Model of the Proposed ANN

The network of the studied artificial neurons is of the feed-forward type, it consists of an input layer, an output layer and a hidden layer with a variable number of neurons (Fig. 2). Information is propagated through the layers from the input layer to the output layer. The response of the network is interpreted from the activation value of its output neurons, including the output vector. At the end of this process, the network should be able to generate the right solutions for examples that have not been seen before. That is the goal of the generalization phase. This process consists in generalizing the output results of the network for entries not belonging to the learning base.

It is very difficult to know which learning algorithm will be the fastest for a given problem. It depends on many factors, including the complexity of the problem, the number of input nodes in the learning set, the number of weights and polarizations in the network, the error value and the use of the neural network in pattern recognition or function approximation. There are several algorithms for learning neural networks that are applied in various applications. During this research, for the best learning algorithm, we tried to apply some algorithms that converge rapidly, and that produce a very

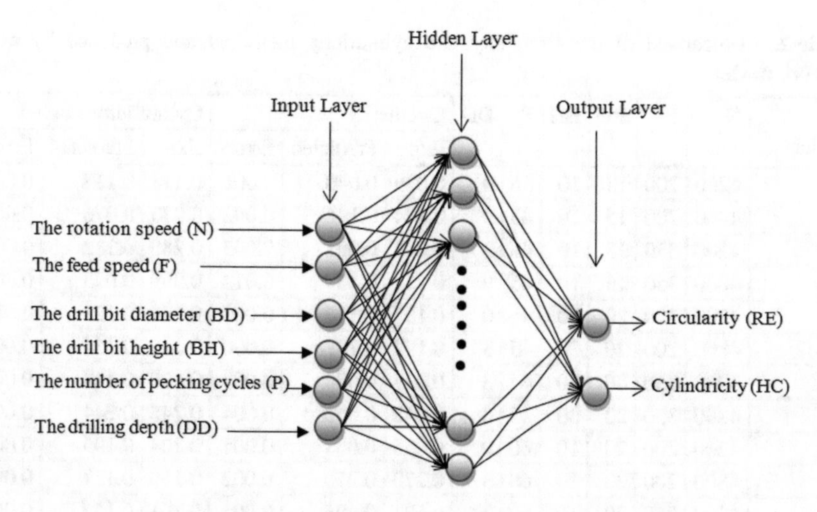

Fig. 2. Neural Network Architecture Used

low mean squared error of learning. The Levenberg-Marquardt (LM) algorithm has the fastest convergence. A neuron in the network produces its input by processing the net input through an activation (transfer) function which is usually non linear. There are several types of activation functions used for BP. However, the sigmoidal activation function is the most utilized. Three types of sigmoid functions are usually used, as follows:

$$f(x) = \frac{1}{1 + e^{-x}} \qquad \text{range } (0, 1) \tag{1}$$

or

$$f(x) = \frac{2}{1 + e^{-x}} - 1 \qquad \text{range } (-1, 1) \tag{2}$$

or

$$f(x) = \frac{e^{x} - e^{-x}}{e^{x} + e^{-x}} \qquad \text{range}(-1, 1) \tag{3}$$

The general architecture of the studied network is "6-X-2" illustrated in Fig. 2 with x is the number of neurons in the hidden layer ranging from 4 to 20.

For the learning phase, 16 examples of inputs/outputs were used (Table 2). The linear regression report of the test (R-test) was be used as an indicator of performance during the training of the network. The obtained results show that the number of neurons in the hidden layer strongly influences the efficiency of the network.

Table 3 shows the variation in MSE results. Data belonging to the 16 trials were used for training 12 data set, 70% of total trials, validating 2 data set, 15% of total trials and testing 2data set, 15% of total trials of ANNs.

Table 2. Comparison of the circularity and cylindricity measured and predicted by neural network model

Std. Order	N	F	BD	BH	P	DD	Circularity			Cylindricity		
							Exp	Predicted	Errors	Exp	Predicted	Errors
1	4200	200	15	10	45	0	0.039	0.040	0.001	0.113	0.112	0.001
2	4800	200	15	10	45	3	0.144	0.146	0.002	0.073	0.076	0.003
3	4200	350	15	10	45	3	0.094	0.091	0.003	0.280	0.282	0.002
4	4800	350	15	10	45	0	0.127	0.139	0.012	0.099	0.093	0.006
5	4200	200	20	30	45	0	0.195	0.186	0.009	0.104	0.121	0.017
6	4800	200	20	30	45	3	0.197	0.197	0.000	0.155	0.154	0.001
7	4200	350	20	30	45	3	0.027	0.024	0.003	0.120	0.118	0.002
8	4800	350	20	30	45	0	0.042	0.046	0.004	0.243	0.244	0.001
9	4200	200	20	10	60	0	0.053	0.058	0.005	0.204	0.199	0.005
10	4800	200	20	10	60	3	0.272	0.270	0.002	0.150	0.151	0.001
11	4200	350	20	10	60	3	0.501	0.496	0.005	0.161	0.157	0.004
12	4800	350	20	10	60	0	0.175	0.187	0.012	0.176	0.166	0.010
13	4200	200	15	30	60	0	0.05	0.075	0.025	0.076	0.053	0.024
14	4800	200	15	30	60	3	0.463	0.461	0.002	0.292	0.293	0.001
15	4200	350	15	30	60	3	0.172	0.195	0.023	0.139	0.128	0.011
16	4200	200	15	10	45	0	0.039	0.054	0.015	0.113	0.123	0.010

Table 3. Results obtained for different architectures

Structure	R Training	R Test	MSE training	MSE Test
6-4-2	0.571	0.603	$6.8474e^{-3}$	$2.2682e^{-2}$
6-6-1	0.742	0.703	$5.0436e^{-3}$	$1.6111e^{-2}$
6-8-2	0.802	0.892	$3.6067e^{-3}$	$1.3552e^{-2}$
6-10-2	0.958	0.728	$1.2339e^{-3}$	$1.9654e^{-2}$
6-12-2	**0.999**	**0.968**	$\mathbf{1.4559e^{-5}}$	$\mathbf{6.0252e^{-3}}$
6-14-2	1	0.700	$1.4425e^{-7}$	$2.7207e^{-2}$
6-16-2	1	0.703	$2.1624e^{-5}$	$2.0351e^{-2}$
6-18-2	1	0.702	$2.4131e^{-5}$	$4.0096e^{-2}$
6-20-2	1	0.667	$1.2053e^{-22}$	$3.7673e^{-2}$

The input layer is associated with factors N, F, BD, HD, BH and P, and 12 PEs are assigned to the hidden layer based on the Kolmogorov theorem. The output layer received two PEs as for the circularity (RE) and cylindricity (HC) responses to achieve the best roundness values for the objective function. It was executed under Matlab.

Circularity error decreases by increasing cutting speed and decreasing feed rate as shown in Fig. 3.

The comparison of response values (RE and HC) of both experimental values and ANN predicted values are shown in Fig. 3(a), (b), respectively. It is observed that the predicted values are close to the experimentally determined values. The results of

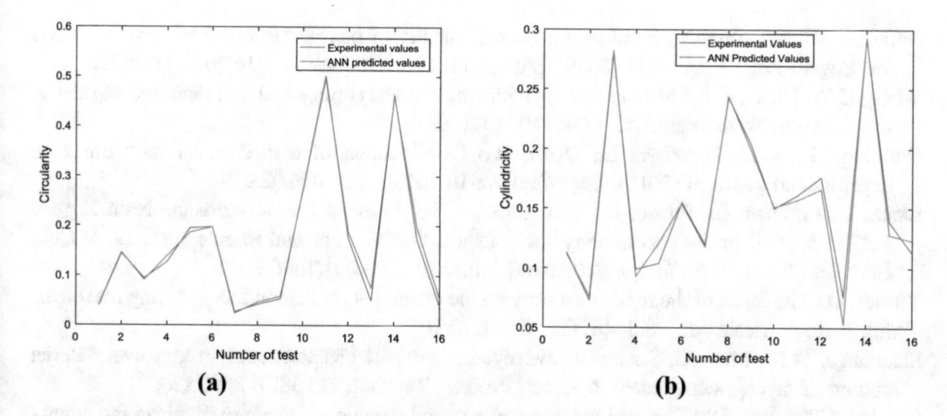

Fig. 3. Results of predicted values of structure 6-12-2 compared to experimental data.

simulations with a 6-12-2 architecture obtained clearly show the performance of the proposed neural network configuration.

4 Conclusion

A computational model based on neural networks has been developed to predict the circularity and cylindricity of the holes generated by a cutting operation. The studied network is a multilayer perceptron feed-forward with three-layer error gradient back-propagation. An architecture comprising six inputs (N, F, BD, HD, BH and P) and two outputs (circularity and cylindricity) was used. A sigmoid activation function in the hidden layer and a linear function in the output layer have been used.

The analysis of the different architectures, based on MSE as a function of the number of neurons in the hidden layer and on the R training performance indicator of the network, clearly show the performance architecture of the chosen net-work. The developed ANN model can be successfully deployed, without experiments in industrial settings, to accurately predict responses considering the negligible error rate achieved.

In order to determine planning production of marble blocks as well as cost estimation, cutting parameters have to be known. We propose, in the future works, to optimize machining costs using optimal cutting parameters.

Acknowledgements. The authors are grateful to MARBLE TUNIS-CARTHAGE Company for providing the Ishikawa dossier and their assistance throughout the experimental study.

References

El-Gammal, M.I., Ibrahim, M.S., El-Sayed, A.B., Asker, S.A., El-Galad, N.M.: Health risk assessment of marble dust at marble workshops. Nat. Sci. **9**, 144–154 (2011)

Gunaydin, O., Kahraman, S., Fener, M.: Sawability prediction of carbonate rocks from brittleness indexes. J. S. Afr. Inst. Min. Metall. **104**, 239–244 (2004)

Abdullah, R., Mahrous, A., Barakat, A.: Surface quality of marble machined by abrasive water jet. Cogent Eng. **3**, 239–244 (2016). https://doi.org/10.1080/23311916.2016.1178626

Wang, C.Y., Clausen, R.: Marble cutting with single point cutting tool and diamond segments. Int. J. Mach. Tools Manuf **42**, 1045–1054 (2002)

Qiuming, G., Xiuli, D., Zhen, L., Qixin, W.: Development of a mechanical rock breakage experimental platform (2016). https://doi.org/10.1016/j.tust.2016.02.019

Elena, T., Lorenzo, L., Renzo, G., Fabrizio, A.: The decay of the polysiloxane resin Sogesil XR893 applied in the past century for consolidating monumental marble surfaces. J. Cult. Heritage (2017). https://doi.org/10.1016/j.culher.2017.03.001/1296-2074

Yilmaz, O.: The effect of the rock anisotropy on the efficiency of diamond wire cutting machines. Int. J. Rock Mech. Min. Sci. **48**, 626–636 (2011)

Shanmuga, D.K., Masood, S.H.: An investigation on kerf characteristics in abrasive waterjet cutting of layered composites. J. Mater. Process. Technol. **20**, 3887–3893 (2009)

Lin, H.L.: The use of the Taguchi method and a neural-genetic approach to optimize the quality of a pulsed Nd: YAG laser welding process. Exp. Tech. (2012). https://doi.org/10.1111/j.1747-1567.2012.00849.x

Karataş, C., Sozen, A., Dulek, E.: Modelling of residual stresses in the shot penned material C-1020 by artificial neural network. Expert Syst. Appl. **36**, 3514–3521 (2009)

Experimental Effect of Cutting Parameters and Tool Geometry in Drilling Woven CFRP

Amani Mahdi[1](✉), Yosra Turki[1], Zoubeir Bouaziz[1], Malek Habak[2], and Souhail El Bouami[2]

[1] Laboratory of Applied Fluid Mechanics, Environment and Process Engineering, National School of Engineers of Sfax, Route Soukra km 3.5, 3035 Sfax, Tunisia
amani.mahdi@enis.tn, yosra.turki@hotmail.fr, zoubeir.bouaziz@enis.rnu.tn

[2] Laboratory of Innovative Technology, IUT of Amiens, GMP Department, University of Picardie Jules Verne, Avenue des Facultés Le Bailly, 80001 Amiens Cedex 1, France
{malek.habak, souhail.el.bouami}@u-picardie.fr

Abstract. This paper focuses on the effect of drill geometry and cutting parameters on the drilling of 4 shaft satin weave carbon fiber and epoxy matrix. Moreover, two different geometry of drilling tool are selected in this study. Surface quality was evaluated in terms of delamination and superficial defects. It was found that the increasing value of thrust force enhanced serious risk of delamination. The augmentation of thrust force is related to feed rate and drill geometry.

Keywords: Woven CFRP · Drilling · Delamination · Thrust force

1 Introduction

Nowadays, carbon fiber reinforced polymer composite (CFRP) is commonly used in many fields due to its specific proprieties such as low weight, heigh strength and stiffness, excellent fatigue and corrosion resistance and low thermal expansion coefficient (Nagaraja et al. 2013). Within this family of materials, woven CFRP has been used in aerospace and aeronautic application especially in hot parts of structures such as rocket engines because of its high thermal protection (Gornet 2008). Nevertheless, the drilling of such material causes defects. The composite is exposed to generate damage during processing due to delamination phenomenon. These damages are related to machining parameters and drill geometry (Phadnis et al. 2013).

Several researchers have studied analytically, numerically and experimentally the process of drilling CFRP. Various authors have studied drilling CFRP unidirectional composites. Qui et al. (2018) have studied the influence of machining parameters and tool structure on cutting force and hole wall damage in drilling unidirectional CFRP with twist and stepped drill. All the fiber in this case were in the same direction. They have been noted that low feed rate and the use of stepped drill reduce the risk of delamination. The influence of feed rate at drilling defects has been also treated by

A. Benamara et al. (Eds.): CoTuMe 2018, LNME, pp. 135–142, 2019.
https://doi.org/10.1007/978-3-030-19781-0_17

Li et al. (2018). They have noted that the lower feed rate was applied, the more defects were disappeared. The study of Turki et al. (2014a, b) has been conducted to understand the influence of drilling on carbon/epoxy unidirectional composite [(0°/ + 45°/90°/-45°)$_3$]$_s$. They have noted that reducing thrust force is the most effective way to decrease the risk of damage. Grilo et al. (2013) have studied the influence of different types of tool geometry in the case of drilling unidirectional CFRP [0°/90°]$_{13}$. Spur drill bit gave the best results compared to twist and four-flute drill. It caused small damage extension at the hole entrance. However, the twist drill presented higher delamination compered to another tool geometry. Similar tests were carried out on unidirectional CFRP in many other papers such as Turki et al. (2014a, b).

Drilling operations of bidirectional carbon fiber reinforced epoxy composite (BCFREC) have also motivated the development of different studies. This dilemma has been proposed by aerospace industry. There are many types of woven applied to BCFREC. The common used one in aerospace industry is woven CFRP composite which is based on 4 shaft satin weave (AS-4) carbon fiber and epoxy matrix. Feito et al. (2017) have developed in their manuscript a rapid estimation of delamination factor and thrust force via numerical analysis of step drill bit performance when drilling AS-4 woven CFRP. They have also noted, in previous work published in 2015, that reamer drill showed the best results in terms of delamination compared to brad and stepped drill. Stepped drill presented an entry delamination higher than exit one. However, brad drill was the worst one compared to other geometry. In the literature, the study of drilling defects of woven CFRP have been poorly developed. So, it is still a challenge to advance the comprehension of the effect of tool geometry and the level of wear.

The objective of this paper is to understand the influence of drilling parameters on BCFREC and the impact of tool geometry at defects. Two different type of drilling tool were utilized: twist and spur drills. Resultant thrust force has been also evaluated together with surface integrity and analyzed in terms of delamination at the entrance and the exit of the hole.

2 Experimental Work

2.1 Workpiece Material and Drills

The material studied in this work is a AS-4 woven CFRP composite based on 4 shaft satin weave carbon fiber and epoxy matrix. It was manufactured by Hexcel composites and it was made in plate of 100 mm × 100 mm × 4.6 mm thick by Zodiac aerospace. It is composed of 18 plies with the same orientation.

Uncoated twist and spur carbide drills were used in this study (Fig. 1). The twist drill has two sides. Its nominal diameter is equal to 6.3 mm. It has 150° tip angle, 5° cutting angle, 11° draft angle and 30° helix angle. However, spur drill has 56° tip angle, 12° cutting angle, 6° draft angle and 30° helix angle. Its nominal diameter is equal to 6 mm.

Fig. 1. Drills used in machining tests

2.2 Machining Tests

The drilling operation was performed on a numerical controlled Charlyrobot CPR0705 machine. During the drilling process, the woven CFRP plate was fixed in particular way to have hole without support (Fig. 2). The used drilling mode in this study was without lubrification. That's why the use of a vacuum cleaner is essential. The support shown in Fig. 2 was fixed on a Kistler 9257B plate. This dynamometer is used to measure the thrust force Fz.

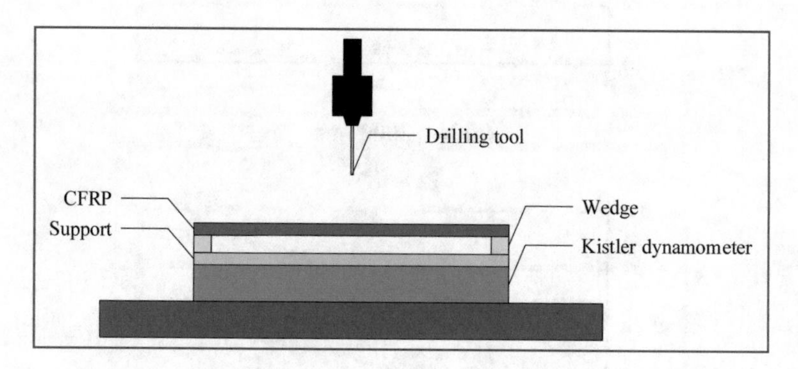

Fig. 2. Experimental device for drilling test

Concerning the cutting conditions, the cutting parameters were summarized in Table 1. Macroscopic observations of the drilling holes were made with a Leica binocular loupe. The position of the drilling hole related to the textile float may influence the delamination phenomenon. However, Feito et al. (2014) have indicated that the influence of this parameter is very low.

Table 1. Cutting parameters used in drilling tests.

f (mm/rev)	N (rpm)
0.05	2000
0.2	3500
0.35	5000

3 Results and Discussion

3.1 Thrust Force

The evolution of the thrust force was recorded using the dynamometer. Figure 3 shows the influence of feed rate f on Fz at various spindle speed N. It can be noted that the increase of Fz is proportional to the increase of f. This result is logic because the

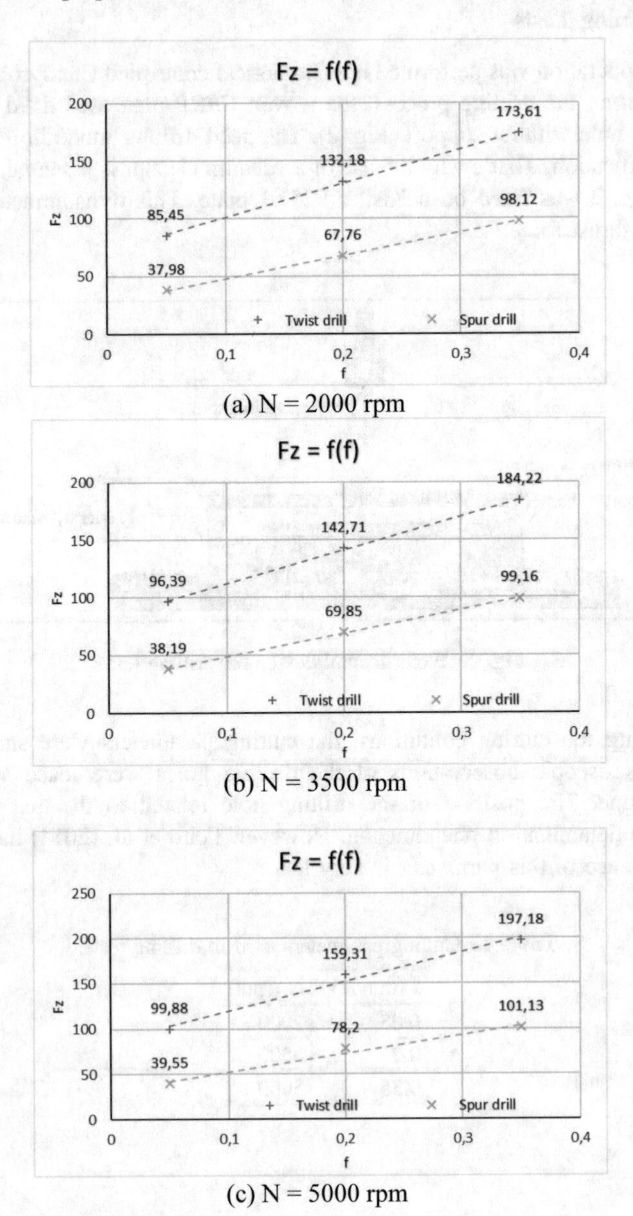

Fig. 3. Fz = f(f) at various spindle speed N

increase of Fz is related to the rise of the thickness to drill per revolution. However, the feed rate has a greater influence on the thrust force than the spindle speed. The increase of N does not have a remarkable effect. It causes a slight increase of Fz value. This result is similar to results obtained in the case of drilling unidirectional CFRP composite (Qui et al. 2018). The thrust force varies also with the type of tool geometry. Fz obtained with spur drill are practically the half of Fz obtained with twist drill.

The main reason of the relation between Fz and tool geometry is that the elevated feed rate lead to an increasing in cutting depth per revolution and the drill bit is required to cut off more materials volumes per revolutions and to overcome much high drilling resistance. The relation between thrust force and tool geometry is caused by the design of the drill bit. Each drilling tool have a specific drill bit.

3.2 Surface Quality

The hole quality was evaluated in terms of hole diameter and visual observation of delamination phenomenon at the entrance and the exit of the hole. The diameter was measured using a micrometer 3P15 (Swiss made). The values of obtained diameter show a good quality of holes using twist or spur tool. Indeed, the error is in order to 0.5% for twist drill and 0.1% for spur drill. However, the delamination phenomenon at the entrance and the exit of the hole increases with f. This is mainly because of high shear stress caused by thrust force and degradation of the material. The severity of this phenomenon is low when f = 0.05 mm/rev and it is more common at the exit of the hole. But the increase of N does not have a remarkable effect. Figure 4 shows macroscopic observations of the delamination phenomenon at N = 2000 rpm using a scale of 2 mm. The choice of the tool geometry impacts the drilling defects. The observed delamination is higher with twist tool at the entrance and the exit of the hole than spur tool.

The use of the twist drill causes more drilling damage than the spur drill. Furthermore, it causes spalling (delamination accompanied by tearing of a piece of ply) at the entrance of the hole and uncut fibers at the exit of the hole with height value of f. However, drilling with spur drill provides a better hole and reduces the effect of cutting parameters. It provides an optimum cutting of hole propriety and delays the delamination onset. The best quality obtained of the hole corresponds at spur drill and f = 0.05 mm/rev.

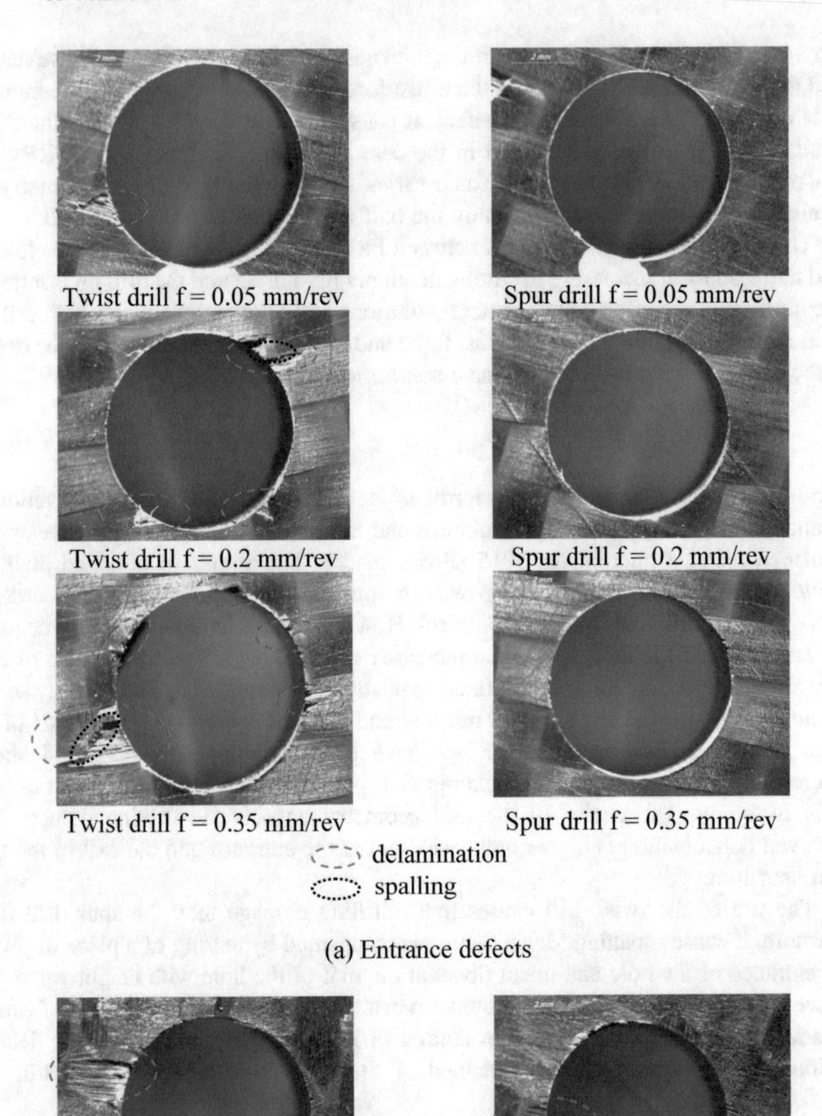

Fig. 4. Macroscopic observations of the delamination phenomenon

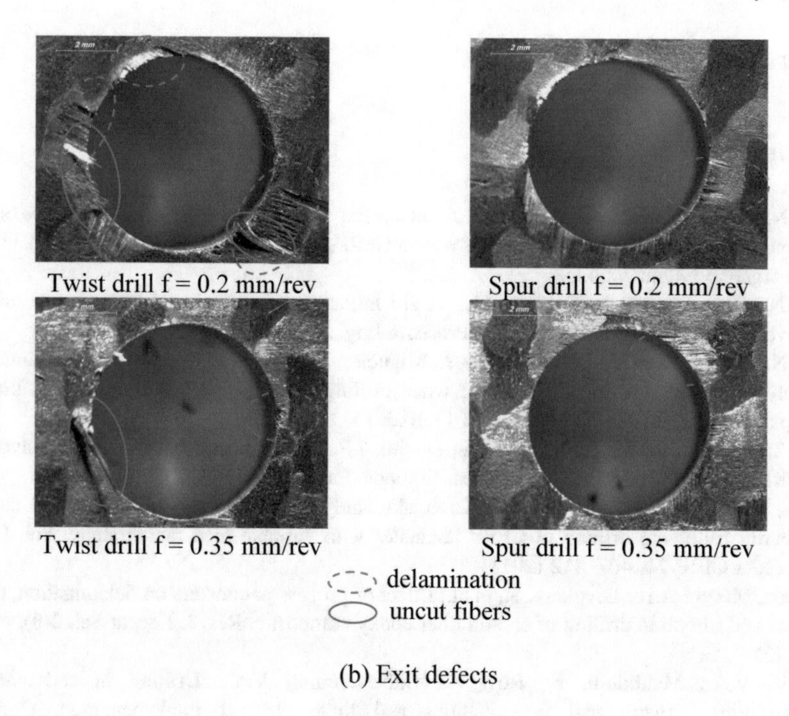

(b) Exit defects

Fig. 4. (*continued*)

4 Conclusion

In this paper, the cutting parameters and the tool geometry in drilling of woven CFRP have been analyzed. The following conclusions can be drawn. The increased value of thrust force is related to enhanced risk of delamination. The tool geometry and the feed rate f have been the most responsible ones of thrust force augmentation and delamination phenomenon. So, the increased value of thrust force should be related to drilling defects. In the literature, delamination in unidirectional CFRP is related to the same causes obtained in this study. However, it seems very important to calculate the delamination factor and to compare delamination in CFRP and BCFREC. A study of defects inside holes is also necessary.

Acknowledgements. This work is partially supported by Laboratory of Applied Fluid Mechanics. The authors also gratefully acknowledge the helpful comments and suggestions of the reviewers, which have improved the presentation.

References

Journal article

Feito, N., Diaz-Alvarez, J., Diaz-Alvarez, A., et al.: Experimental analysis of the influence of drill point angle and wear on the drilling of woven CFRPs. Materials **7**, 4258–4271 (2014). https://doi.org/10.3390/ma7064258

Feito, N., Diaz-Alvarez, J., Cantero, J.L., et al.: Influence of special tool geometry in drilling woven CFRPs materials. Sci. Direct Procedia Eng. **132**, 632–638 (2015)

Feito, N., Diaz-Alvarez, J., Lopez-Puento, J., Miguelez, M.H., et al.: Experimental and numerical analysis of step drill bit performance when drilling woven CFRPs. Comp. Struct. (2017). https://doi.org/10.1016/j.compstruct.2017.10.061

Grilo, T.J., Paulo, R.M.F., Silva, C.R.M., Davim, J.P.: Experimental delamination analyses of CFRPs using different drill geometries. Compos. Part B **45**, 1344–1350 (2013)

Li, M., Leung Soo, S., Aspinwall, D.K., et al.: Study on tool wear and work piece surface integrity following drilling of CFRP laminates with variable feed rate strategy. Sci. Direct Procedia CIRP **71**, 407–412 (2018)

Nagaraja, Mervin A.H., Divokara, S., et al.: Effect of process parameters on delamination, thrust force and torque in drilling of carbon fiber epoxy composite. Res. J. Recent Sci. **2**(8), 47–51 (2013)

Phadnis, V.A., Makhdum, F., Roy, A., Silberschmidt, V.V.: Drilling in carbon/epoxy composites: experimental investigations and finite element implementation. Compos. Part A **47**, 41–51 (2013)

Qui, X., Li, P., Niu, Q., et al.: Influence of machining parameters and tool structure on cutting force and hole wall damage in drilling CFRP with stepped drills. Int. J. Adv. Manuf. Technol. (2018). https://doi.org/10.1007/s00170-018-1981-2

Turki, Y., Habac, M., Velasco, R., et al.: Experimental investigation of drilling damage and stitching effects on the mechanical behavior of carbon/epoxy composites. Int. J. Mach. Manuf. **87**, 61–72 (2014a)

Journal article only by DOI

Turki, Y., Hbak, M., Velasco, K., et al.: Influence of cutting forces and damage on 2D and 3D carbon/epoxy composites in drilling (2014b). https://doi.org/10.4028/www.scientific.net/KEM.611-612.1217

Online document

Gornet, L.: Généralités sur les matériaux composites. Ecole d'ingénieurs (2008). < cel-00470296v1>

Effects of the Tool Bending on the Cutting Force in Ball End Milling

Rami Belguith[1,2,3(✉)], Hassen Khlifi[2], Lotfi Sai[2], Maher Baili[3], Gilles Dessein[3], and Wassila Bouzid[2]

[1] Institut Supérieur des Etudes Technologiques de Gabès, 6011 Teboulbou, Gabes, Tunisia
rami.belguith@gmail.com
[2] Unité de Génie de Production Mécanique et Matériaux, UGPMM, ENIS, 3038 Sfax, Tunisia
hassenreal@hotmail.fr, lotfi_sai@yahoo.com, wassilabouzid@yahoo.fr
[3] Laboratoire de Génie de Production, LGP, ENIT, 65000 Tarbes, France
{maher.baili,gilles.dessein}@enit.fr

Abstract. In this paper, the influence of the tool bending during a ball end milling on the cutting force magnitude is investigated. A thermomechanical model of the cutting force using the cutter workpiece engagement region (CWE) method to determine the actual cutting geometry is developed. The tool bending was considered in calculating the equivalent radius of each discretized element of the tool. The rotational center shifts according to the bending direction. A decrease in the cutting force and a good agreement between the modeled and measured forces is observed.

Keywords: Ball end milling · Cutting forces · Equivalent radius · Bending · Thermomechanical model

1 Introduction

The ball end milling process is frequently used in the manufacturing of molds and the aerospace industries. The surface integrity is highly dependent on the cutting forces. Thereby, the thermomechanical oblique modeling of the cutting forces is used. (Ben Said et al. 2009) and (Abdellaoui and Bouzid 2016) confirmed that this method is considered as convenient to model the cutting forces in high speed machining. Many works have been carried out to develop an accurate thermomechanical model. (Fontaine et al. 2006, 2007) proposed a predictive model of cutting force in ball end milling process based on their previous work (Fontaine 2004) and on the thermomechanical oblique approach outlined by (Dudzinski and Molinari 1997; Moufki et al. 1998, 2000, 2004). Bearing in mind that the uncut chip thickness is non-uniform, the used method consists in discretizing the immersed zone of the tool in the workpiece to elementary cutting edges. The cutting forces are computed as a function of the machining parameters and the thermomechanical behavior of the tool/workpiece materials. One of the main errors affecting the machined surface is the tool bending

A. Benamara et al. (Eds.): CoTuMe 2018, LNME, pp. 143–151, 2019.
https://doi.org/10.1007/978-3-030-19781-0_18

generated by the cutting forces. Therefore, many researchers involved this error to enhance their models. (Sim and Yang 1993) affirmed that the cutter deflection is the main factor which affects the machining stability in ball end milling process. (Kim et al. 2003) analyzed the three dimensional form error due to the elastic behavior of the tool. It was considered as a beam divided into two parts, the shank and the flutes. The contact area tool/workpiece was computed using the Z-map method to calculate the forces and bending.

In the present study, a predictive thermomechanical model of cutting forces in 3-axes machining with ball end mill is established. This model is based on the oblique cutting approach and considering the variation of the material behavior according to the cutting temperature. The engagement region of the immersed zone of the tool in workpiece is predicted using the CWE region method developed by (Sai et al. 2016, 2018) in order to have a good precision of the removed material. The effect of bending is considered where the equivalent radius of the tool due to bending is calculated from the shift of the rotational center. This equivalent radius will affect the geometrical parameters of the tool and its engaged region in the workpiece. The cutting forces and the tool bending are modeled according to the diagram (Fig. 1) based on the equivalent radius. In this stage, we will neglect the runout, tool wear and vibrations. Experiments are carried out in order to validate the developed model in 3-axes machining of the AISI4142 steel with ball end mill.

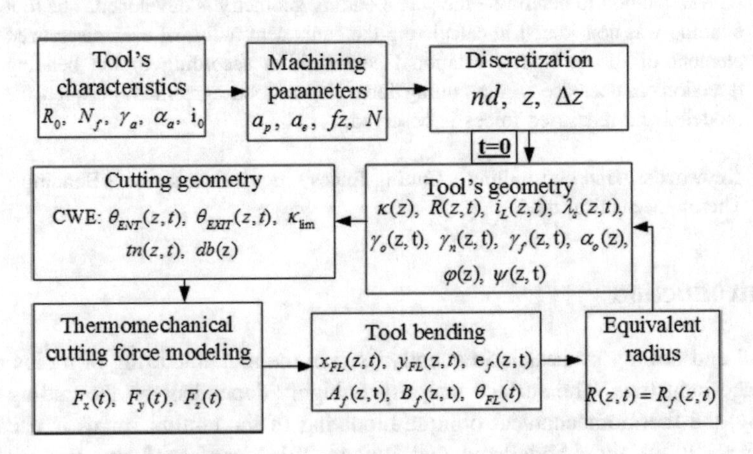

Fig. 1. Cutting force and bending calculation diagram

2 Geometry of the Tool

The geometrical parameters of the ball end-mill are defined in Fig. 2 according to the previous work of (Ben Said et al. 2009). $(C, \overrightarrow{X_S}, \overrightarrow{Y_S}, \overrightarrow{Z_S})$ present the local coordinate system attached to the spindle rotation center C. The immersed depth of the cutting edges in the workpiece is decomposed axially into a series of n_z elements equidistant of Δz.

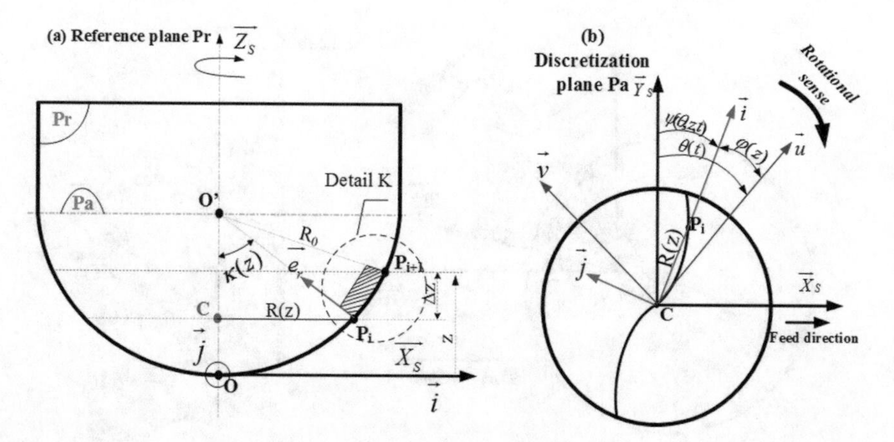

Fig. 2. Geometry of the ball end mill (a) in the plane Pr (b) in the plane Pa

3 Effect of the Bending on the Tool Geometry

The tool bending is the primary defect caused by the cutting forces during machining in ball end milling process. It is calculated in this study in the static conditions considering the small displacements theory. The tool displacements according to X and Y are considered. Whereas displacement due to the vertical force is neglected because of its low value.

The deflections $x_{FL}(z, t)$ and $y_{FL}(z,$ ols Manu$t)$ are calculated as a function of the time t and the position z in the free length L(z) as follows:

$$x_{FL}(z,t) = \frac{F_x(t)L^3(z)}{3EI_{Gz}} \text{ and } y_{FL}(z,t) = \frac{F_y(t)L^3(z)}{3EI_{Gz}} \tag{1}$$

$F_x(t)$ and $F_y(t)$ are the cutting force components in X and Y directions, $L(z)$ presents the free length of the tool, E the young modulus and I_{Gz} the second moment of the circular area defining the tool section.

The effective equivalent radius (Fig. 3) considering the tool bending is calculated as follows:

$$R_f(z,t) = \sqrt{B_f^2(z,t) + (A_f(z,t) - e_f(z,t))^2} \tag{2}$$

$$A_f(z,t) = R(z)\cos(\varphi(z) + \theta_f(z,t)) \text{ and } B_f(z,t) = \sqrt{R^2(z) - A_f^2(z,t)} \tag{3}$$

$$e_f(z,t) = \sqrt{(x_{FL}(z,t))^2 + (y_{FL}(z,t))^2} \tag{4}$$

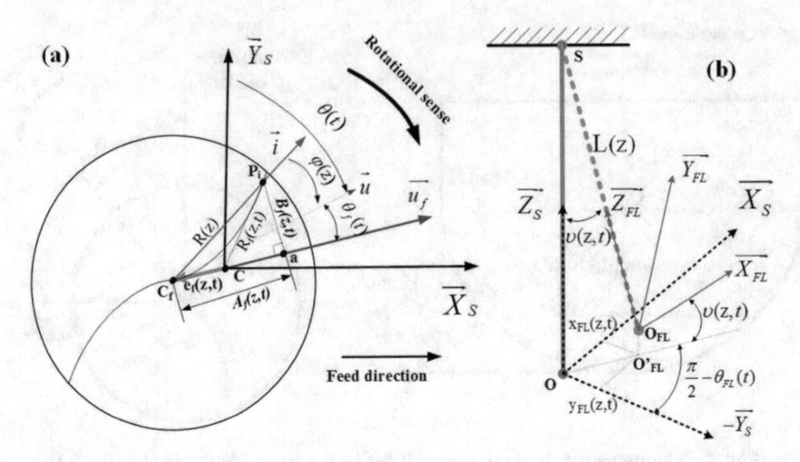

Fig. 3. Bending parameters (a) Equivalent radius in Pa, (b) Displacements of the bent tool

4 Thermomechanical Cutting Force Modeling

The cutting forces, which are a function of the elementary shear force dFs, are calculated using the thermomechanical approach. It is determined as follows:

$$dF_s(z,t) = -\tau_h(z,t)dS_{cs}(z,t) \tag{5}$$

τ_h presents the shear stress modeled using the thermomechanical behavior of the workpiece material (Johnson-Cook law) (Fontaine et al. 2007). The elementary section of cut is calculated from the uncut chip thickness and the chip width by:

$$dS_{cs}(z,t) = \frac{db(z)}{\cos(\lambda_s(z,t))} \frac{t_n(\theta,z,t)}{\sin(\phi_n(z,t))} \tag{6}$$

$$t_n(\theta,z,t) = f_z(t)\sin(\psi(\theta,z,t))sin(\kappa(z)) \quad for \begin{cases} 0 \leq \kappa(z) \leq \kappa_{\lim} \\ ENT \leq \psi(\theta,z,t) \leq EXIT \end{cases} \tag{7}$$

The tool bending is implicitly introduced in the uncut chip thickness t_n when calculating the entry and exit angles with replacing R(z) by the equivalent radius of the bent tool $R_f(z, t)$. The CWE region calculation is detailed in previous works (Sai et al. 2018).

The elementary normal force at the exit of the primary shear zone is:

$$dN_s(z,t) = \frac{\tan(\phi_n(z,t) - \gamma_n(z,t)) + tan(\beta_n(z,t))cos(\eta_c(z,t))}{tan(\beta_n(z,t))cos(\eta_c(z,t))\tan(\phi_n(z,t) - \gamma_n(z,t)) - 1}\cos(\eta_s(z,t))dF_s(z,t) \tag{8}$$

The elementary oblique force components dFr, dFa and dFt are projected on the machine axes (X, Y, Z) to dFx, dFy and dFz as follows:

$$\begin{cases} dF_r(z,t) = -dF_s(z,t)\cos(\eta_s(z,t))\sin(\phi_n(z,t)) - dN_s(z,t)\cos(\phi_n(z,t)) \\ dF_a(z,t) = dF_s(z,t)\cos(\eta_s(z,t))[tan(\eta_s(z,t))\cos(\lambda_s(z,t)) + \cos(\phi_n(z,t)) \\ sin(\lambda_s(z,t))] - dN_s(z,t)sin(\phi_n(z,t))sin(\lambda_s(z,t)) \\ dF_t(z,t) = -dF_s(z,t)\cos(\eta_s(z,t))[tan(\eta_s(z,t))sin(\lambda_s(z,t)) - \cos(\phi_n(z,t)) \\ cos(\lambda_s(z,t))] - dN_s(z,t)sin(\phi_n(z,t))cos(\lambda_s(z,t)) \end{cases} \tag{9}$$

$$\begin{pmatrix} dF_x(z,t) \\ dF_y(z,t) \\ dF_z(z,t) \end{pmatrix} = \begin{bmatrix} \sin(\psi(z,t))sin(\kappa(z,t)) & \sin(\psi(z,t))cos(\kappa(z,t)) & cos(\psi(z,t)) \\ cos(\psi(z,t))sin(\kappa(z,t)) & cos(\psi(z,t))\cos(\kappa(z,t)) & -\sin(\psi(z,t)) \\ -\cos(\kappa(z,t)) & \sin(\kappa(z,t)) & 0 \end{bmatrix} \times \begin{pmatrix} dF_r(z,t) \\ dF_a(z,t) \\ dF_t(z,t) \end{pmatrix} \tag{10}$$

The cutting force components Fx, Fy and Fz are calculated by summing the discretized elements in the engaged region as follows:

$$\begin{pmatrix} F_x(t) \\ F_y(t) \\ F_z(t) \end{pmatrix} = \begin{pmatrix} \int_0^z dF_x(z,t) \\ \int_0^z dF_y(z,t) \\ \int_0^z dF_z(z,t) \end{pmatrix} \tag{11}$$

5 Experimental Work

Experiments were conducted in dry three axis high speed machining using the milling center Huron KX10 (Fig. 4).

The ball end mill is a coated tungsten carbide tool for dry milling. Their principle parameters are: R = 5 mm, L = 100 mm, $i_0 = 30°$, $\gamma_a = 20°$ and $\alpha_a = 12.5°$. The workpiece material is the AISI4142 Steel. A piezoelectric force dynamometer Kistler 9257 B and a charge amplifier 5019A were used to measure the cutting forces.

Fig. 4. Experimental setup (a) Huron center KX10 (b) Tool/Workpiece/Kistler dynamometer

6 Results and Discussion

6.1 Cutting Force Results

Figure 5 shows the predicted variation of the tool bending as a function of the time according to X and Y. The maximum displacement values of the cutter is computed in the tool tip (z = 0) for different free length. It's noticed that the bending takes its maximum in the feed rate direction and rise when increasing the tool's free length.

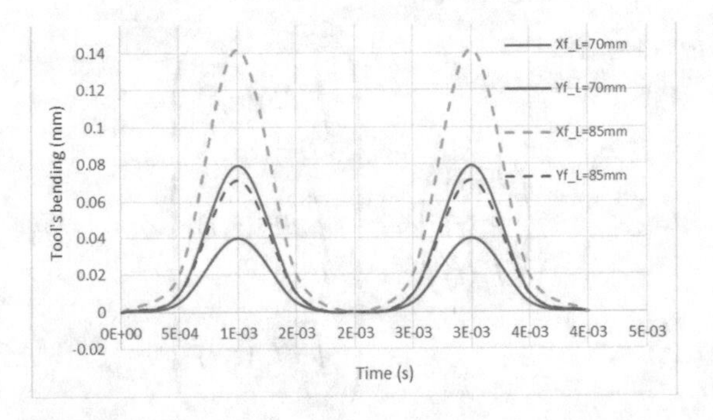

Fig. 5. Predicted tool bending for different free lengths (Nr = 15000 rpm, Vf = 2500 mm/min, ap = 1 mm)

Figure 6 illustrates the measured and the predicted cutting forces considering the tool bending for a free length L = 70 mm.

The global magnitudes stay in a good agreement, but there are some discrepancies especially between the first and the second tooth caused by the tool runout and vibrations not considered in this model.

6.2 Tool Radius

The calculated equivalent effective radius in the upper discretized element considering the tool bending is presented in the Fig. 7.

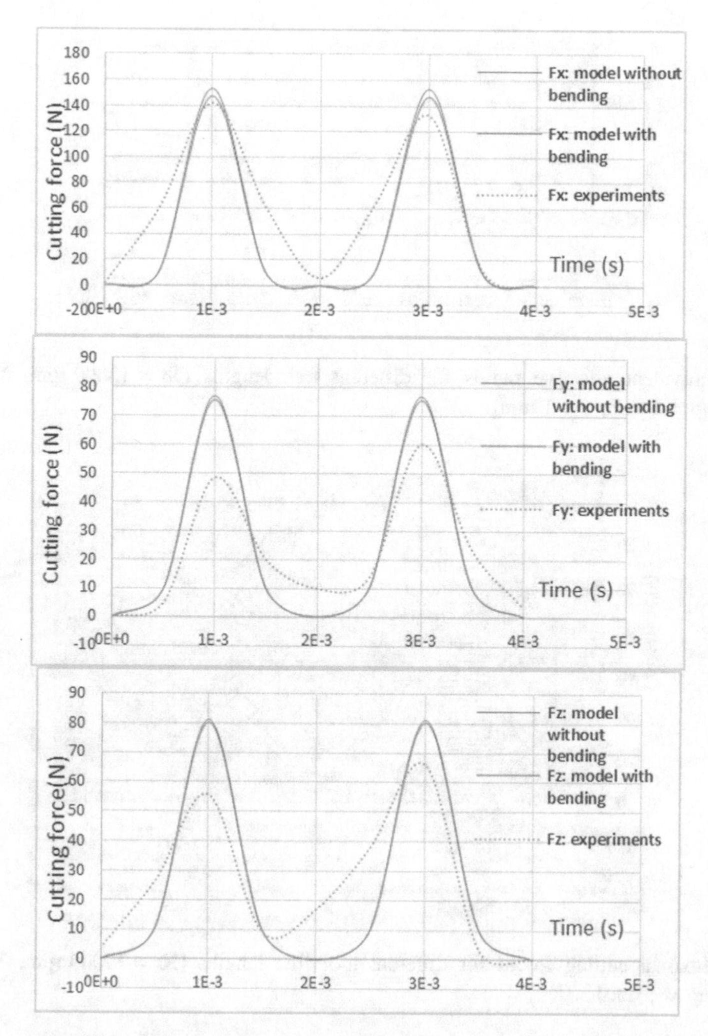

Fig. 6. Modeled and measured forces (a) Fx (b) Fy (c) Fz (Nr = 15000 rpm, Vf = 2500 mm/min, ap = 1 mm)

The elementary radius decreases from its initial value $R(z = 1) = 3\ mm$ in terms of the free length and bending growth.

The maximum cutting force for the three components Fx, Fy and Fz decrease by increasing the tool free length which produce a growth of the bending of the tool (Fig. 8). This is caused by the diminution of the equivalent radius (Fig. 7) and the uncut chip thickness when the cutter shifts inversely to the direction of the forces.

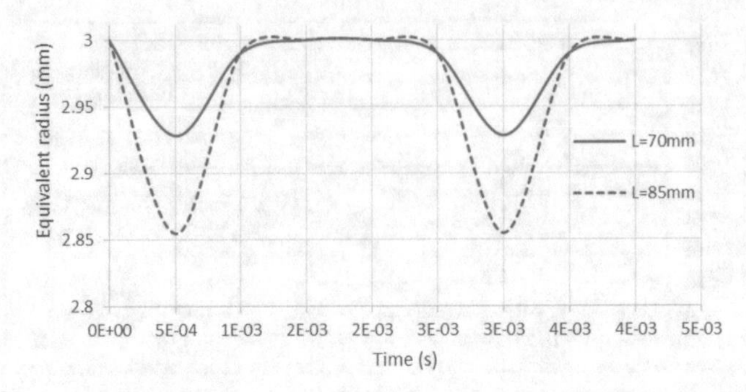

Fig. 7. Equivalent effective radius for different free lengths (Nr = 15000 rpm, Vf = 2500 mm/min, ap = 1 mm, z = 1 mm)

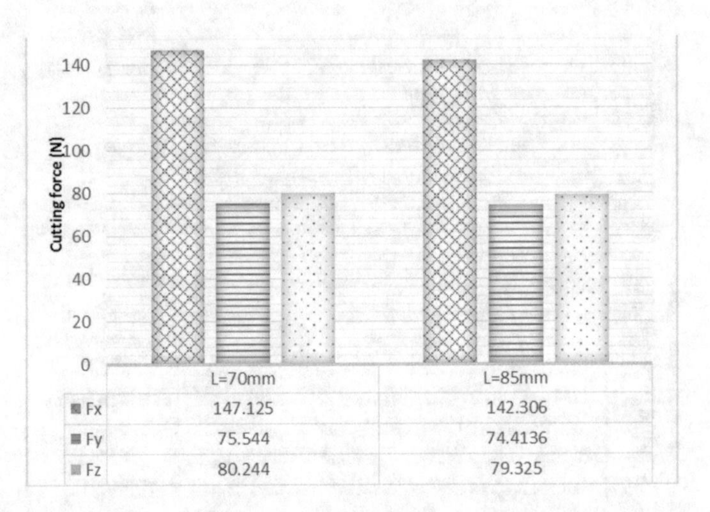

Fig. 8. Maximal cutting forces for different tool free lengths (Nr = 15000 rpm, Vf = 2500 mm/min, ap = 1 mm)

7 Conclusion

In this paper, a new method was developed based on the calculation of the equivalent radius to include the tool bending in the thermomechanical modeling of cutting force. The method of CWE region was used in order to have an accurate uncut chip thickness. The influence of cutting parameters and tool bending on the cutting forces was investigated. The uncut thickness decreases due to the diminution of the equivalent radius, while increasing the tool bending. To have a more accurate model the effect of runout, wear and vibrations on the cutting forces will be considered in future works.

Acknowledgments. The work is carried out thanks to the support and funding allocated to the Unit of Mechanical and Materials Production Engineering (UGPMM/UR17ES43) by the Tunisian Ministry of Higher Education and Scientific Research.

References

Abdellaoui, L., Bouzid, W.: Thermomechanical approach for the modeling of oblique machining with a single cutting edge. Mach. Sci. Technol. **20**, 655–680 (2016). https://doi.org/10.1080/10910344.2016.1224020

Ben Said, M., Saï, K., Bouzid Saï, W.: An investigation of cutting forces in machining with worn ball-end mill. J. Mater. Process. Technol. **209**, 3198–3217 (2009). https://doi.org/10.1016/j.jmatprotec.2008.07.028

Dudzinski, D., Molinari, A.: A modelling of cutting for viscoplastic materials. Int. J. Mech. Sci. **39**, 369–389 (1997). https://doi.org/10.1016/S0020-7403(96)00043-4

Fontaine, M.: Modélisation thermomécanique du fraisage de forme et validation expérimentale. Metz (2004)

Fontaine, M., Devillez, A., Moufki, A., Dudzinski, D.: Predictive force model for ball-end milling and experimental validation with a wavelike form machining test. Int. J. Mach. Tools Manuf. **46**, 367–380 (2006). https://doi.org/10.1016/j.ijmachtools.2005.05.011

Fontaine, M., Moufki, A., Devillez, A., Dudzinski, D.: Modelling of cutting forces in ball-end milling with tool–surface inclination: part I: predictive force model and experimental validation. J. Mater. Process. Technol. **189**, 73–84 (2007). https://doi.org/10.1016/j.jmatprotec.2007.01.006

Kim, G.M., Kim, B.H., Chu, C.N.: Estimation of cutter deflection and form error in ball-end milling processes. Int. J. Mach. Tools Manuf. **43**, 917–924 (2003). https://doi.org/10.1016/S0890-6955(03)00056-

Moufki, A., Devillez, A., Dudzinski, D., Molinari, A.: Thermomechanical modelling of oblique cutting and experimental validation. Int. J. Mach. Tools Manuf. **44**, 971–989 (2004). https://doi.org/10.1016/j.ijmachtools.2004.01.018

Moufki, A., Dudzinski, D., Molinari, A., Rausch, M.: Thermoviscoplastic modelling of oblique cutting: forces and chip flow predictions. Int. J. Mech. Sci. **42**, 1205–1232 (2000). https://doi.org/10.1016/S0020-7403(99)00036-3

Moufki, A., Molinari, A., Dudzinski, D.: Modelling of orthogonal cutting with a temperature dependent friction law. J. Mech. Phys. Solids **46**, 2103–2138 (1998). https://doi.org/10.1016/S0022-5096(98)00032-5

Sai, L., Bouzid, W., Dessein, G.: Cutter workpiece engagement region and surface topography prediction in five-axis ball-end milling. Mach. Sci. Technol. **22**, 181–202 (2016). https://doi.org/10.1080/10910344.2017.1337131

Sai, L., Belguith, R., Baili, M., et al.: An approach to modeling the chip thickness and cutter workpiece engagement region in 3 and 5 axis ball end milling. J. Manuf. Process. **34**, 7–17 (2018). https://doi.org/10.1016/j.jmapro.2018.05.018. Part A

Sim, C., Yang, M.: The prediction of the cutting force in ball-end milling with a flexible cutter. Int. J. Mach. Tools Manuf **33**, 267–284 (1993). https://doi.org/10.1016/0890-6955(93)90079-A

Influence of the Nose Radius on the Cutting Forces During Turning

Hassen Khlifi[1,2](✉), Lefi Abdellaoui[1], Hedi Hamdi[3], and Wassila Bouzid[1]

[1] Unité de Génie de Production Mécanique et Matériaux, ENIS, Route Soukra Km 3, 5, B.P. 1173-3038 Sfax, Tunisia
hassenreal@hotmail.fr, Lefiabdellaoui@yahoo.fr, wassilabouzid@yahoo.fr

[2] Institut Supérieur des Etudes Technologiques de Gafsa, Campus Universitaire Sidi Ahmed Zarrouk, 2112 Gafsa, Tunisia

[3] Laboratoire de Tribologie et Dynamique des Systèmes UMR 5513 CNRS/ECL/ENISE, 58 Rue Jean Parot, 42023 Saint Etienne Cedex 2, France
hedi.hamdi@enise.fr

Abstract. The primary goal of this paper consists of developing a cutting force model in oblique longitudinal turning. A new method, based on the equivalent geometry of cutting tool, is proposed and included in a 3D thermomechanical model of cutting. Roughing and finishing turning operations were considered for validation. The predicted results are in line with experimental data over a wide range of cutting conditions. Besides, it is underlined that the proposed method represents a robust model taking into account tool-workpiece behavior, friction and it leads to reaching cutting force components quickly.

Keywords: Nose radius · Cutting forces · Thermomechanical modeling · Turning · Johnson-Cook behavior

1 Introduction

Controlling cutting forces in the machining process may outcome an enhanced tool life, better surface finish and less chatter vibrations (Campocasso et al. 2014). Hence, the prediction of cutting forces has always been key to manufacturing fields. Different cutting conditions and geometrical parameters influence the machining forces directly. Among these parameters, the effect of the tool's nose radius on the workpiece's roughness, the residual stresses, and the tool wear has been noted in many researches (Beauchamp and Thomas 1996; Endres and Kountanya 2002; Childs et al. 2008; Meyer et al. 2012; Orra and Choudhury 2018). This significant fact justifies the requirement of including the nose radius in the cutting force models. Several studies have been conducted on predicting the machining forces using tools with a nose radius (Imani and Yussefian 2008; Campocasso et al. 2014; Abdellaoui and Bouzid 2016a). The tool nose radius causes a non-uniform distribution of the chip thickness to which the cutting forces are directly related. Therefore, the chip formation becomes a complicated 3D process.

A. Benamara et al. (Eds.): CoTuMe 2018, LNME, pp. 152–159, 2019.
https://doi.org/10.1007/978-3-030-19781-0_19

Empirical approaches are used to estimate cutting forces (Tang et al. 2014). The empirical coefficients are obtained for a specific pair of tool/workpiece and for a limited range of cutting conditions. These limitations lead some authors to develop an analytical approach for computing cutting forces. Imani and Yussefian (2008) have considered each element of the tool's rounded corner as a single straight cutting edge, where a local cutting-edge direction angle was defined, and the oblique theory (Ben Said et al. 2009) is applied to get the local cutting forces. Likewise, in the work of (Abdellaoui and Bouzid 2016a), a new thermomechanical approach was developed for each element of the discretized geometry to deal with the non-uniform uncut chip area. An elementary edge direction angle, an elementary depth of cut, and an average uncut chip thickness were determined for each discretized element. However, the analytical methods based on geometry discretization were confined to some limitations such as the complexity of the cutting forces computation, the dependence of the results accuracy on the number of discretized elements which highly increase the time computation. Some authors have proposed a solution which takes advantage of the analytical thermomechanical models with the existence of the nose radius, and it may overcome the drawbacks of discretization. The concept of the equivalent cutting edge was developed by (Arsecularatne et al. 1995). They discretize the cross section into infinitesimal elements; for each element the friction force is estimated, assuming it collinear with the local direction of flow. They finally summed over the entire section of the chip to find the friction resultant and the corresponding equivalent chip flow direction.

In this paper, a new original method to define an equivalent geometry is proposed with the goal to overcome the limitations mentioned above. The effect of the nose radius on the cutting forces has been studied using a thermochemical cutting forces model.

2 Tool Geometry Modeling

As considered in the mechanistic approaches, simply, the total cutting force can be obtained by multiplying the area of the cross-section of cutting by the specific cutting force. One can conclude that, for given cutting conditions, the resultant cutting force is still the same if the cross-section of cutting S_{real} is still the same as well. (Figure 1).

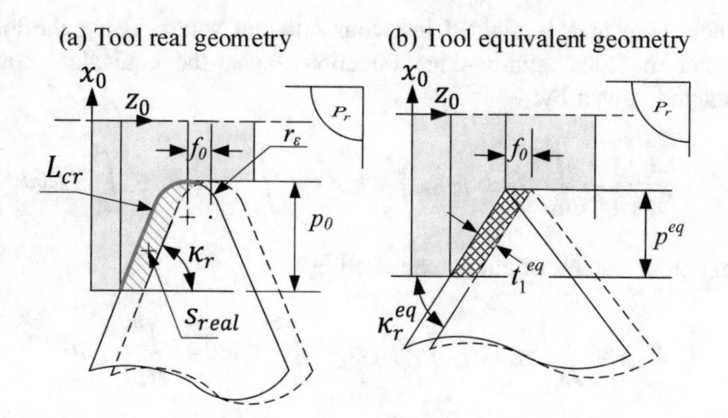

Fig. 1. Definition of the equivalent tool geometry during turning operation

Hence, for the new equivalent geometry, this cross-section equality is well-maintained, and the equivalent mean uncut chip thickness is expressed by:

$$t_1^{eq} = f_0 \sin\left(\kappa_r^{eq}\right) \quad (1)$$

Where the challenge is to find the corresponding equivalent cutting-edge direction angle κ_r^{eq}. To accomplish this, and knowing the tool nose radius (r_ε), the cutting-edge direction angle (κ_r), the depth of cut (p_0) and the feed (f_0), three cases may be distinguished according to (Abdellaoui and Bouzid 2016a). In this work, only the two cases of roughing and finishing are studied (Fig. 2). The uncut chip area is divided into zones and an equivalent cutting edge is defined as the uncut chip area is the same for the real geometry and the equivalent one.

The roughing case is considered when the depth of cut is large enough until the rounded corner is fully engaged in the workpiece (Fig. 2a).

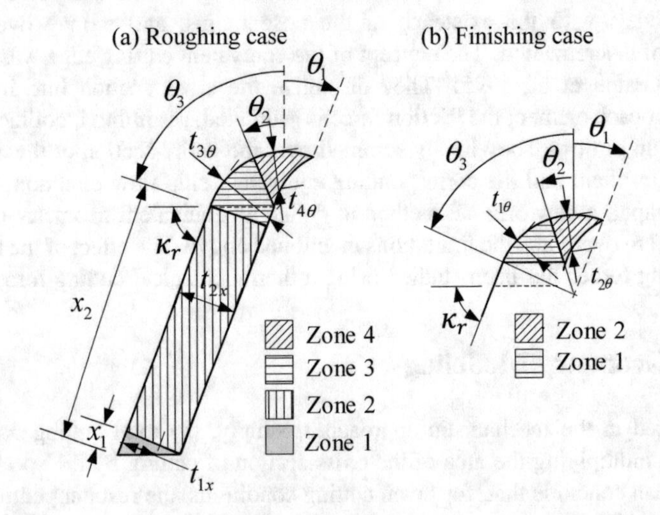

Fig. 2. Equivalent tool geometry for (a) roughing case (b) finishing case during turning

The uncut chip area is divided into four different zones. Using the local uncut thickness and the local cutting-edge direction angle, the equivalent cutting-edge direction angle is given by:

$$\kappa_r^{eq} = \frac{1}{S_{real}}\left\{\kappa_r \int_0^{x1} t_{1x}dx + \kappa_r \int_{x_1}^{x_2} t_{2x}dx + \int_{\theta_2}^{\theta_3} t_{3\theta}\theta d\theta + \int_{\theta_1}^{\theta_2} t_{4\theta}\theta d\theta\right\} \quad (2)$$

The real cross-section of cut is expressed by:

$$S_{real} = \int_0^{x1} t_{1x}dx + \int_{x_1}^{x_2} t_{2x}dx + \int_{\theta_2}^{\theta_3} t_{3\theta}d\theta + \int_{\theta_1}^{\theta_2} t_{4\theta}d\theta \quad (3)$$

The mean uncut thickness t_{1m} in the real geometry and the equivalent depth of cut in the equivalent geometry are deduced as follows:

$$S_{real} = t_{1m} L_{cr} = f_0 p^{eq} \tag{4}$$

The length of the engaged cutting edge in the workpiece is calculated by:

$$L_{cr} = \frac{p_0 - r_\varepsilon(1 - \cos(\kappa_r))}{\sin(\kappa_r)} + r_\varepsilon(\kappa_r + \theta_1) \tag{5}$$

On the other hand, the finishing case is considered when the depth of cut is as small as the rounded cutting edge is not entirely engaged in the workpiece (Fig. 2b). The uncut chip area is divided into two different zones. The equivalent cutting-edge direction angle is given by:

$$\kappa_r^{eq} = \frac{1}{L_{cr} t_{1m}} \left\{ \int_{\theta_2}^{\theta_3} t_{1\theta} \theta \, d\theta + \int_{\theta_1}^{\theta_2} t_{2\theta} \theta \, d\theta \right\} \text{ where } L_{cr} = r_\varepsilon(\theta_3 + \theta_1) \tag{6}$$

3 Cutting Forces Modeling

A Johnson-Cook law is used to describe the thermomechanical behavior of the chip flow in the primary shear zone. The adopted friction law is dependent on the mean temperature at the tool-chip interface. This law is established by (Abdellaoui and Bouzid 2016b) with ($\alpha_1 = 0.5$; $\alpha_2 = 0.32$ *and* $\alpha_3 = 0.12$):

$$\mu = \tan(\lambda) = \alpha_1 \left(1 - \left(\frac{T_{\text{int}}}{T_m} \right)^{\frac{\alpha_2}{t_1^{eq}}} \right) + \alpha_3 \tag{7}$$

Once the equivalent geometry is defined with the equivalent parameters (t_1^{eq}, κ_r^{eq} and p^{eq}), an analytical model is used to predict the cutting forces and the tool-chip contact temperature. The resultant force exerted by the workpiece on the tool is resolved into three components, F_x, F_y, and F_z with respect to the machine's 3-axis coordinates (Fig. 3). Their expressions are given by:

$$\begin{cases} F_x = -R_{T/C} \cos\lambda \cos\kappa_r^{eq} [\cos\gamma_n (\sin\lambda_s \tan\kappa_r^{eq} + \tan\gamma_n) \\ \quad + \mu \cos\eta_c (\cos\lambda_s \tan\kappa_r^{eq} (\sin\gamma_n \tan\lambda_s - \tan\eta_c) - \cos\gamma_n)] \\ F_y = R_{T/C} \cos\lambda \cos\lambda_s (\cos\gamma_n + \mu \cos\eta_c (\sin\gamma_n + \tan\eta_c \tan\lambda_s)) \\ F_z = R_{T/C} \cos\lambda \cos\kappa_r^{eq} [\cos\gamma_n (\sin\lambda_s - \tan\kappa_r^{eq} \tan\gamma_n) \\ \quad + \mu \cos\eta_c (\cos\gamma_n (\tan\kappa_r^{eq} + \tan\gamma_n \sin\lambda_s) - \cos\lambda_s \tan\eta_c)] \end{cases} \tag{8}$$

The resultant force $R_{T/C}$ is expressed by the following:

$$R_{T/C} = \frac{F_{sh}\cos(\eta_{sh})}{\cos(\lambda)\cos(\phi_n - \gamma_n) - \cos(\eta_c)\sin(\lambda)sin(\phi_n - \gamma_n)} \tag{9}$$

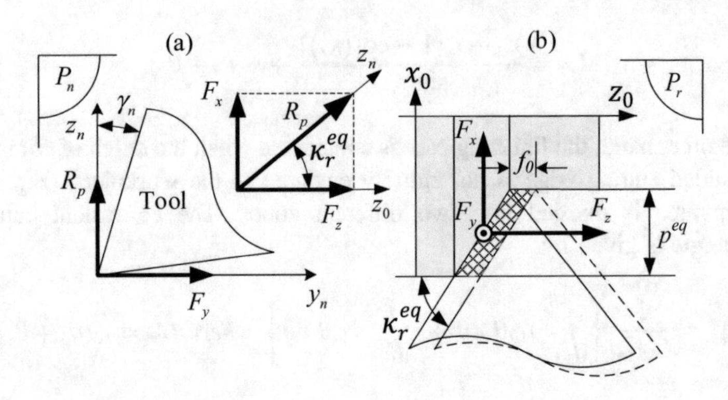

Fig. 3. Cutting force components in the (a) normal plane P_n (b) reference plane P_r

4 Results and Discussion

For the validation of the equivalent cutting-edge direction angle, an experimental data for oblique turning with 304 stainless steel conducted by (Abdellaoui and Bouzid 2016a), are used. The used tool holder is PCBNL2525M12 ($\kappa_r = 75°$, $\gamma_n = -6°$, $\lambda_s = -6°$). The cutting inserts are referenced CNMG120408 (r_ε = 0.8 mm). The cutting force components were registered on TRANSMAB 450TD lathe machine using a triaxial force dynamometer KISTLER 9257A. Experimental cutting conditions were conducted using Taguchi array L9, as shown in Table 1.

Table 1. Cutting conditions used in the experiments

Tests	1	2	3	4	5	6	7	8	9
V_c (m/min)	180	180	180	250	250	250	400	400	400
f_0 (mm/rev)	0.1	0.15	0.2	0.1	0.15	0.2	0.1	0.15	0.2
p_0 (mm)	0.5	1	2	1	2	0.5	2	0.5	1

A comparison between the experimental data, the discretization and the equivalent methods for modeling of cutting forces is carried out for the three cutting forces components: the cutting force (Fig. 4a), the feed force (Fig. 4b) and the radial force (Fig. 4c).

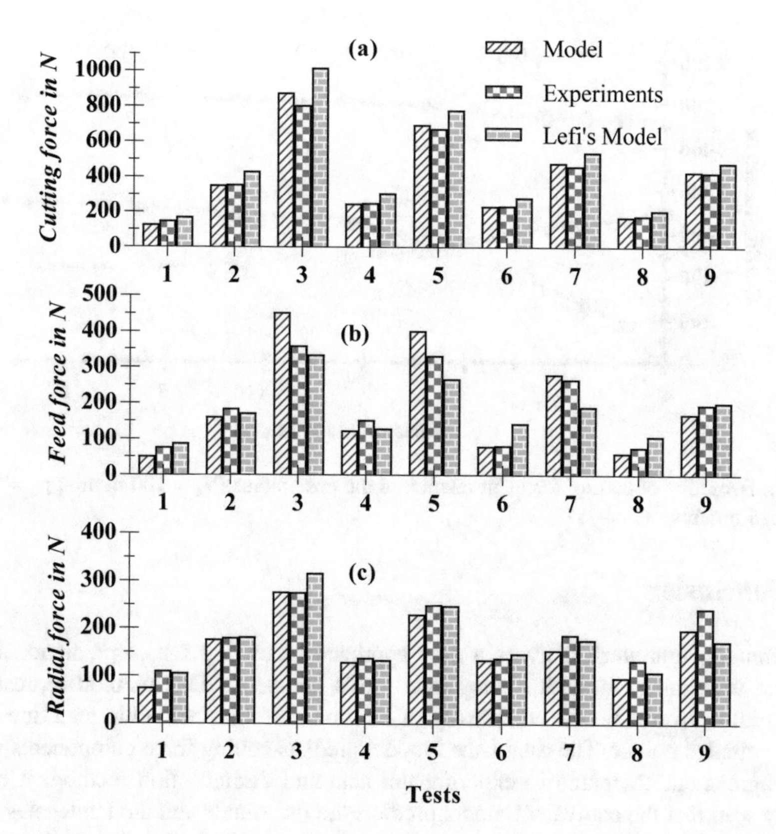

Fig. 4. The discretization, the equivalent models, and the experimental data for (a) cutting forces (b) feed forces and (c) radial forces (nose radius $r_\varepsilon = 0.8\,\text{mm}$)

The two cases of roughing and finishing are considered. The tests (1, 6 and 8) represent the finishing case, and the tests (2, 3, 4, 5, 7 and 9) are included in the roughing case.

A good agreement was found for the equivalent model compared to the experimental data for both cases. Especially for the prediction of the main cutting edge F_y, a high accuracy is registered. Despites of some errors in the test number 3, the equivalent model gives an accurate prediction of the feed and radial forces. This approach based on using the average value of the equivalent uncut chip thickness along the cutting edge provides a satisfactory prediction of the cutting force components even for small depth of cut comparing to nose radius.

Figure 5 shows the evolution of the cutting forces as a function of the nose radius. One can note a decrease in feed force with the radius while the radial component increases instead. However, a slight increase of the cutting force is noted. This result goes in line with the literature (Tang et al. 2014; Abdellaoui and Bouzid 2016a).

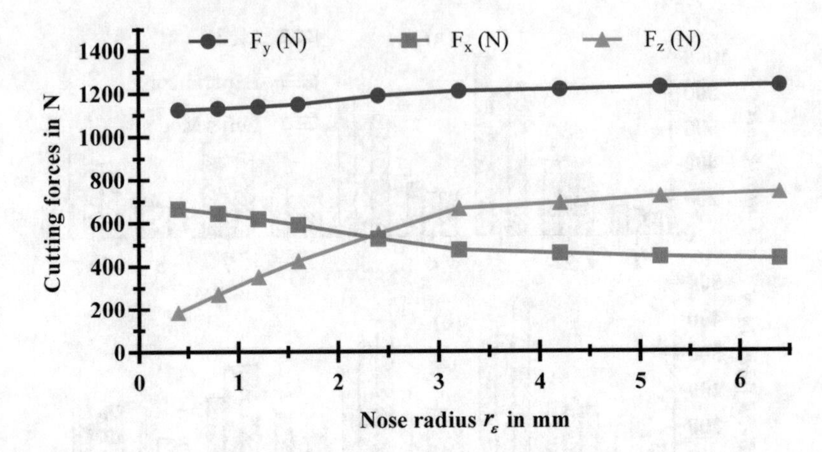

Fig. 5. Evolution of cutting forces in relation to the nose radius (V_c = 180 m/min; p_0 = 3 mm; f_0 = 0.15 mm/rev; κ_r = 75)

5 Conclusion

In summary, this work delivers a new approach of cutting forces prediction in 3D oblique machining, through a simplified tool's geometry. This method represents a good trial to overcome the complexity of the non-uniform uncut chip area due to the tool's rounded corner. The established model predicts cutting force components with a good agreement compared to experimental data and discretization method. It can be clearly seen that the equivalent model predicts the magnitude and the tendencies of the cutting forces under several cutting conditions quite accurately. Further studies, which take edge radius, chip breaker and tool wear into account, will need to be performed based on the model developed in this paper.

Acknowledgments. This work is carried out thanks to the support and funding allocated to the Unit of Mechanical and Materials Production Engineering (UGPMM/UR17ES43) by the Tunisian Ministry of Higher Education and Scientific Research.

The authors gratefully thank the Laboratory of Tribology and Dynamics of Systems, ENISE, France for their support in developing the experiments.

References

Abdellaoui, L., Bouzid, W.: Thermomechanical modeling of oblique turning in relation to tool-nose radius. Mach. Sci. Technol. **20**, 586–614 (2016a). https://doi.org/10.1080/10910344.2016.1224017

Abdellaoui, L., Bouzid, W.: Thermomechanical approach for the modeling of oblique machining with a single cutting edge. Mach. Sci. Technol. **20**, 655–680 (2016b). https://doi.org/10.1080/10910344.2016.1224020

Arsecularatne, J.A., Mathew, P., Oxley, P.L.B.: Prediction of chip flow direction and cutting forces in oblique machining with nose radius tools. Proc. Inst. Mech. Eng. Part B J. Eng. Manuf. **209**, 305–315 (1995). https://doi.org/10.1243/PIMEPROC_1995_209_087_02

Beauchamp, Y., Thomas, M.: Investigation of cutting parameter effects on surface roughness in lathe boring operation by use of a full factorial design. Comput. Ind. **31**, 645–651 (1996). https://doi.org/10.1016/s0360-8352(96)00234-3

Ben Said, M., Saï, K., Bouzid Saï, W.: An investigation of cutting forces in machining with worn ball-end mill. J. Mater. Process. Technol. **209**, 3198–3217 (2009). https://doi.org/10.1016/j.jmatprotec.2008.07.028

Campocasso, S., Poulachon, G., Costes, J.P., Bissey-Breton, S.: An innovative experimental study of corner radius effect on cutting forces. CIRP Ann. Manuf. Technol. **63**, 121–124 (2014). https://doi.org/10.1016/j.cirp.2014.03.076

Childs, T.H.C., Sekiya, K., Tezuka, R., et al.: Surface finishes from turning and facing with round nosed tools. CIRP Ann. **57**, 89–92 (2008). https://doi.org/10.1016/j.cirp.2008.03.121

Endres, W.J., Kountanya, R.K.: The effects of corner radius and edge radius on tool flank wear. J. Manuf. Process. **4**, 89–96 (2002). https://doi.org/10.1016/S1526-6125(02)70135-7

Imani, B.M., Yussefian, N.Z.: Cutting force simulation of machining with nose radius tools. In: 2008 International Conference on Smart Manufacturing Application, pp. 19–23. IEEE (2008)

Meyer, R., Köhler, J., Denkena, B.: Influence of the tool corner radius on the tool wear and process forces during hard turning. Int. J. Adv. Manuf. Technol. **58**, 933–940 (2012). https://doi.org/10.1007/s00170-011-3451-y

Orra, K., Choudhury, S.K.: Mechanistic modelling for predicting cutting forces in machining considering effect of tool nose radius on chip formation and tool wear land. Int. J. Mech. Sci. **142–143**, 255–268 (2018). https://doi.org/10.1016/j.ijmecsci.2018.05.004

Tang, L., Cheng, Z., Huang, J., et al.: Empirical models for cutting forces in finish dry hard turning of hardened tool steel at different hardness levels. Int. J. Adv. Manuf. Technol. **76**, 691–703 (2014). https://doi.org/10.1007/s00170-014-6291-8

Effect of the Interpolator Properties During the Multi-Point Hydroforming Process (MPHF)

Mohamed Amen Gahbiche, Safa Boudhaouia(✉), and Wacef Ben Salem

Mechanical Engineering Laboratory, National Engineering School of Monastir, University of Monastir, Monastir, Tunisia
amen.gahbiche@gmail.com, sa.boudhaouia@gmail.com, Wacef.bensalem@gmail.com

Abstract. Multi-Point Hydroforming (MPHF) is a recently developed hybrid process. An experimental set up has been designed and created by the authors in order to further investigate this new process's potentialities for creating complex parts and different geometries using the same tools. Moreover, finite element simulations have been performed to determine the influence of the process parameters, namely the effect of shapes and the density of the reconfigurable die's punches and the contribution of the insertion of an elastomeric membrane (commonly called interpolator) between the die and the blank sheet. The FE analysis and the experimental results have shown that MPHF is a very promising process: As a matter of fact, several parts with different shapes have been successfully obtained using the same tools. In addition, the use of the elastomeric interpolator has considerably improved the quality of the formed parts by eliminating the geometric defect called 'dimples' which is caused by the tips of the reconfigurable die's punches and thus generating a better distribution of residual stresses in the manufactured parts.

Keywords: Hydroforming · Flexible sheet forming · MPHF · Finite element simulation · Interpolator properties

1 Introduction

Conventional forming techniques such as stamping and deep drawing are very costly and not recommendable for small series and prototyping. As an alternative, much more flexible techniques have been developed like the hydroforming and the multipoint forming. Recently, there have been attempts to combine these two different forming processes in order to create a new hybrid forming technique called Multi-Point Hydroforming (MPHF) which takes advantages of both original processes. Selmi and BelHadjSalah (2013) were the first to test the feasibility of this process. In their work, a doubly curved shell product was successfully obtained using the experimental set up that they developed. In their apparatus, the fluid chamber is adopted on one side and on the other side, the rigid die (of the conventional hydroforming or drawing processes) is substituted by a reconfigurable one so that multitude shapes could be manufactured

A. Benamara et al. (Eds.): CoTuMe 2018, LNME, pp. 160–166, 2019.
https://doi.org/10.1007/978-3-030-19781-0_20

using a different configuration of the same multipoint die each time. They concluded that using a metallic medium between the blank sheet and the reconfigurable die is a more efficient way to eliminate dimples and edge buckling than the use of an elastomeric interpolator especially when manufacturing thin aluminum alloys materials. In a more recent study (Selmi and BelHadjSalah 2017) found that the quality of doubly curved shell obtained by the flexible hydroforming with segmented rigid tool is relatively enhanced which confirms the extended potentialities of this process in flexibility and accuracy for thin and thick shells. On the other hand (Liu et al. 2016) investigated a similar version of this new hybrid process but in their study the reconfigurable multipoint tool was moving during the experiments and thus it was considered as a punch rather than a die and the liquid chamber acted as a die. In their study, a half-ellipsoidal component was successfully manufactured and the authors reconfirmed that the increase of the thickness of metallic cover sheet, inserted under multipoint punch, reduces stress concentrations and dimpling severities and also brings the geometrical profile of the obtained part closer to the desired geometry. In this study, this new hybrid process (MPHF) is investigated further. An experimental set up was developed and some of the manufactured parts will be presented. The process is modeled using finite element analysis and the effect of the geometry and density of the multipoint die pins will be analyzed. Moreover, the influence of the insertion of an elastomeric interpolator and a cover sheet (called a martyr sheet) between the blank sheet and the reconfigurable die's pins will be presented.

2 Description of the Experimental Set Up, Materials and Finite Element Model

The proposed MPHF device is composed of three parts: a reconfigurable upper die, a blank holder and a fluid chamber which substitutes the punch of the classical stamping process. The multipoint die carries 37 cylindrical pins having a 10 mm diameter with a 5 mm radius spherical end tip. Each pin's height is controlled by a corresponding lower screw. The blank sheet (respectively the interpolator and the cover sheet) is clamped between two blank holders. Figure 1 shows the experimental setup and its different components.

Using the manufactured MPHF device, different shapes have been successfully formed. A sample of the die configurations and the corresponding manufactured parts is shown in Fig. 2a. In fact, despite the geometric dissimilarities of these two parts, they were both obtained using the same reconfigurable die only by adjusting the height of the reconfigurable die's pins accordingly to the CAD model. Therefore, no specific expensive dies were required, which highlights the flexibility of this new process.

In this study, all the forming experiments were carried out on AISI 304 steel sheets having a thickness of 1 mm. It has a Young modulus $E = 193$ GPa and a Poisson's ratio $\upsilon = 0.3$. This material exhibits a good ultimate strength and a high elongation percent at break which is very interesting for high strained forming processes. The corresponding stress-strain curve obtained after a tensile test on this material along the rolling direction is presented in Fig. 2b. To overcome the dimpling effect caused by discrete pins, both rubber sheet and a metallic cover sheet were used in the experimental studies. The used interpolator is a polyurethane rubber with Shore hardness of 50. As for the metallic cover sheet, the same material as the blank sheet was used with the same thickness.

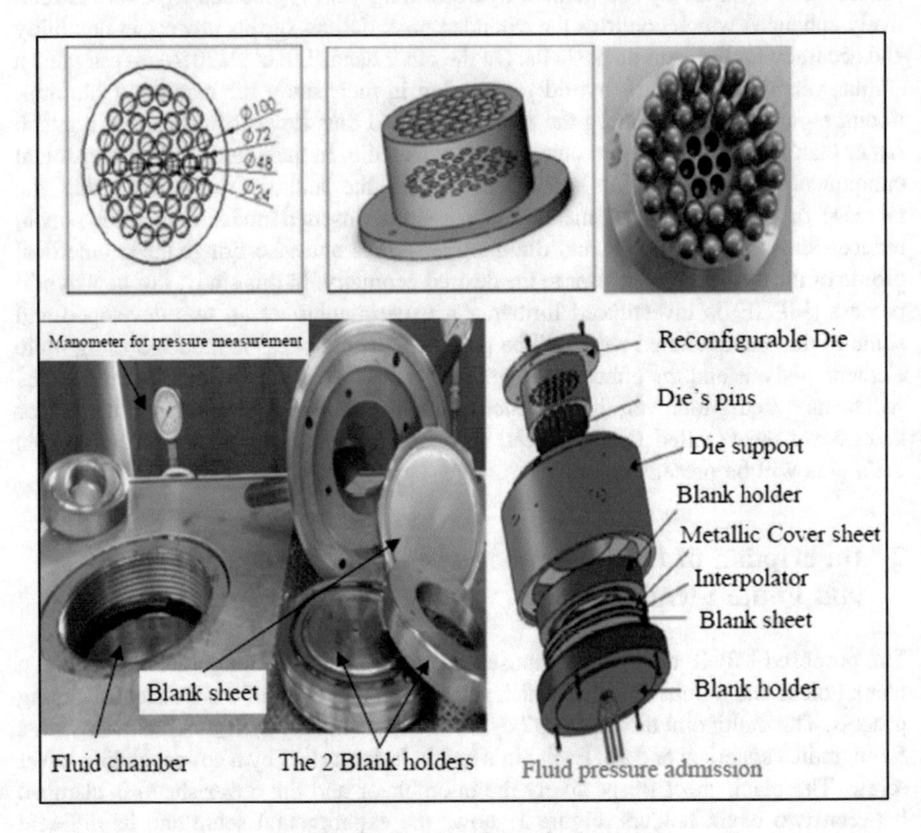

Fig. 1. The experimental set up

In order to investigate the MPHF process, a finite element model was developed using the finite element analysis software Abaqus/Explicit. The blank sheet, the cover sheet and the interpolator were all meshed using 8-node linear bricks with reduced integration (C3D8R elements). An elastoplastic behavior introduced as a tabulated flow rule was assigned to the first two instances. As for the interpolator, its hyper-elastic behavior was described using the Mooney–Rivlin constitutive law. The corresponding material parameters have been identified by means of an uniaxial compression tests from the work of (Zhang et al. 2006). The rest of the instances (reconfigurable die, the blank holders, the die support) were assumed to be rigid bodies. A surface to surface contact interaction with a coulomb friction law were used for the interaction between reconfigurable die (master surface) and the deformable blank sheet (slave surface). A global friction coefficient of $\mu 1 = 0.1$ is chosen for this contact property (Fig. 3).

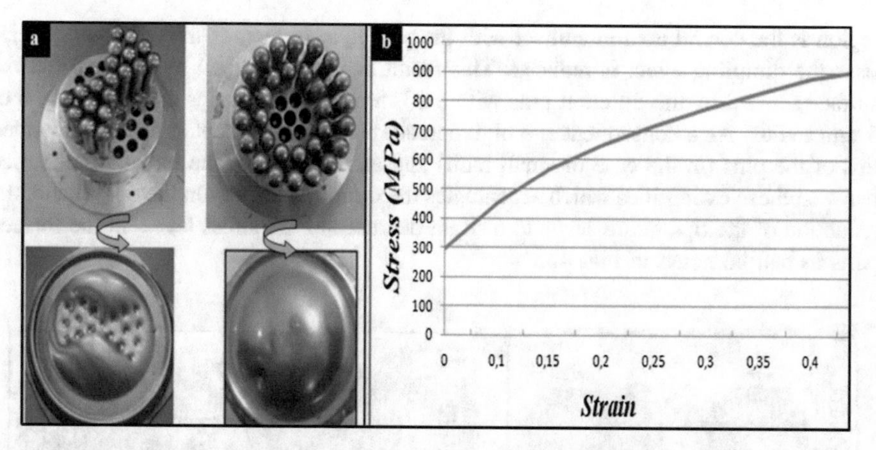

Fig. 2. (a) Sample of the manufactured parts using the MPHF device and the corresponding die configurations (b) Stress–strain curve of AISI 304 steel (Gahbiche 2015).

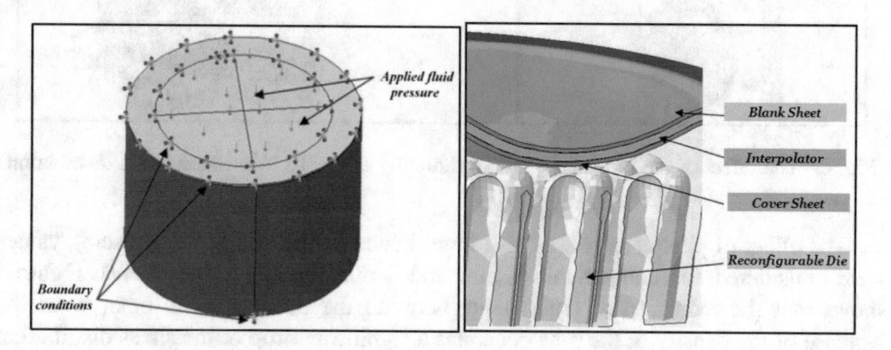

Fig. 3. The numerical model for the MPHF process

3 Results and Discussion

3.1 The Effect of the Pins Geometry and Density of the Final Product Quality

Three different pins geometries were considered and their potential effect on the dimpling effect and the surface homogeneity was investigated. As a matter of fact, for the three case studies, the same pin diameter was considered (so that the die configuration is kept the same for all of the simulation) and only the radii of the tip was modified (R1 = 5 mm, R2 = 10 mm and R3 = 20 mm).

As can be remarked in Fig. 4a, the profile of the different formed parts is not quite affected by the change of the pins geometry (almost the same height of the formed domes is witnessed in the three cases). However, the dimpling effect which is related to the regularity and smoothness of the product surface and as a consequence the quality of the formed part, is highly affected by this later parameter and the most affected region is the central area in contact with the pins tips. In fact, as the radius of the tips rises the dimpling effect is reduced. This result can be explained by the fact that the surface formed by the different pins with a 20 mm radius is planer than the surface of 5 mm radius. As a consequence, a high contact pressure is excepted near the rounded tips of the pins (in the case of small radii) and an overflow of the material is trapped between these extremities which accentuates the dimpling effect. On the other hand, the reduction of the tips radius leads to a slight decrease of the stress levels in the formed parts as can be noted in Fig. 4b.

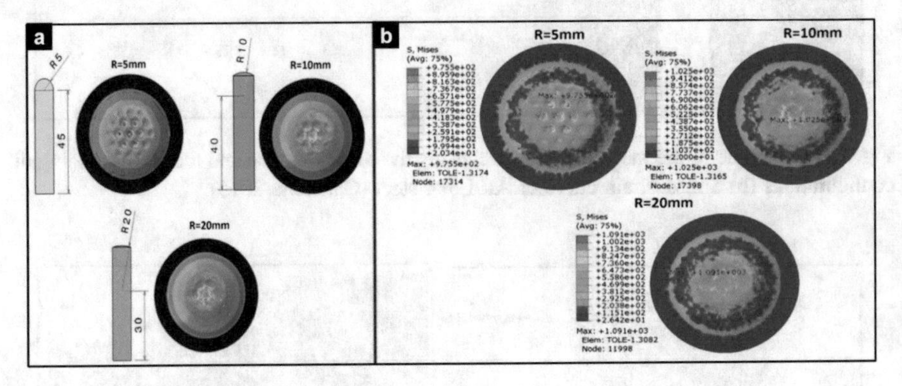

Fig. 4. The effect of the pins geometry (a) dimpling effect (b) Von Mises stress distribution

The effect of pins density was also investigated in this study. Two spacing values were considered for comparison (1 mm and 2 mm spacing respectively). Figure 5 shows that the reduction of the spacing between the punches or in other word the increase of the density of the pins generates a significant drop of the stress distribution in the formed part. The increase of the density also decreases the height of the dimples on the formed parts. This finding was also remarked in the case of conventional multipoint forming process: In the work of (Cai et al. 2008) the dimpling effect is significantly reduced and even eliminated when a larger pins density is considered (i.e. when small sized pins are used). However, the reduction of the pins dimensions will complicate the die assembly and the pins adjustment and positioning and also will imply a less economic tooling.

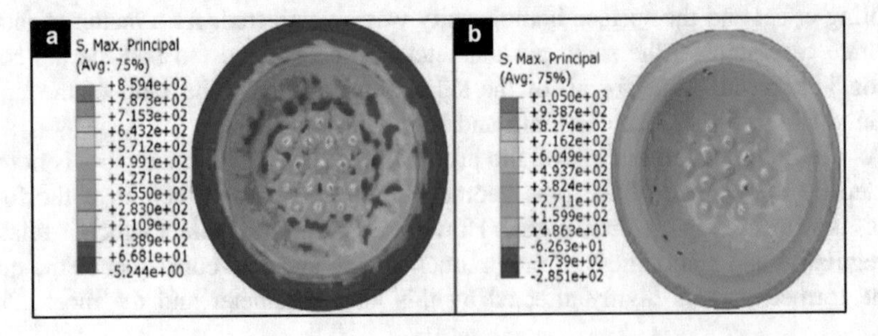

Fig. 5. The effect of the pins density (a) 1 mm spacing (b) 2 mm spacing

3.2 The Effect of an Interpolator and a Cover Sheet Insertion

Although the studied new process seems very interesting and promising thanks to its high flexibility, the presence of the dimples on the formed parts could limit the efficiency of this proposed technique and its use for industrial parts. It has been shown in many conventional Multi Point forming process studies that the use of an interpolator and/or a metallic cover sheet improves the contact between the blank and the discrete die and thus reduces and even eliminates the dimpling phenomena (Cai et al. 2008; Selmi and BelHadjSalah 2013). This idea was tested in our case and a 2 mm polyurethane interpolator with a Shore hardness of 50 and a cover sheet identical to the blank sheet used in experiments were inserted on the top of the reconfigurable die.

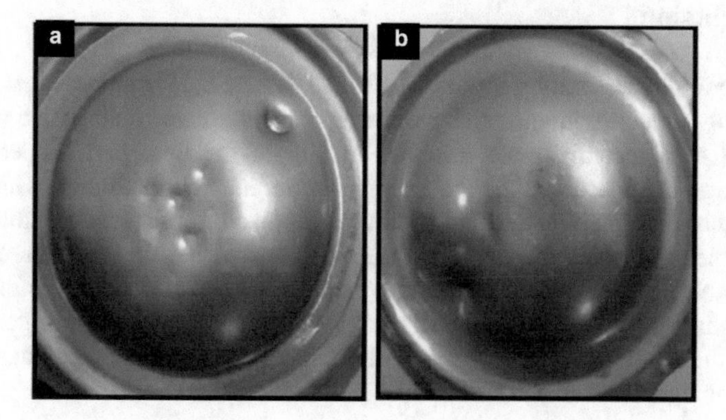

Fig. 6. (a) Cover sheet (b) blank sheet

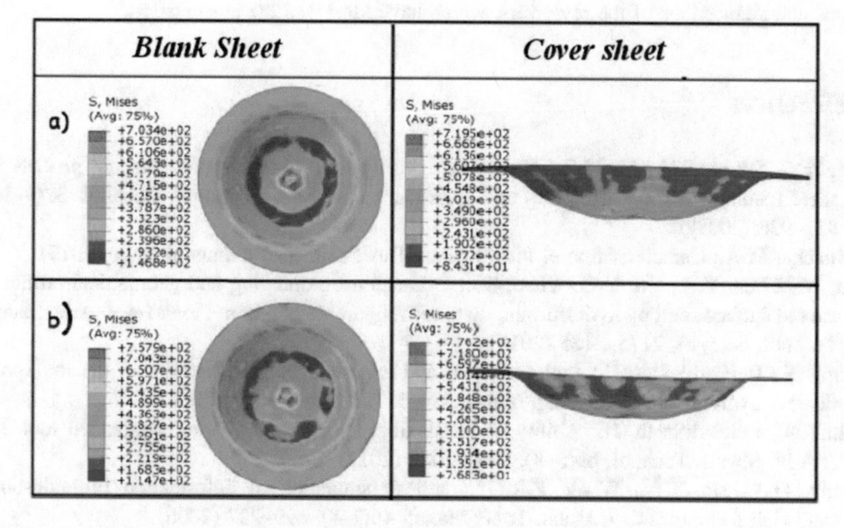

Fig. 7. The effect of the interpolator thickness (a) 2 mm thickness (b) 1 mm thickness

As can be seen in Fig. 6, after the insertion of these two instances, the quality of the formed part have been considerably improved and the totality of the dimples has been eliminated from the final part. In fact, since the cover sheet becomes in direct contact with the pins, it has absorbed all of the contact pressure applied by the pins tips. As a result, the blank sheet was not in contact with a discrete surface anymore. Moreover the presence of the interpolator prohibited the material overflow of the blank sheet. The presence of the interpolator has also participated in the establish of a more homogeneous stress distribution in both the deformed cover sheet and the formed product. In addition, the increase of the interpolator thickness has led to the reduction of the stress levels in both sheets as presented in Fig. 7.

4 Conclusion

The multipoint hydroforming process (MPHF) is a new hybrid process that allows the obtaining of products with different geometries using the same tooling each time thanks to the use of a reconfigurable die. The quality of the final parts depends on the geometry and the density of the die's pins. Despite its high flexibility, which make it quite adequate for small series and prototype production, this process suffers from a major problem which is the dimples defect caused by the impact of the tips of the reconfigurable die on the blank sheet. This geometrical defect was successfully reduced and even eliminated after the insertion of a metallic cover sheet and an interpolator on the top of the discrete die which has greatly improved the parts surface quality and led to more homogenous stress distribution.

Acknowledgements. This work is partially supported by the 'Ministère de la Recherche Scientifique' (SERST), LAB-MA 05. The authors also gratefully acknowledge the helpful comments and suggestions of the reviewers, which have improved the presentation.

References

Cai, Z.Y., Wang, S.H., Li, M.Z.: Numerical investigation of multi-point forming process for sheet metal: wrinkling, dimpling and springback. Int. J. Adv. Manuf. Technol. **37**(9–10), 927–936 (2008)

Gahbiche, M.A.: Caracterisation et Identification Des Paramètres d'emboutissage (2015)

Liu, W., Chen, Y.Z., Xu, Y.C., Yuan, S.J.: Evaluation on dimpling and geometrical profile of curved surface shell by hydroforming with reconfigurable multipoint tool. Int. J. Adv. Manuf. Technol. **86**(5–8), 2175–2185 (2016)

Selmi, N., BelHadjSalah, H.: Finite element and experimental investigation of the multipoint flexible hydroforming. Key Eng. Mater. **554–557**, 1290–1297 (2013)

Selmi, N., BelHadjSalah, H.: Ability of the flexible hydroforming using segmented tool. Int. J. Adv. Manuf. Technol. **89**(5–8), 1431–1442 (2017)

Zhang, Q., Dean, T.A., Wang, Z.R.: Numerical simulation of deformation in multi-point sandwich forming. Int. J. Mach. Tools Manuf. **46**(7–8), 699–707 (2006)

Materials: Mechanical Behaviour and Structure

Probabilistic Fatigue Life Prediction of Parabolic Leaf Spring Based on Latin Hypercube Simulation Method

Akram Atig[1,2,3], Rabï Ben Sghaier[1,2(✉)], and Raouf Fathallah[1,3]

[1] Unité de Génie de Production Mécanique et Matériaux (UGPM2), Ecole Nationale d'Ingénieurs de Sfax (ENIS), Université de Sfax, Sfax, Tunisia
rabibensghaier@gmail.com
[2] Institut supérieur des sciences appliquées et de technologie de Sousse, Université de Sousse, Rue Tahar Ben Achour Sousse, Sousse, Tunisia
[3] Ecole nationale d'ingénieur de Sousse, Université de Sousse, BP 264, Sousse Erriadh, Tunisia

Abstract. Fatigue phenomenon is one of the main causes of parabolic leaf spring failure. Therefore, fatigue life assessment and prediction represent an important aspect during parabolic leaf spring design stage. Nevertheless, the estimation of fatigue life is usually affected by many inherent uncertainties which must be considered in a fatigue design approach. In this work, a stochastic approach based on Latin hypercube simulation method has been performed to predict the fatigue life of parabolic leaf spring. The strain based approach and Morrow fatigue criterion have been used to compute the number of cycles to failure. The proposed approach has been applied on a finite element and a response surface model of parabolic leaf spring. The dispersion of geometrical dimensions, materials properties and cyclic loading parameters have been taken into consideration. The number of cycles to failure distribution has been presented and characterized. The effects of probabilistic variables on the fatigue life results have been studied in order to enhance the fatigue behavior of parabolic leaf spring.

Keywords: Parabolic leaf spring · Probabilistic fatigue life prediction · Finite element analysis · Response surface method

1 Introduction

Leaf spring presents an important flexible component of light and heavy vehicle suspension systems. It can be used to absorb and release shocks due to road surface irregularities (Reimpell et al. 2001; Atig et al. 2018a, b). Moreover, leaf spring is utilized as a connecting element. It links the axle to the vehicle chassis (Heißing and Ersoy 2011; Atig et al. 2018a, b). It is well Known that Fatigue phenomenon is one of the principal reasons of fracture of Parabolic Leaf Spring (PLS). Commonly, leaf spring fatigue behavior is characterized by random experimental results. Svensson (1997) joined the random fatigue results with the uncertainty of the cyclic load, the model error, material and geometrical properties of the structure. Many probabilistic

A. Benamara et al. (Eds.): CoTuMe 2018, LNME, pp. 169–176, 2019.
https://doi.org/10.1007/978-3-030-19781-0_21

simulation methods can be adopted to estimate the fatigue life distribution (Ben Sghaier et al. 2010; McKay et al. 1979). With a similar efficiency and accuracy, the required number of simulations Latin hypercube sampling method is lower than the number used by the standard Monte Carlo simulation method (Olsson et al. 2003). Hence, Latin hypercube sampling method can be used to reduce the extra-computational time.

In this works, finite element model of PLS has been implemented using Ansys workbench® software. The PLS fatigue life has been computed by using Morrow fatigue criterion. A probabilistic study based on Latin hypercube sampling method has been performed to estimate the fatigue life distribution.

2 Materials and Methods

A Three dimensional parametrical finite element model of single asymmetric parabolic leaf spring was developed using ANSYS commercial finite element software. For the assignment of material in fatigue analysis, the SAE 5160 steel, typically employed in the leaf spring manufacturing, is selected (Yamada 2007). The elastic plastic model of Ramberg-Osgood (1943), given by Eq. (1), is adopted.

$$\varepsilon_a = \frac{\sigma_a}{E} + \left(\frac{\sigma_a}{K}\right)^{\frac{1}{n}} \tag{1}$$

Where ε_a is the total strain amplitude, σ_a equivalent stress amplitude, E is the Young modulus, K is the cyclic strength coefficient and n is the cyclic strain hardening exponent.

As boundary conditions of PLS, the two leaf spring eyes are allowed only to rotate in X-axis and translate in Y-axis. The applied loading is assimilated as a uniformly distributed force applied on the seat surface of PLS. The cyclic loading has a sinusoidal form with a maximum loading and a load ratio equal to 5500 N and 1/3 respectively. Boundary conditions and loading are schematized in Fig. 1. For solid meshing, 8-node hexahedra elements are used with a maximal length of 5 mm. Consequently, 11649 elements with 59886 nodes are generated.

In this work, the morrow fatigue criterion is used to assess the PLS fatigue life (Morrow 1968). This criterion takes into account the mean stress effect and covers both low and high cycle fatigue regions (Stephens et al. 2001). The relation describing the Morrow fatigue criterion is defined as:

$$\varepsilon_a = \frac{\sigma_f' - \sigma_m}{E}(2N_f)^b + \varepsilon_f'(2N_f)^c \tag{2}$$

where ε_a is the total strain amplitude, E is the Young modulus, σ_f' is the fatigue strength coefficient, σ_m is the mean stress, b is the fatigue strength exponent, ε_f' is the fatigue ductility coefficient, c is the fatigue ductility exponent and N_f is the number of cycles to failure. The mechanical proprieties of SAE 5160 steel are illustrated in Table 1 (Boardman 1982).

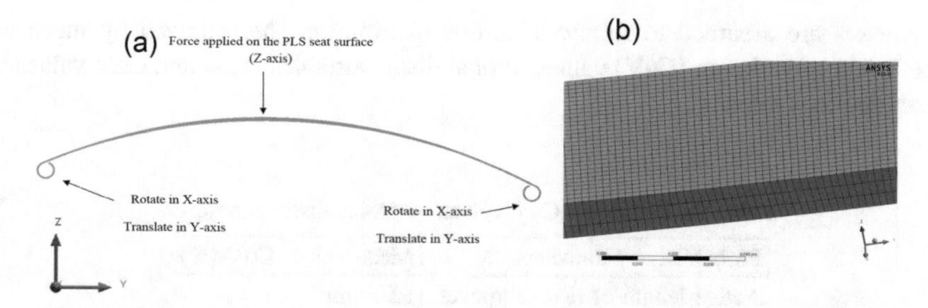

Fig. 1. (a) Loading, boundary conditions (b) mesh of PLS of loading conditions applied on the SAPLS

The number of cycles to failure is computed using the Morrow strain-life relation. The strain amplitude is calculated through the cyclic stress-strain relation of Ramberg-Ostgood. The equivalent stress amplitude is obtained from finite element analysis. The flowchart of PLS the fatigue life analysis is depicted in Fig. 2.

Table 1. Mechanical properties of SAE 5160.

Mechanical properties of SAE 5160 steel	Value
Ultimate tensile strength (MPa)	1584
Yield strength (MPa)	1487
Young modulus (GPa)	207
Fatigue strength coefficient (MPa)	2063
Fatigue strength exponent	−0.08
Fatigue ductility coefficient	9.56
Fatigue ductility exponent	−1.05
Cyclic strength coefficient (MPa)	2432
Cyclic strain hardening exponent	0,15

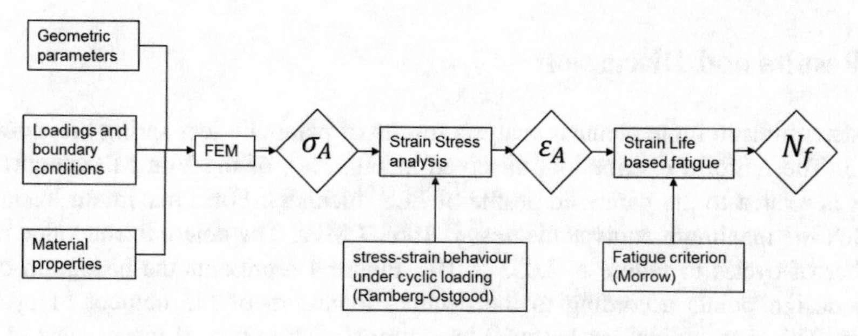

Fig. 2. Flow chart of parabolic leaf spring fatigue life analysis

In order to analyze the effect of design parameter dispersion on the PLS fatigue life, ten input parameters are considered as probabilistic variables. These probabilistic input parameters are assumed to follow a normal distribution characterized by mean and Coefficient of variation (CoV) values. Probabilistic variables mean and CoV values are illustrated in Table 2.

Table 2. Mean and CoV values of probabilistic parameters.

Probabilistic parameters	Mean value	CoV (%)
Active length of rear cantilever	850 mm	1
Active length of front cantilever	600 mm	1
Width	63 mm	1
Maximum thickness	13.2 mm	1
Maximum loading	5500 N	5
Young modulus	207 GPa	5
Fatigue strength coefficient	2063 MPa	5
Fatigue strength exponent	−0.08	5
Fatigue ductility coefficient	9.56	5
Fatigue ductility exponent	−1.05	5

The fatigue life results are affected by the dispersion of the input parameters. To evaluate the probabilistic distribution of the number of cycle to failure, a probabilistic approach, based on Latin hypercube simulation method, was developed. 1000 random values of each probabilistic input variable are generated by using the Latin Hypercube sampling method and 1000 Morrow fatigue life results are computed for the studied case. Generally, the generation of a big number of Finite element simulations is a time consuming process. In order to reduce the computational time, Finite element model can be substituted by second order polynomial model using response surface method (Myers and Montgomery 2002; Atig et al. 2017). For response surface model construction, a central composite design of experiments of ten factors with two levels is implemented (Montgomery 1991).

3 Results and Discussion

The deterministic finite element analysis results of parabolic leaf spring are shown in Fig. 3. The uniform distribution, observed in Fig. 3(a), of the Von Mises equivalent stress is related to the parabolic profile of PLS thickness. For a maximum loading of 5500 N the maximum equivalent stress is 1063.7 MPa. The deterministic value of the number of cycles to failure is 2.053×10^7. Figure 4 represents the histogram of the 1000 design points according to the decimal logarithm of the number of cycle to failure. This distribution can be fitted to a normal distribution of mean value of 7.32 and a CoV of 9.1%.

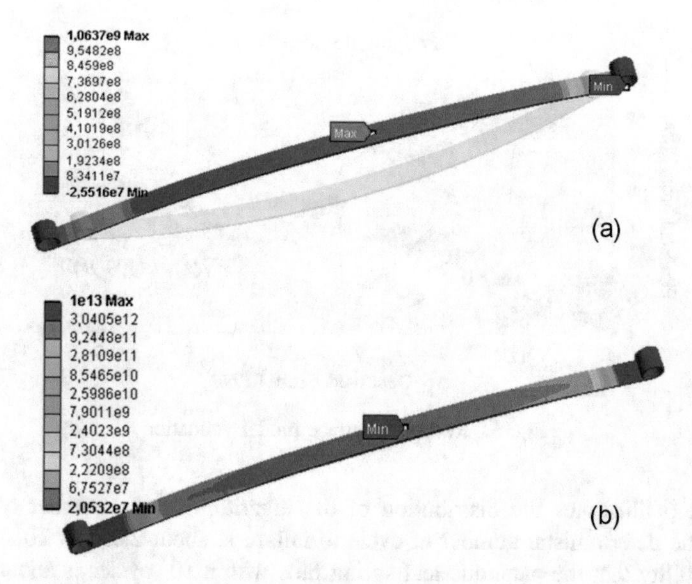

Fig. 3. (a) Equivalent stress (Pa) and (b) fatigue life distribution along parabolic leaf spring

To reduce the extra-computational time, related to the calling a big number of times the finite element model of parabolic leaf spring, response surface method is adopted. To evaluate the regression coefficients of second order polynomial response surface model, a central composite design of experiments is implemented. Afterwards the validation of the response surface model of parabolic leaf spring is performed by applying thirty verification points.

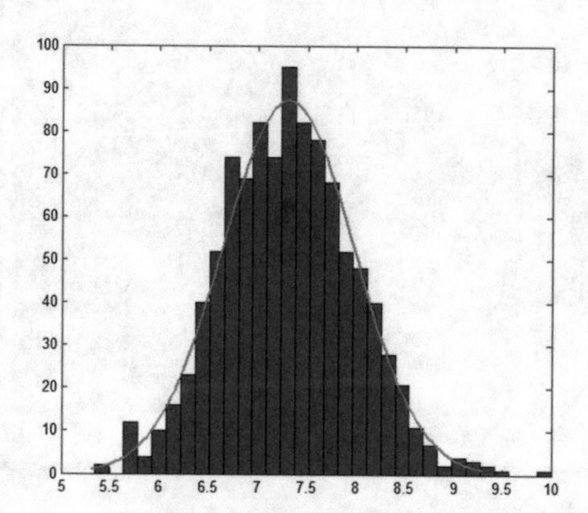

Fig. 4. Distribution of the decimal logarithm of the number of cycle to failure

As shown in Fig. 5, a good coefficient of determination level of 99% is attained. To compute the fatigue life distribution using the response surface model, 10^5 Latin Hypercube simulations are executed.

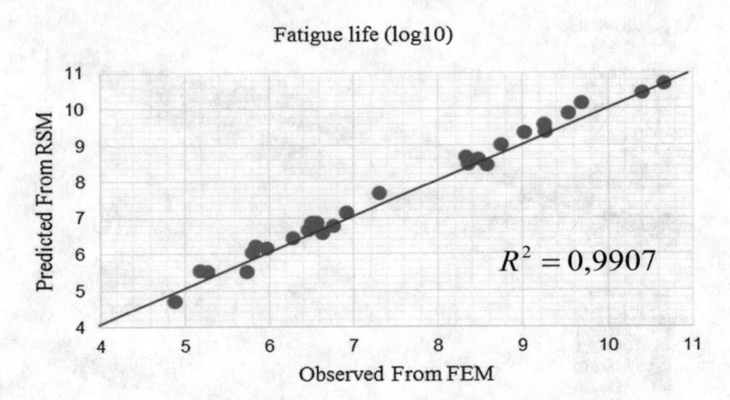

Fig. 5. Response surface model validation

Figure 6 illustrates the distribution of the logarithm of the number of cycle to failure. The deterministic number of cycle to failure is about 2.053×10^7. However, the probability that the parabolic leaf spring fails within 10^7 cycles is about 33.85%.

Figure 7 shows the effect of design parameters on the Morrow fatigue life. It is observed that the fatigue ductility coefficient and fatigue ductility exponent have no effect fatigue life of parabolic leaf spring. Therefore, the plastic term of morrow criterion can be neglected in the fatigue life prediction of parabolic leaf spring. The variation of 1% of the maximum thickness has the most important effect on the fatigue life. The variations of the maximum load and the two active lengths have a negative effect on fatigue life.

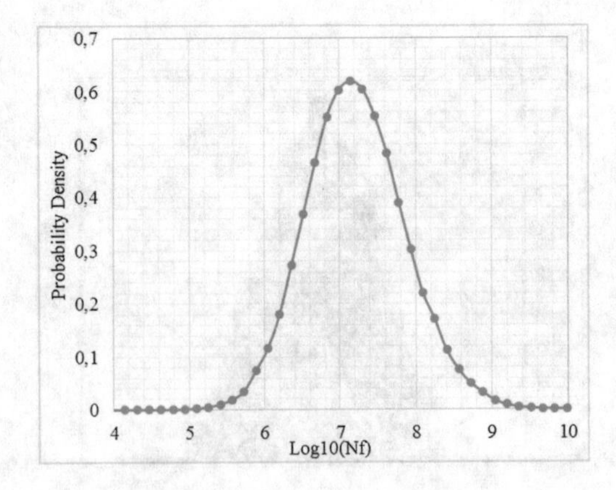

Fig. 6. Fatigue life distribution obtained from response surface model

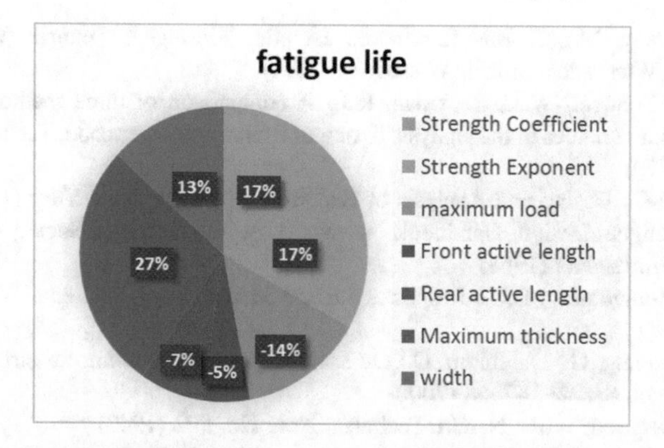

Fig. 7. Effect of design parameters on the Morrow fatigue life

4 Conclusion

A probabilistic approach based on morrow criterion and Latin hypercube simulation method has been presented to predict the fatigue behavior of parabolic leaf spring. It takes into account the dispersion of geometric, loading and material design parameters. The uncertainty effect has been evaluated through a sensitivity analysis. The variation of maximum thickness has the most important effect on the number of cycles to failure. However, the fatigue ductility exponent and coefficient have no effect on the fatigue life of parabolic leaf spring. The probabilistic fatigue life prevision approach can evaluate with precision the fatigue failure probability and solves the oversizing design problems. Moreover, it is very important to make a probabilistic optimization of design parameters by taking into account the mass, the stiffness and the manufacturing cost of parabolic leaf spring.

References

Atig, A., Ben Sghaier, R., Seddik, R., Fathallah, R.: Reliability-based high cycle fatigue design approach of parabolic leaf spring. Proc. Inst. Mech. Eng. Part L J. Mater. Des. Appl. (2017). https://doi.org/10.1177/1464420716680499

Ben Sghaier, R., Bouraoui, C., Fathallah, R., Degallaix, G.: Probabilistic prediction of high cycle fatigue reliability of high strength steel Butt-welded joints. Fatigue Fract. Eng. Mater. Struct. **33**, 575–594 (2010)

Atig, A., Sghaier, R.B., Seddik, R., Fathallah, R.: Probabilistic methodology for predicting the dispersion of residual stresses and Almen intensity considering shot peening process uncertainties. Int. J. Adv. Manuf. Technol. **94**(5–8), 2125–2136 (2018a)

Atig, A., Ben Sghaier, R., Seddik, R., Fathallah, R.: A simple analytical bending stress model of parabolic leaf spring. Proc. Inst. Mech. Eng. Part C J. Mech. Eng. Sci. **232**(10), 1838–1850 (2018b)

Boardman, B.: Crack initiation fatigue—data, analysis, trends and estimation. SAE Technical Paper; 820682 (1982)

Heißing, B., Ersoy, M.: Chassis Handbook, 1st edn. Vieweg & Teubner Verlag, Springer Fachmedien Wiesbaden GmbH, Wiesbaden (2011)

McKay, M.D., Conover, W.J., Beckman, R.J.: A comparison of three methods for selecting values of input variables in the analysis of output from a computer code. Technometrics **21**(2), 239–245 (1979)

Montgomery, D.C.: Design and Analysis of Experiments. Wiley, New York (1991)

Morrow, J.: Fatigue Design Handbook Advances in Engineering. Society of Automotive Engineers, Warrandale (1968)

Myers, R.H., Montgomery, D.C.: Response Surface Methodology, 2nd edn. Wiley, New York (2002)

Olsson, A., Sandberg, G., Dahlblom, O.: On Latin hypercube sampling for structural reliability analysis. Struct. Saf. **25**, 47–68 (2003)

Ramberg, W., Osgood, W.R.: NACA Technical Note No. 902 (1943)

Reimpell, J., Stoll, H., Betzler, J.W.: The Automotive Chassis: Engineering Principles, 2nd edn. Butterworth-Heinemann, Oxford (2001)

Stephens, R.I., Fatemi, A., Stephens, R.P., Fuchs, H.O.: Metal Fatigue in Engineering, 2nd edn. Wiley, New York (2001)

Svensson, T.: Prediction uncertainties at variable amplitude fatigue. Int. J. Fatigue **19**(1), 295–302 (1997)

Yamada, Y.: Materials for Springs. Springer, Berlin (2007)

The Relation Between R Phase Presence Level and Stress-Temperature Diagram of an Aged NiTi Shape Memory Alloy

Boutheina Ben Fraj[1(✉)], Slim Zghal[2], Amen Gahbiche[3], and Zoubeir Tourki[1]

[1] Mechanical Laboratory of Sousse (LMS), National Engineering School of Sousse ENISo, University of Sousse, 4023 Sousse, Tunisia
bfb.ingmec@gmail.com, zbrtourki@gmail.com

[2] Laboratory of Multifunctional Materials and Applications (LaMMA), Faculty of Sciences of Sfax, University of Sfax, BP 1171, 3000 Sfax, Tunisia
salim_zghal@yahoo.fr

[3] Laboratoire de Génie Mécanique LGM, Ecole Nationale d'Ingénieurs de Monastir ENIM, Université de Monastir, 5019 Monastir, Tunisie
amen.gahbiche@gmail.com

Abstract. This paper aims to study the effect of the R phase on the phase transformation behavior of an aged Ni-rich NiTi Shape Memory Alloy (SMA). The thermal analysis at zero stress was performed through Differential Scanning Calorimetry (DSC) technique. It was reported that the NiTi thermal behavior is strongly influenced by the R phase presence level. The calorimetric findings were supported by X-Ray Diffraction (XRD) analysis which reveals the relationship between R phase and Ni_4Ti_3 precipitates. In order to characterize the stress-induced martensitic transformation, an isothermal compression tests were performed. Based on the critical stress-temperature diagram, two different transformation processes, with and without R phase, were selected to study the global NiTi thermomechanical behavior. The obtained results can be useful to predict the domain where the effects of the Ni-rich NiTi alloys can be occurred.

Keywords: NiTi alloys · Phase transformation · Aging temperature · R phase · Stress-temperature diagram

1 Introduction

In the investigation study of the NiTi thermomechanical behavior, usually, both transformation temperatures and stresses play a key role since they define the critical stress-temperature diagram from which the relationship is clearly described and the evolution of the overall NiTi behavior is easily understood (Tanaka 1986; Brinson 1993; Brocca et al. 2002). Thus, in order to identify these main properties, the experimental thermal characterization remains unavoidable.

A. Benamara et al. (Eds.): CoTuMe 2018, LNME, pp. 177–185, 2019.
https://doi.org/10.1007/978-3-030-19781-0_22

Based on previous experimental studies, it has been reported that the thermomechanical properties of the Ni-rich NiTi SMA are often sensitive to several factors such as the aging temperature and duration (Ben Fraj et al. 2017a, b; Sadiq et al. 2010; Khaleghi et al. 2013; Wang et al. 2015; Eggeler et al. 2005), cooling media and cooling/heating rate during the thermal transformation process (Motemani et al. 2009; Ben Fraj et al. 2016; Kaya 2017; Ben Fraj et al. 2017a, b), strain rate (Kazemi Choobi et al. 2014), test temperature (Duerig and Bhattacharya 2015; Churchill et al. 2009; Laplanche et al. 2017). Under specific aging conditions which lead to the formation of Ni_4Ti_3 precipitates in the NiTi matrix, the forward transformation can be carried out through a rhombohedral phase known as R phase, where a multistage transformation takes place during the cooling process (Duerig and Bhattacharya 2015).

The purpose of this study is to present an experimental analyze on the influence of the R phase presence level on the thermomechanical properties of an aged Ni-rich NiTi SMA. The evolution of both thermal and microstructural behavior as function of aging temperature is investigated through a Differential Scanning Calorimetry (DSC) and a X-Ray Diffraction (XRD) techniques, respectively. Moreover, a mechanical analysis is performed based on an isothermal compressive tests in order to investigate the effect of the aging temperature on the stress induced martensitic transformation. The global thermomechanical behavior of the NiTi SMA and the critical stress-temperature diagram are finally discussed as a function of the R phase presence level.

2 Material and Experimental Method

The NiTi SMA was supplied by Baoji Rare Titanium-Nickel Co. Ltd., China. The corresponding chemical composition is given in Table 1. The temperature A_f and M_f values are about 33 °C and less than −60 °C, respectively.

Table 1. The chemical composition of the studied NiTi alloy/wt%

Ni	H	N	O	C	Ti
55.89	0.001	0.002	0.048	0.041	Balance

2.1 Thermal Analysis

All samples were cut from the same NiTi hot rolled bar and then aged at various temperatures, from 450 °C to 650 °C, for 1 h and cooled at ambient temperature. Thereafter, these samples were polished using waterproof sand paper and rinsed with ethanol. The characterization of the transformation behavior was performed with a differential scanning calorimeter DSC 4000/Perkin Elmer with an accuracy of ±0.1 °C. The samples were encapsulated in an aluminum specimen pans. As a reference, a matching empty specimen pan with crimped lid was used. The imposed cooling/heating rate was of 5 °C/min with a constant nitrogen flow rate of 20 ml/min. All the DSC tests have consisted of a complete cooling and heating cycle. The cycle was repeated twice, for each sample, in order to check the repeatability of the material response.

2.2 Microstructural Analysis

The phase formations of the samples were analyzed at ambient temperature using XRD with D8 Advance diffractometer and Cu Kα radiation (λCu = 1.5418 *Å*). The XRD samples were cut into disks with a diameter of 10 mm and height of 3.5 mm. The diffraction patterns were recorded over 2θ ranging from 30° to 90°, during 45 min, by continuous scan with tube voltage of 40 kV and tube current of 40 mA.

2.3 Mechanical Analysis

The NiTi samples were cut into cylinder forms with a diameter of 10 mm and a height of 20 mm. The NiTi cylindrical samples were aged at various temperatures, from 500 °C to 650 °C, for 1 h and then cooled at ambient temperature.

The compression tests were performed using 30-ton Shimadzu press at constant loading rate of 0.1 mm/min. As shown in Fig. 1, the compression plate admits a greatly higher rigidity compared to that of the tested sample. The experiments were performed at a temperature of 37 °C controlled using a digital thermocouple positioned closely beside the sample. The compression test was repeated twice, for each sample, to ensure the repeatability of the material response.

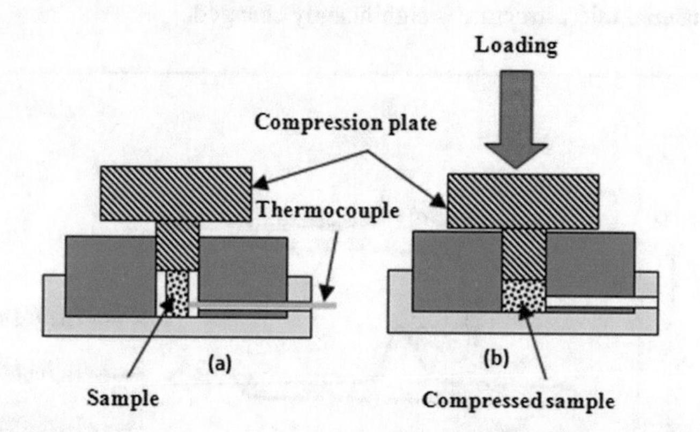

Fig. 1. Schematic drawing of the compression test: (a) sample positioning; (b) compression loading.

3 Results and Discussion

3.1 DSC Analysis

The DSC thermogrammes at various aging temperatures are displayed in Fig. 2. It can be clearly observed that the increase of the aging temperature has a significant effect on both forward (A → M) and reverse (M → A) transformations during cooling and heating processes, respectively.

From the DSC analysis (Fig. 3), three sequences of phase transformation can be revealed as a function of aging temperature:

1. For the 450 °C and 500 °C aged samples: A symmetric R phase transformation, A → R → M on cooling and M → R → A on heating is observed.
2. For the 550 °C and 600 °C aged samples: An asymmetric R phase transformation (A → R → M on cooling and M → A on heating) is observed.
3. For the 650 °C aged sample: A direct (A ↔ M) transformation is taken place during cooling and heating paths. The R phase is completely absent in this case.

In the Ni-rich NiTi alloys, the R phase behavior is directly linked to the kinetic of the Ni_4Ti_3 precipitates which is mainly dependent to the aging temperature (Wang et al. 2015, Jiang et al. 2015). By increasing the aging temperature, the gradual disappearance of the R phase can be attributed to the gradual dissolution of the Ni_4Ti_3 precipitates in the matrix (Otsuka and Ren 2005). Accordingly, the total superposition of the two transformation peaks (M → R) and (R → A), observed during heating at 550 °C indicates the partial dissolution of these precipitates. During the cooling path, the two peaks (A → R) and (R → M) are progressively merged until reaching the total superposition at an aging temperature of 650 °C which is indicative of the total dissolution of the Ni_4Ti_3 precipitates. Thus, 600 °C–650 °C may be considered as the critical aging temperature range in which the material microstructure is significantly changed.

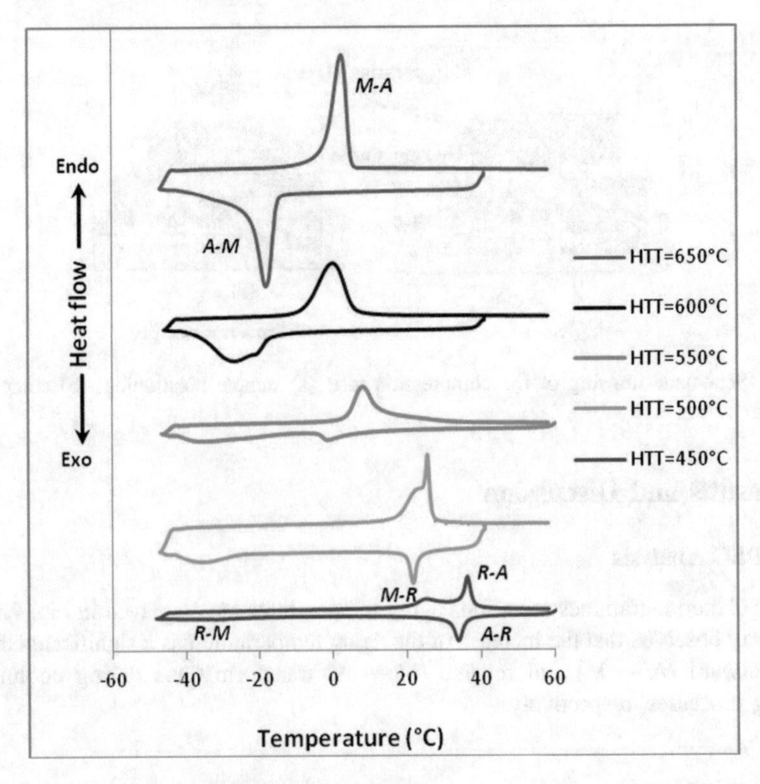

Fig. 2. DSC curves of Ni-rich NiTi SMA aged at different temperatures: (a) 450 °C; (b) 500 °C; (c) 550 °C; (d) 600 °C; (e) 650 °C. (1 h/air cooling)

3.2 XRD Analysis

For the identification of the NiTi SMA phases, mainly the Ni_4Ti_3 phase, a XRD analysis was performed. The XRD patterns recorded for the aged samples are presented in Fig. 3. A gradual evolution of the microstructure is clearly visible. According to the DSC curve of the 450 °C aged sample, at ambient temperature (~25 °C), the NiTi alloy should be completely in R phase. Consequently, in Fig. 3, the corresponding XRD pattern shows two principal superimposed peaks, R(112) and R(030). Furthermore, a small P(20-22) and P(21-32) peaks are revealed, which indicate the presence of the Ni_4Ti_3 phase formed during the aging process (P denotes Precipitates). The low intensity of these peaks proves that the precipitated Ni_4Ti_3 particles are very fine. The precipitation peaks are also visible in the other XRD patterns in addition to the austenitic peaks and tend to disappear with the increase of the aging temperature from 500 °C to 650 °C.

For the aged samples at the range of 500 °C–650 °C, the integral intensity of the A (110) peak decreases by increasing the aging temperature which reveals that the lower aging temperature can promotes the grains to preferentially growing up. The main peak A(110) and the minor peak A(211) are the main and the characteristic peaks of the austenitic phase (Motemani et al. 2009).

It should be noted that a new diffraction peak of the parent phase A(200) is detected at 650 °C which can be interpreted by a change in the material texture at this aging temperature.

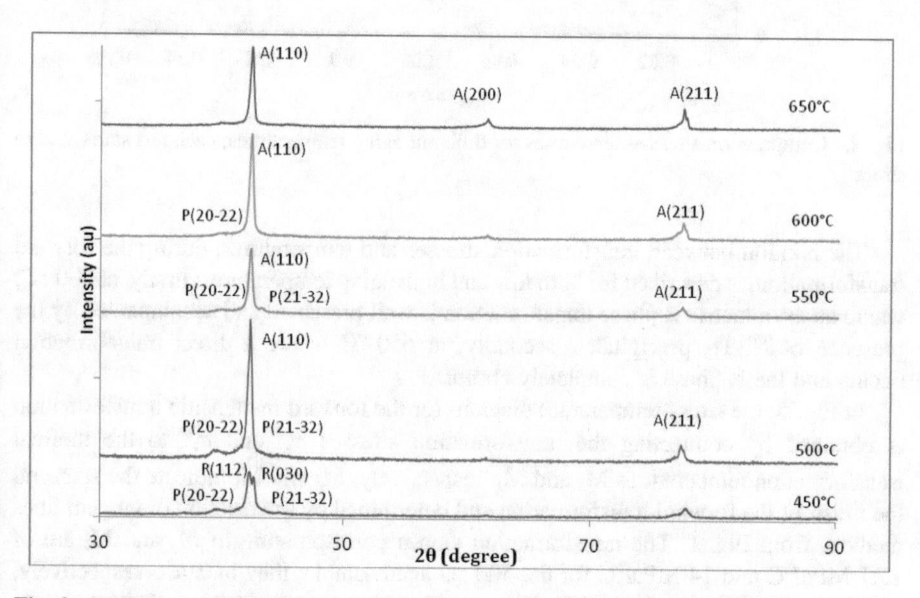

Fig. 3. XRD patterns of the Ni-rich NiTi SMA heat treated at various temperatures. A: austenite, M: martensite and P: precipitates.

3.3 Compression Tests and Stress-Temperature Diagram

The samples aged at 500 °C and 650 °C are subjected to an isothermal compression loading at a test temperature of 37 °C. It should be noted that the aging temperature of 450 °C was excluded from the mechanical study since the correspondent DSC thermogram reveals that the presented phase is not purely austenitic at 37 °C.

The obtained compressive stress-strain curves for the selected aging temperatures are depicted in Fig. 4. The transformation stresses σ_s^M and σ_f^M which characterize the start and the finish of the stress-induced transformation (A → M), respectively, are determined from the stress-strain curves by the intersection of tangent lines method.

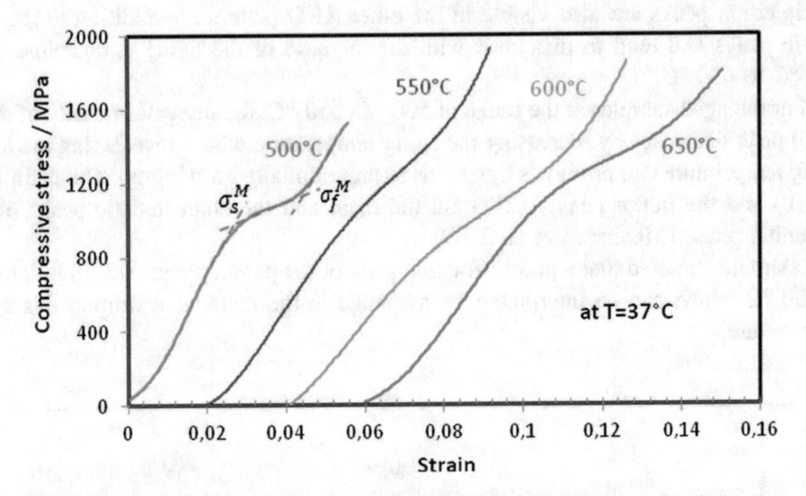

Fig. 4. Compression stress-strain curves for different aging temperatures; each test starts at zero strain.

The relation between transformation stresses and temperatures, during the forward transformation, is described for both low and high aging temperatures: firstly, at 500 °C, where an asymmetric R phase transformation is well presented and accompanied by the presence of Ni_4Ti_3 precipitates; secondly, at 650 °C where a direct transformation occurs and the R phase is completely absent.

In Fig. 5, the stress-temperature diagram for the forward martensitic transformation is obtained by connecting the transformation stresses σ_s^M and σ_f^M to the thermal transformation temperatures M_s and M_f, respectively. M_s and M_f indicate the start and the finish of the forward transformation and determined by intersection of tangent lines method, from Fig. 1. The transformation slopes corresponding to M_s and M_f are of 17.7 MPa/°C and 14 MPa/°C for the 500 °C aged sample; they increase, respectively, to 22.36 MPa/°C and 23.98 MPa/°C at 650 °C aging temperature. The evolution tendency of these transformation slopes as function of aging temperature can be attributed to the R phase presence level which affects the transformation temperatures M_s and M_f. Regarding the transformation process, it is observed that, for the 500 °C aged sample, the difference between M_s and M_f is remarkably decreased by the increase of the applied stress which reduces the martensitic transformation window width. Consequently, in the sample containing initially the Ni_4Ti_3 precipitates at zero stress, we can deduce that the martensitic transformation rate increases by increasing the compressive stress. For the 650 °C aged sample, the stress/temperature relationship exhibits a slight increase in the difference between M_s and M_f. This tendency may be explained, then, by the creation of dislocations by deformation which act as a resistance to the martensitic transformation (Zheng et al. 2008).

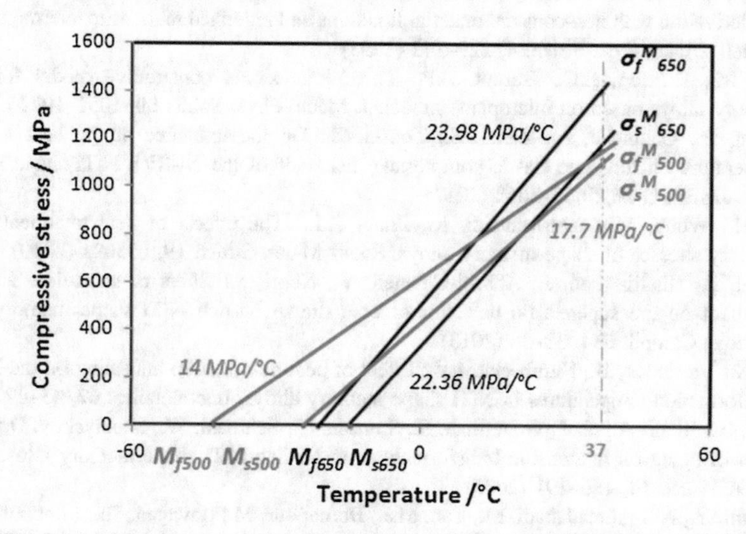

Fig. 5. Critical stress-temperature diagram for the martensitic transformation of 500 °C-1 h and 650 °C-1 h aged Ni-rich NiTi samples and compressed at 37 °C.

4 Conclusion

In this work, an experimental investigation was performed to characterize the phase transformation behavior of an aged Ni-rich NiTi shape memory alloy. The thermal and microstructural analyzes was carried out at zero stress using both DSC and XRD techniques. It was reported that the R phase presence level, revealed through the calorimetric measurements, has a great effect on the NiTi thermal properties. By increasing the aging temperature, the gradual disappearance of the R phase is attributed to the dissolution of the Ni_4Ti_3 precipitates.

The characterization of the aged NiTi mechanical behavior was performed through an isothermal compression tests and the overall thermomechanical behavior was discussed based on the critical stress-temperature diagram for different transformation processes. It was proved that the lower transformation slopes are obtained where the transformation process exhibits the R phase during the thermal transformation at zero stress. For the transformation process free from the R phase, the martensitic transformation kinetic seems to be not very sensible to the thermomechanical loading.

References

Tanaka, K.: A thermomechanical sketch of shape memory effect: one dimensional tensile behaviour. Res. Mech. **18**, 251–263 (1986)

Brinson, L.C.: One-dimensional constitutive behavior of shape memory alloys: thermomechanical derivation with non-constant material functions and redefined martensite internal variable. J. Intell. Mater. Syst. Struct. **4**, 229–242 (1993)

Brocca, M., Brinson, L.C., Bazant, Z.P.: Three-dimensional constitutive model for shape memory alloys based on microplane model. J. Mech. Phys. Solids **50**, 1051–1077 (2002)

Ben Fraj, B., Gahbiche, A., Zghal, S., Tourki, Z.: On the influence of the heat treatment temperature on the superelastic compressive behavior of the Ni-Rich NiTi shape memory alloy. JMEPEG **26**, 5660–5668 (2017)

Sadiq, H., Wong, M.B., Al-Mahaidi, R., Zhao, X.L.: The effects of heat treatment on the recovery stresses of shape memory alloys. Smart Mater. Struct. **19**, 035021 (2010)

Khaleghi, F., Khalil-Allafi, J., Abbasi-Chianeh, V., Noori, S.: Effect of short-time annealing treatment on the superelastic behavior of cold drawn Ni-rich NiTi shape memory wires. J. Alloys Compd. **554**, 32–38 (2013)

Wang, X., Verlinden, B., Humbeeck, J.V.: Effect of post-deformation annealing on the R-phase transformation temperatures in NiTi shape memory alloys. Intermetallics **62**, 43–49 (2015)

Eggeler, G., Khalil-Allafi, J., Gollerthan, S., Somsen, C., Schmahl, W., Sheptyakov, D.: On the effect of aging on martensitic transformations in Ni-rich NiTi shape memory alloys. Smart Mater. Struct. **14**, 186–191 (2005)

Motemani, Y., Nili-Ahmadabadi, M., Tan, M.J., Bornapour, M., Rayagan, Sh.: Effect of cooling rate on the phase transformation behavior and mechanical properties of Ni-rich NiTi shape memory alloy. J. Alloys Compd. **469**, 164–168 (2009)

Ben Fraj, B., Tourki, Z.: In the 23rd International Federation of Heat Treatment and Surface Engineering Congress, Savannah, Georgia, USA, pp. 481–486 (2016)

Kaya, I.: Effect of cooling rate on the shape memory behavior of Ti-54at.%Ni alloys. Anadolu Univ. J. Sci. Technol. A-Appl. Sci. Eng. **18**(2), 535–542 (2017)

Ben Fraj, B., Zghal, S., Tourki, Z.: DSC investigation on entropy and enthalpy changes in Ni-rich NiTi shape memory alloy at various cooling/heating rates. Springer book: Design and Modeling of Mechanical Systems-III, pp. 631–640 (2017b)

Kazemi-Choobi, K., Khalil-Allafi, J., Abbasi-Chianeh, V.: Influence of recrystallization and subsequent aging treatment on superelastic behavior and martensitic transformation of Ni50.9Ti wires. J. Alloys Compd. **582**, 348–354 (2014)

Duerig, T.W., Bhattacharya, K.: The Influence of the R-phase on the Superelastic Behavior of NiTi. Shape Mem. Superelasticity **1**, 153–161 (2015)

Churchill, C.B., Shaw, J.A., Iadicola, M.A.: Tips and tricks for characterizing shape memory alloy wire: Part 2-fundamental isothermal responses. Tech. Exp. (2009). https://doi.org/10.1111/j.1747-1567.2008.00460.x

Laplanche, G., Birk, T., Schneider, S., Frenzel, J., Eggeler, G.: Effect of temperature and texture on the reorientation of martensite variants in NiTi shape memory alloys. Acta Mater (2017). https://doi.org/10.1016/j.actamat.2017.01.023

Jiang, S., Zhang, Z., Zhao, Y., Liu, S., Hu, L., Zhao, C.: Influence of Ni4Ti3 precipitates on phase transformation of NiTi shape memory alloy. Trans. Nonferrous Met. Soc. China **25**, 4063–4071 (2015)

Otsuka, K., Ren, X.: Physical metallurgy of Ti–Ni-based shape memory alloys. Prog. Mater Sci. **50**, 511–678 (2005)

Zheng, Y., Jiang, F., Li, L., Yang, H., Liu, Y.: Effect of ageing treatment on the transformation behaviour of Ti-50.9 at.% Ni alloy. Acta Mater **56**, 736–745 (2008)

FTIR Spectroscopy Characterization and Numerical Simulation of Cyclic Loading of Carbon Black Filled SBR

Amina Dinari(✉), Makram Chaabane, and Tarek Benameur

Mechanical Engineering Lab, LR99ES32 ENIM,
Monastir University, 5000 Monastir, Tunisia
dinariamina@gmail.com,
{makram.chaabane, t.benameur}@enim.rnu.tn

Abstract. This contribution presents a combined approach including experimental observation at both macroscopic and microscopic scales of carbon black Styrene butadiene rubber (SBR) in form of thick sample after cyclic loading using FTIR. Due to fatigue, the rubber network contains intact and altered network portions, which is introducing the scission/rotation mechanism. The chain scission explains the dramatic stress decrease under fatigue loading. The decrease of hydrocarbon area confirms the rotation in some chemical group. The breakdown/rebound mechanism of inter-aggregate links in the rubber-filler material system is also responsible of the relaxation response at the macro scale. FTIR spectroscopy characterization is carried out in order to clarify the effect of stress concentration, filler amount on SBR softening do to cyclic loading. The spectroscopic changes are measured in two different zones the median and the upper section of the specimen. Numerical simulation is developed. The response at the macro-scale after cyclic loading is predicted, of thick carbon-filled SBR samples. The predictive capability of the FE model is tested by comparison with experimental data.

Keywords: SBR rubber · Cyclic loading ·
Stress softening FTIR-ATR spectroscopy · FE simulation

1 Introduction

Firstly studied in the 1960s by Mullin and coworkers, the fatigue effects in rubbers have a long research history. The authors experimentally observed an appreciable change on material response, the stress lower for the same applied strain from the first extension (Mullins 1969), becomes negligible after several cycles to reach eventually a stationary state with constant stress amplitude, and stabilized hysteresis loop. Moreover, they investigated the fillers particles effects on the stiffness and the strength of rubber compound. Two types of microstructural network rearrangements take place in rubbers after mechanical fatigue: a recoverable viscoelastic rearrangement due to the original network stretching (e.g. entanglement slippage) and unrecoverable rearrangements via scission reactions of bonds. These mechanisms are dependent on the frequency, filler content and stretch amplitude. Many researchers resort to

A. Benamara et al. (Eds.): CoTuMe 2018, LNME, pp. 186–194, 2019.
https://doi.org/10.1007/978-3-030-19781-0_23

spectroscopic techniques to reveal the molecular changes of polymeric material (Celina et al. 1998; Choi et al. 2014; Diaz et al. 2014; He et al. 2017). During cyclic loading, in both filled and unfilled rubber, stress softening is detected which known as Mullin effect. From a physical point of view, plenty interpretations are proposed, we review some of them: matrix damage including cavities appearance, chain scission, disentanglement, change of number of chemical crosslinks, filler network alteration and rubber-filler interface change (Guo et al. 2017). In our contribution, we present an investigation about this phenomena in both macroscopic (mechanical) and microscopic (molecular) scales. In the microscopic scale, we use Infrared absorption spectroscopy with ATR technique to evaluate the structural degradation of styrene butadiene rubbers containing three different amount of carbon black particles. FTIR analysis is done in two different regions: the sample upper and median section to figure out the effect of heat buildup, geometry and fillers on the network alteration. Since the finite element, computation is essential to predict the macroscopic behavior of SBRs under fatigue and to highlights the several causes of microscopic changes, we resorts to FE simulation model. A subroutine describes the hyper viscoelastic model is implemented into MSC Marc. The predictive capability of the FE model is tested by comparison with experimental data.

2 Experimental

2.1 Sample Geometry and Materials

Cylindrical diabolo specimens with curvature radius of 42 mm (referred to as AE42mm) are used in this study (Fig. 1). The reduced section in the center of the sample allows locating the highest increase of stress and temperature. The specimens material is a sulphur vulcanized Styrene Butadiene Rubber (SBR) with three Carbon Black filler content, 15, 25 and 43 phr (part per hundred of rubber in weight).

2.2 Mechanical Fatigue Characterization

Fatigue tests are performed using an electro-pulse testing machine Instron-5500 and defined by four different loading. A first load inducing stabilization of the cyclic response is applied before each fatigue test. This is to enfranchise the Mullins effect (Mullins 1969). Two types of series cyclic tests are made. Fatigue test at imposed stretch level and fatigue test at imposed frequency level. The stretches ramp to maximum value of 2 mm and 14 mm, respectively for the first and the second series of tests. After defining the maximum loading cycle number, load and displacement versus time is recorded during the cyclic test. The same absolute axial frequency 2 Hz is imposed to the loading and unloading paths. For the second type of fatigue test, the same stretch level rises to maximum value at 10 mm and the imposed frequency is 0.4 Hz for the first tests and 2.8 Hz for the second tests.

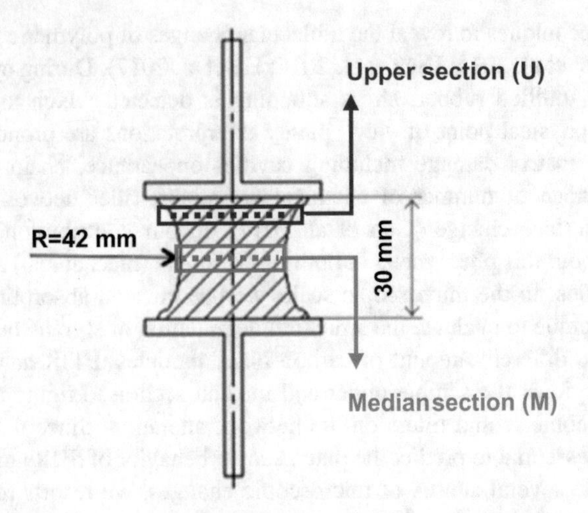

Fig. 1. Sample geometry with different areas of cutting for FTIR characterization

2.3 Fourier Transform Infrared Characterization

Infrared spectra were measured with a FT-IR spectrometer (Perkin Elmer Spectrum Two Fourier Transform Infrared (L120 2057)). Thin sections of SBRs samples (about 100-µ thickness) are cut from two regions: one from the median section (M) and the other one was taken from the upper side (U) of the specimen. Then, the disc is placed on top of ATR- crystal (Diamond) and exposed to an infrared beam across a range of wavelengths from 2.5 µm to 25 µm (wavenumber range 4000 cm^{-1} to 400 cm^{-1}) that passes through the diamond crystal forming an evanescent wave that penetrates and reflects off the crystal surface. The pressure force applied to all specimens is 50 ± 1 N. In order to minimize experimental errors and to ensure our experiments reproducibility, FTIR tests are repeated three times and the spectrums are fitted using Omnic software. The experimental protocol is presented in Fig. 2.

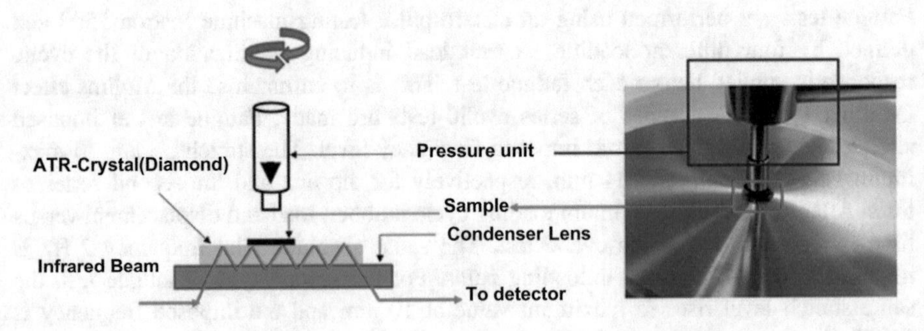

Fig. 2. Infrared spectroscopy: Beam path of ATR accessory and FTIR Sample techniques: Attenuated Total Reflection (ATR)

3 Results and Discussion

3.1 Fillers Softening Effects

Figure 3 shows Fatigue cyclic loading induced stress softening at two stretch amplitudes for the three SBR filler content. The curves provides the stress evolution during 400 cycles for 2 mm imposed stretch (Fig. 3a) and 14 mm imposed stretch (Fig. 3b) at the same given frequency 2 Hz. We are interested on the rapid decrease in stress observed in the first 100 cycles. Following this substantial decrease in stress, the SBR presents less and less softening and tends to a stable stress state for important cycles. The magnitude of softening stress depends on three factors, the number of cycles, the imposed stretch level and carbon black filler rate. Rapid decay of the stress initially is often associated with irreversible state of macromolecular network (Mullins 1960; Göktepe and Miehe 2005) and damage caused by the breakage chains during the first cycles (Ayoub et al. 2011b, 2014b) of non-recoverable network (Guo et al. 2017). Because of their time-dependent nature, the inelastic phenomenon in elastomers can recover during unloading time. In this case, the initial stiffness of the material may be recoverable. This recoverability depends on the magnitude of stress and the temperature (Mullins 1969). It is reported in the review of (Guo et al. 2017) that cyclic load applied to SBR containing different amounts of carbon-black induce two types of macromolecular network arrangement in the rubber-filler material system. The first one is the recoverable network present a viscoelastic part and unrecoverable network damage mechanisms part (Dinari et al. 2017) induces softening. The decrease of stress attributed to the unrecoverable network.

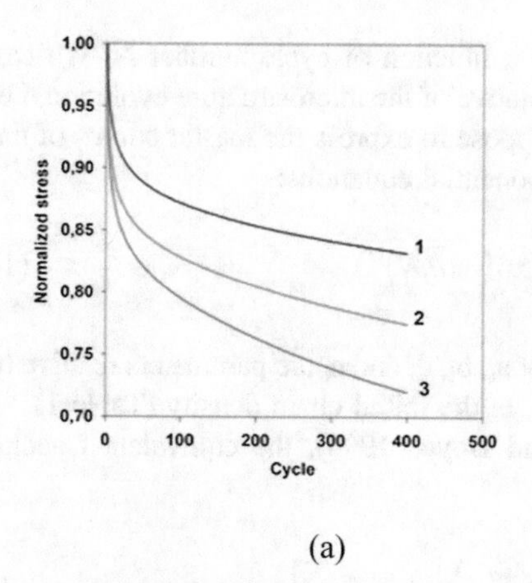

(a)

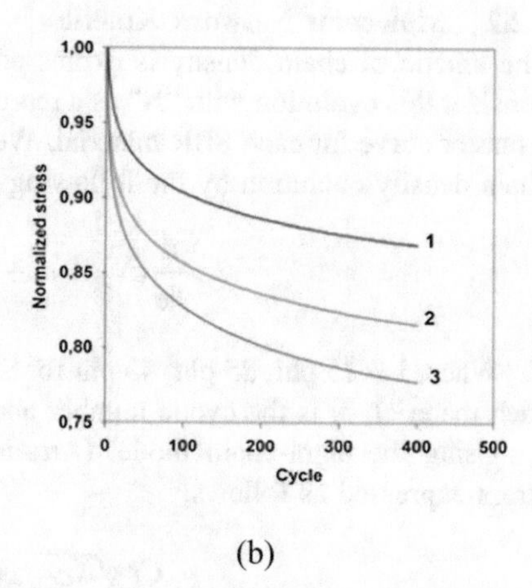

(b)

Fig. 3. Relaxation stress during fatigue loading for: (a) U_{max} = 2 mm; (b) U_{max} = 14 mm (1:15 phr, 2:25 phr, 3:43 phr)

3.2 Fillers Microstructural Effects

3.2.1 ATR-IR Analysis

Two specimens sections are examined by FTIR spectroscopy, the median section (M) which has highest stress and the upper section, the least solicited zone (U). Results of the FTIR spectra of the rubber with three Carbon Black filler content from all previous sections are reported. We present the FTIR result of SBR15phr and SBR43phr in Fig. 4. In addition, two peaks are observed in 3000 cm^{-1} and 3070 cm^{-1} related to phenyl groups (Choi et al. 2014; Biao et al. 2018). As previously mentioned by He and co-worked (He et al. 2017), the absorption peaks at 1250 cm^{-1}, 800 cm^{-1} in the spectra of SBR are attributed to C-O (carbonyl) stretching and bending vibrations. Indeed, these carbonyl groups are observed in our spectra for the three SBRs, however there is not a noticeable difference in the both zones (the median section (M)/the upper side of sample (U)). The histograms of characteristic absorption area presented below are measured from the collected data of three materials spectra. The area variation of peaks assigned to butadiene group follow the same trend of change in SBR 15 phr in both zone. For the less filled SBR, the cis 1.4, 1.2 and trans 1.4 butadiene area are correspondingly to 2.6, 0.26 and 1.4 in the Upper side, while they are equal to 1.03, 0.16 and 0.98 in the median section (Histogram 5.a). On the contrary, the characteristic absorption peaks of SBR43 phr remain the same in both zones. Therefore, the areas are equal to 2.11, 0.25 and 2.85 for cis 1.4, cis 1.2 and Trans 1.4 butadiene, respectively (Histogram 5.b). The strong diminution of hydrocarbon group in SBR 15 phr is explained by the loss of conformational chains ordering and the rotation of double bond position (Figs. 4 and 5).

3.2.2 Molecular Network Kinetic

The kinetic of chain density is expressed as a function of cycle number N. We can consider this evolution with “N” as a representative of the microstructure evolution, i.e. a master curve for each SBR material. We propose to express the master curves of the chain density evolution by the following exponential equations:

$$\frac{n}{n_0}(N) = a_i + b_i \exp(-(d_i N)^{ci}) \tag{1}$$

Where i = 15 phr, 25 phr, 43 phr for SBR_i: a_i, b_i, d_i and c_i are parameters relative to each material; N is the cyclic number and n_0 in the initial chain density (Table 1).

Using the eight-chain model (Arruda and Boyce 1993), the equivalent Cauchy stress expressed as follows:

$$\underline{\underline{\sigma_{eq}}} = \frac{Cr}{3}\frac{\sqrt{N_{Cr}}}{\lambda_{eq}} L^{-1}\left(\frac{\lambda_{eq}}{\sqrt{N_{Cr}}}\right)\left[\underline{\underline{B_e}} - \lambda_{eq}\underline{\underline{I}}\right] \tag{2}$$

$Cr = nKT$ is the shear modulus, where k is the Boltzmann constant, T is the absolute temperature and n is the chain density. $\sqrt{N_{CR}}$ is the number of connected rigid links in a chain. $\underline{\underline{B_e}}$ is the right Cauchy strain tensor, which can be written in the following form: $\underline{\underline{B}}_e = \underline{\underline{F}}_e \underline{\underline{F}}_e^T$.

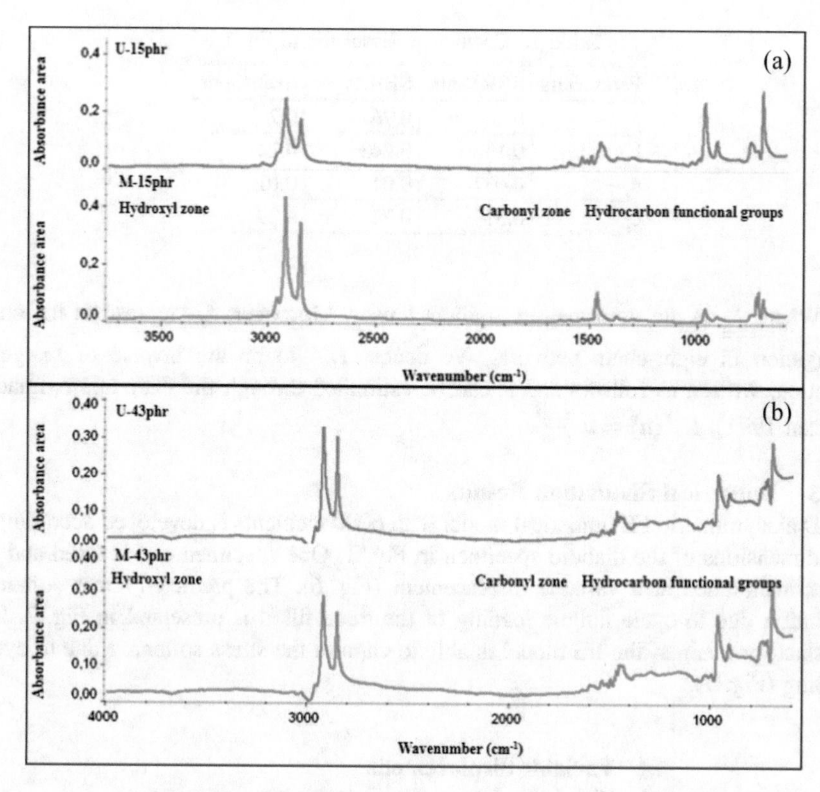

Fig. 4. Infrared spectrum filled SBR after cyclic load in two zone (U-upper section) and (M-median section): SBR15phr (a), SBR43phr (b)

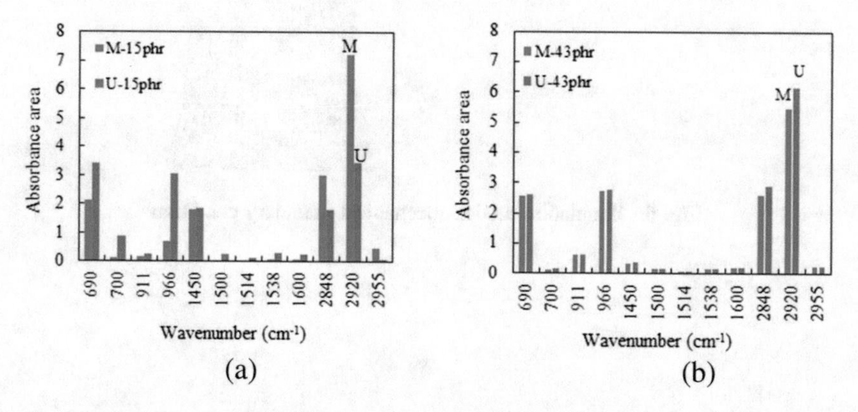

Fig. 5. Absorbance area of chemical functional groups in the median (M) and the upper (U) zones: (a) SBR15phr and (b) SBR43phr

Table 1. Coefficient associated to Eq. 1

Parameters	SBR15phr	SBR25phr	SBR43phr
a_i	0.86	0.76	0.7
b_i	0.136	0.243	0.24
d_i	0.037	0.01	0.01
c_i	0.51	0.72	0.29

Where $\underline{\underline{F_e}}$ is the deformation gradient tensor. Moreover, λ_{eq} represents the chain elongation in eight-chain network. We denote L^{-1} to be the inverse of Langevin function, written as follows and it can be estimated through the pade approximation (Cohen 1991), $L^{-1}(u) = u\,\frac{3-u^2}{1-u^2}$.

3.2.3 Numerical Simulation Results

A 3D axisymmetric FE numerical model with 6000 elements is developed according to the dimensions of the diabolo specimen in Fig. 1. One specimen end is fixed and the other maintained at a variable displacement (Fig. 6). The predicted stress softening evolution due to cycle fatigue loading of the three filled is presented in Fig. 7. In a satisfactory manner, the FE model is able to capture the stress softening due to cyclic loading (Fig. 7).

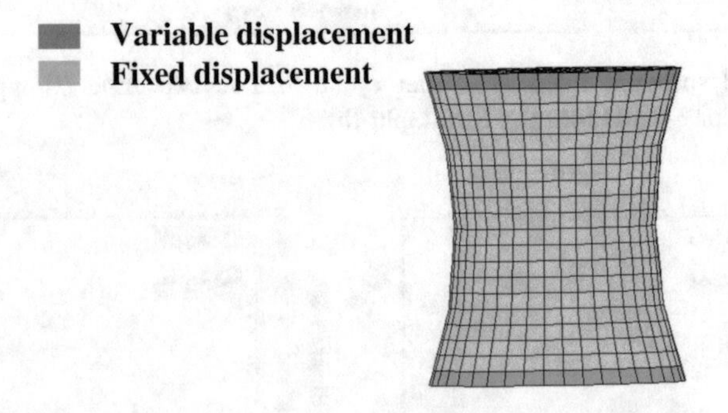

Fig. 6. Simulation model: mechanical boundary condition

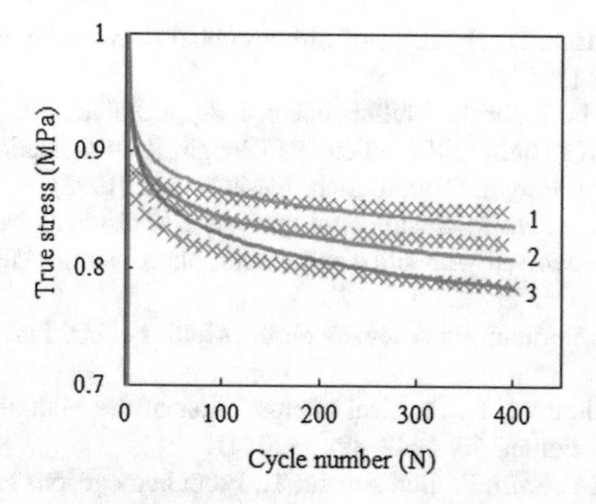

Fig. 7. Predicted and experimental stress softening for 2 mm imposed stretch (1:15 phr; 2:25 phr; 3:43 phr)

4 Conclusion

This contribution is an experimental testing and numerical simulation of the fatigue effects on carbon black filled styrene butadiene rubber. Stress softening induced by cyclic loading observed at both macroscopic and microscopic scale. At macroscopic scale, it has shown that the stress decay after the cyclic loading related to chain breakdown and fillers friction. At microscopic scale, the FTIR ATR spectroscopy analysis show that the decrease of functional group cis and Trans in the specimens median section of SBR25phr and 15 phr. Hence, the functional group of SBR43phr presents more stability. The results show a good agreement between numerical and experimental behavior under cyclic loading.

Acknowledgement. The authors would like to thank Professor Fahmi Zaïri Civil Engineering and geo-Environmental Laboratory, Lille University, (EA 4515 LGCgE) for the sample used in this study.

References

Arruda, E.M., Boyce, M.C.: A three-dimensional constitutive model for the large stretch behavior of rubber elastic materials. J. Mech. Phys. Solids **41**, 389–412 (1993)

Ayoub, G., Zaïri, F., Naït-Abdelaziz, M., Gloaguen, J.M., Kridli, G.: A visco-hyperelastic damage model for cyclic stress-softening, hysteresis and permanent set in rubber using the network alteration theory. Int. J. Plast. **54**, 19–33 (2014)

Ayoub, G., Zaïri, F., Naït-Abdelaziz, M., Gloaguen, J.M.: Modeling the low-cycle fatigue behavior of visco-hyperelastic elastomeric materials using a new network alteration theory: application to styrene-butadiene rubber. J. Mech. Phys. Solids **59**, 473–795 (2011)

Biao, Z., Manabu, I., Daisuke, K., Xinping, W., Ta, K.: Conformational relaxation of poly (styrene-co-butadiene) chains at substrate interface in spin-coated and solvent-cast films. J. Macrom. **51**, 2180–2186 (2018)

Blanchard, A.F., Parkinson, D.: Breakage of carbon-rubber networks by applied stress. Ind. Eng. Chem. **44**, 799–812 (1952)

Bueche, F.: Molecular basis for the Mullins effect. J. Appl. Polym. Sci. **10**, 107–114 (1960)

Celina, M.J., Wise, D.K., Otteseq, K., Gillen, T., Clough, R.L.: Oxidation profiles of thermally aged nitrile rubber. J. Polym. Degrad. Stab. **60**(493), 504 (1998)

Choi, S., Kim, Y., Kwon, H.: Microstructural analysis and cis–trans isomerization of BR and SBR vulcanizates reinforced with silica and carbon black using NMR and IR. RSC Adv. **4**, 31113 (2014)

Diani, J., Fayolle, B., Gilormini, P.: A review on the Mullins effect. Eur. Polym. J. **45**, 601–612 (2009)

Diaz, R., Julie, D., Gilormini, P.: Physical interpretation of the Mullins softening in a carbon black filled SBR. J. Polym. **55**, 4942–4947 (2014)

Dinari, A., Chaabane, M., Zaïri, F., Ben Ameur, T.: From homogenous to hetergenous ageing of filled. Carbon Black Styrene Butadiene Rubber **7**, 144 (2017). CMSM Proceeding/Tunisia

Göktepe, S., Miehe, C.: A micro–macro approach to rubber-like materials. Part III: the micro-sphere model of anisotropic Mullins-type damage. J. Mech. Phys. Solids **53**, 2259–2283 (2005)

Guo, Q., Zaïri, F., Baraket, H., Chaabane, M., Guo, X.: Pre-stretch dependency of the cyclic dissipation in carbon-filled SBR. Eur. Polym. J. **96**, 145–158 (2017)

He, S., Bai, F., Liu, S., Ma, H., Hu, J., Chen, L., Lin, J., Wei, G., Du, X.: Aging properties of styrene-butadiene rubber nanocomposites filled with carbon black and rectorite. J. Polym. Test (2017). https://doi.org/10.1016/olymertesting.2017.09.017

Mullins, L.: Softening of rubber by deformation. Rubber Chem. Technol. **42**, 339–362 (1969)

Shear Bands Behavior in Notched $Cu_{60}Zr_{30}Ti_{10}$ Metallic Glass

Mohamed Ahmedou Senhoury[1(✉)], Bechir Bouzakher[2], Fathi Gharbi[1,3], and Tarek Benameur[1]

[1] Mechanical Engineering Laboratory, LR99ES32, ENIM University of Monastir, Monastir, Tunisia
senhoury90@yahoo.fr, fathi.gharbi@yahoo.fr, t.benameur@enim.rnu.tn

[2] LMMP LR99ES05, Université de Tunis, ENSIT, 1008 Tunis, Tunisia
bechir.bouzakher@gmail.com

[3] Institut Supérieur des Sciences Appliquées et de Technologie de Kairouan, Kairouan, Tunisia

Abstract. A significant role is played by shear bands in controlling plasticity and failure mode in metallic glasses MGs. Atomic force microscopy AFM together with a nanoindender was previously used to observe shear band blocking. In this work, we have developed a correlation between finite element FE analysis and in situ SEM characterization, AFM topographic and frictional measurements. The experiments are performed in notched thin ribbons of $Cu_{60}Zr_{30}Ti_{10}$ MG. Based on Gao's FE formulation integrating Spaepen's micromechanical model and AFM topographic analysis allow us to get further insights on the propagation, intersection and shear band offsets. Different shear band networks emanating at near and far plastic zones from the notch roots are observed as the tensile load was increased. The intersection is found weak between individual shear bands. Simulations and direct AFM frictional tests on pile up of shear bands did not reveal any work hardening that is due to intersection. The conjugate shear bands remains inactive and the propagation direction of shear band that can be deviated by the pre-existing one is not seen.

Keywords: Metallic glasses · Shear band intersection · FE modeling AFM topographic and frictional images

1 Introduction

Bulk metallic glasses BMGs are interesting class of materials both for basic and applied research (Johnson 1999, Inoue 2000). They are useful in various applications such as medical tools, MEMs and NEMs technologies. Despite the extremely high strength, large elastic strain limit combined with relatively high fracture toughness, as well as good wear resistance of BMGs, they suffer a strong tendency for shear localization and fail catastrophically on one dominant shear band in tensile tests at ambient temperature. However, Greer et al. (2013) reported that stable plasticity in BMGs can be achieved by either shear band proliferation with small shear offsets or by introducing heterogeneities into the glass to stimulate or by imposing confinements on the size of the

A. Benamara et al. (Eds.): CoTuMe 2018, LNME, pp. 195–203, 2019.
https://doi.org/10.1007/978-3-030-19781-0_24

shear band and of shear offsets. Moreover, it is reported that more extended plastic strain by shear delocalization crystallization of shear bands operating under compressive applied stress revealed by synchrotron x-ray microscopy in transmission (Yavari et al. 2012). The authors reported that the detected crystallization occurs in a bent $Pd_{40}Cu_{30}Ni_{10}P_{20}$ glassy ribbon only on the compression side of the neutral fiber. Furthermore, and as reported by Bigoni and Dal Corso (2008), SBs is an instability linked to an abrupt loss of homogeneity of deformation occurring in crystalline solid subject to a loading path.

It has been shown through experimental evidence in ductile matrix containing hard inclusions that SBs nucleate at inclusion boundary and grow parallel to them. An important and crucial factor in the understanding of SBs failure mode is the presence of stress raisers such as sharp inclusions (Dal Corso and Bigoni 2009). By creating two symmetrical semi-circular notches, Zhao et al. (2010) indicated that these results are consistent with several reports of plasticity improvement to a high value of 10% under compression tests. Sha et al. (2015) showed a failure mode transition from shear banding to necking induced by the dual effects of large stress gradient at the notch roots and the impingement and subsequent arrest of shear bands emanating from the notch roots. In a different study Wang et al. (2013), reported a densification and a hardening in notched BMGs under multiaxial loading.

The previous studies by Bigoni and Dal Corso (2008), Dal Corso and Bigoni (2009), Zhao et al. (2010), and Sha et al. (2015) all share common event stress raisers and stimulate an important questions: Does SB emerge from the defect tips in BMGs? and why, unlike cracks, do shear bands grow rectilinearly and for 'long distances' under shear loading? In this study, we address these two questions through experimental as well as numerical analysis. The experiments are conducted using both in-situ SEM and AFM characterizations of notched ribbons of $Cu_{60}Zr_{30}Ti_{10}$ MG. The simulations are extended to analyze qualitatively the individual processes of SBs within $Cu_{60}Zr_{30}Ti_{10}$ MG, the intersection between individual SB and between the shear bands and the background stress fields. The numerical simulations are based on Gao's finite element (FE) formulation integrating Spaepen's micromechanical model.

2 Experimental Procedure

$Cu_{60}Zr_{30}Ti_{10}$ alloy is synthesized by melting a mixture of highly pure metals, at least 99.99%. The alloy is then re-melted in quartz crucibles and injected in an inert atmosphere onto a copper wheel. The amorphicity of the as-quenched ribbons (with a thickness of 28 to 30 μm and a width of 0.8 to 4 mm), is examined using a monochromatic x-ray diffractometer. Figure 1 shows the experimental set-up used for applying a gradual tensile elongation. The ribbons are fixed and submitted to a finite and a gradual displacement step; each corresponds to an elastic elongation of 0.13%.

Before applying the incremental elongation to the BMG alloy, a micrometer-sized notch is introduced in the medium region of the ribbon by a pincers. Tensile elongations in the range of 0.13 to 0.52% have been applied at a strain rate of approximately 10^{-4} s^{-1}. The experimental set-up is loaded after each ribbon elongation in the SEM and the AFM's Nanoscope-III Digital microscope chamber for SBs topographic and

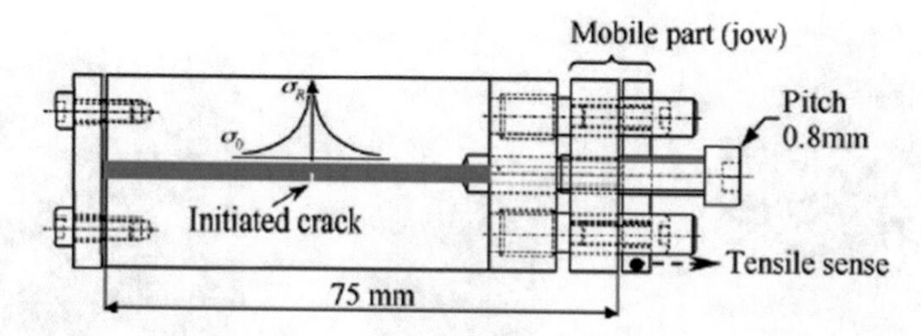

Fig. 1. Experimental set-up used for applying a gradual tensile elongation

frictional characterizations. The SBs are very difficult to study because the underlying instability is often a rapid process of autocatalytic localization of plastic strains; the process is non-steady in time and inhomogeneous in space. However, the in-situ SEM and AFM observations provided significant insights of the mutual interaction of SBs in different monolithic metallic glasses.

3 Results and Discussions

3.1 Experimental Results

Figure 2 shows the SEM images of $Cu_{60}Zr_{30}Ti_{10}$. As illustrated in Fig. 2a, before any loading imposed, some regular shear bands (SBs) initiated on the front of notch and certain shear bands across each other. Obviously, plentiful SBs are formed around the midpoint. From Fig. 2b and c, it is observed that only two major shear bands are developed and bifurcated after 0.39% of elongation (as showed in Fig. 2c). This observation illustrates that shear bands increase with increasing elongations values.

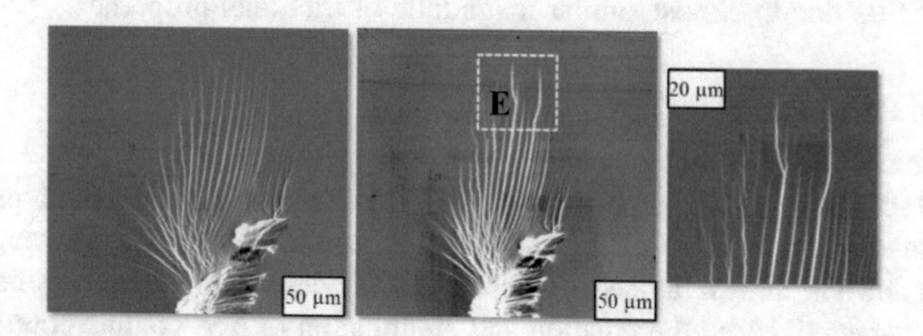

Fig. 2. SEM images of $Cu_{60}Zr_{30}Ti_{10}$ at (a) 0%, (b) 0.39% and (c) the corresponding magnified image of the box E.

Furthermore, the AFM topographic in-situ examination of $Cu_{60}Zr_{30}Ti_{10}$ alloy shows shear bands with heights increases into 400 nm for imposed deformation of 0.39%. In addition, we notice the presence of shear band offsets as illustrated in Fig. 3a and subsidiary thin bands in Fig. 7b. When operating in lateral force mode (friction measurements in AFM contact mode) as shown in Fig. 3a and b (right side), it is

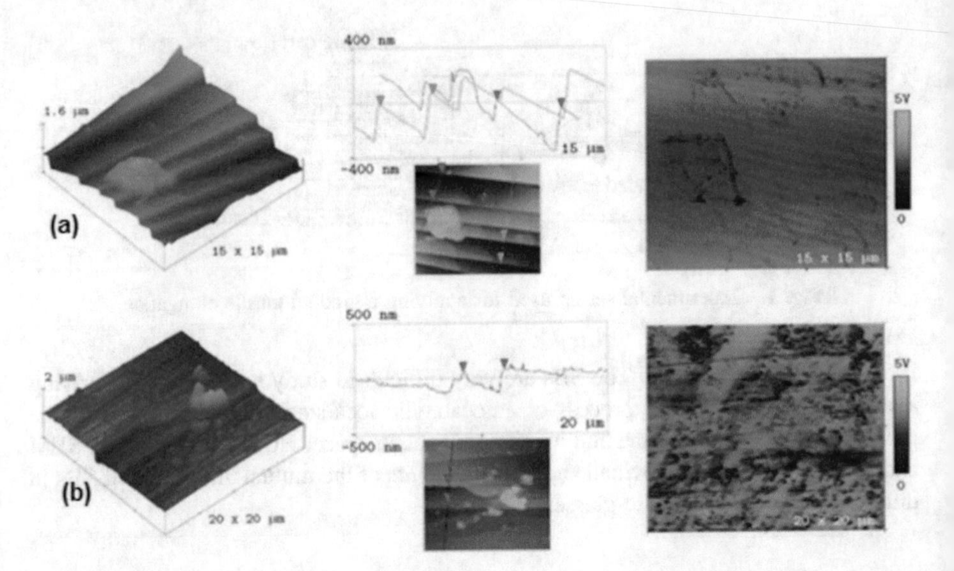

Fig. 3. AFM images of 3D topography, sectional analysis and frictional analysis of shear bands of $Cu_{60}Zr_{30}Ti_{10}$ at (a) 0% and (b) 0.39% elongation.

important to account for the topographic convolution, frequently lateral force images are collected both in the forward (retrace) and reverse (trace) scan direction and then subtracted from one another.

Here, it is worth mentioning that, because the height of the subsidiary bands is less than that of the pre-existing primary one, the suggested blocking mechanism of interactive bands by Wang et al. (2017) is found to be weak and no evidence of the pre-existing primary SB acts as an energy barrier to influence the propagation of the secondary SBs due to almost similar magnitude of frictional properties.

3.2 Numerical Results

3.2.1 Constitutive Model

For metallic glasses, it is reported that the free volume model could predict the inhomogeneous deformation. This framework integrates Spaepen's micromechanical model for BMGs alloys. Indeed, Spaepen (1977) derived an inhomogeneous flow equation characterized by the creation and annihilation of free volume contributing to local shearing deformation. Both the plastic flow equation and the free-volume evolution are driven by shear stress (Steif et al. 1982). Afterwards, Gao (2006) extended this mechanical model to the generalized multiaxial stress state.

It should be noted that the strain rate is decomposed into elastic and plastic parts as indicated below in Eqs. (1a) to (1d), where the elastic strain rate, $\dot{\varepsilon}^e_{ij}$, is generalized by Hooke's law for an elastically isotropic materials. The constant E is the Young's modulus, v is the Poisson's ratio; $\delta_{ij} = 0$ when i $\neq$ j and $\delta_{ij} = 1$ when i = j. σ_e is the Von-Mises effective stress, in which $S_{ij} = \sigma_{ij} - \sigma_{kk}\delta_{ij}/3$ is the deviatoric stress tensor.

$$\dot{\varepsilon}_{ij} = \dot{\varepsilon}_{ij}^{e} + \dot{\varepsilon}_{ij}^{p} \tag{1a}$$

$$\dot{\varepsilon}_{ij}^{e} = \frac{1+v}{E}\left(\dot{\sigma}_{ij} - \frac{v}{1+v}\dot{\sigma}_{kk}\,\delta_{ij}\right) \tag{1b}$$

$$\dot{\varepsilon}_{ij}^{p} = \exp\left(-\frac{1}{\xi}\right)\sinh\left(\frac{\sigma_e}{\sigma_0}\right)\frac{S_{ij}}{\sigma_e} \tag{1c}$$

$$\sigma_e = \sqrt{\frac{3}{2}S_{ij}S_{ij}} \tag{1d}$$

In addition, Eq. (2) below is the free volume evolution equation, where $\sigma_0 = 2K_BT/\Omega$ is the reference stress, in which K_B, T and Ω are Boltzmann constant, absolute temperature and average atomic volume, respectively; $\beta = v^*/\Omega$ and $v_f = v_f/\alpha v^*$ is the normalized free volume. n_D is the number of diffusive jumps necessary to annihilate a free volume as to v^* and is usually taken to be 3–10. ξ is the concentration of free volume defined by $\xi = v_f/v^*$ (v_f is the average free volume per atom and v^* is the critical hard-sphere volume (Huang et al. 2002)).

$$\dot{\xi} = \frac{1}{\alpha}\exp\left(-\frac{1}{\xi}\right)\left\{\frac{3(1-v)}{E}\left(\frac{\sigma_0}{\beta\xi}\right)\left[cosh\left(\frac{\sigma_e}{\sigma_0}\right)-1\right]-\frac{1}{n_D}\right\} \tag{2}$$

This model was implemented into the ABAQUS commercial software through UMAT subroutine. In this study, we investigated uniaxial tensile tests of $Cu_{60}Zr_{30}Ti_{10}$ metallic glass alloy with material parameters listed in Table 1. To further investigate the initiation and propagation of shear bands during the deformation processes, 2D FEM simulations with multiple shear band zones is conducted.

Table 1. The mechanical and physical properties of $Cu_{60}Zr_{30}Ti_{10}$

ρ (g/cm^3)	B (GPa)	μ (GPa)	E (GPa)	v	μ/B	T_g (°K)	T_x (°K)	Ref.
7.408	100.7	36.6	95.5	0.342	0.363	754	745	(Inoue et al. 2001)

3.2.2 Geometrical Model

Figure 4 shows the finite element mesh specimen with loading conditions and boundary. The two-dimensional plane is loaded by displacing the top boundary at constant vertical velocity and the bottom boundary is fixed. A convergence test was conducted for choosing the mesh. The present manuscript develops ABAQUS-8-node plane-strain element (CPE8). The strain rate is 10^{-6} s^{-1}. This numerical model integrates the effects of free volume, in which some elements are defined as the shear-band zones. For this work, two initial free volume, $\xi = 0.05$ and $\xi_0 = 0.0501$, respectively, are defined. A variance of 0.0001 can promote bands to nucleate.

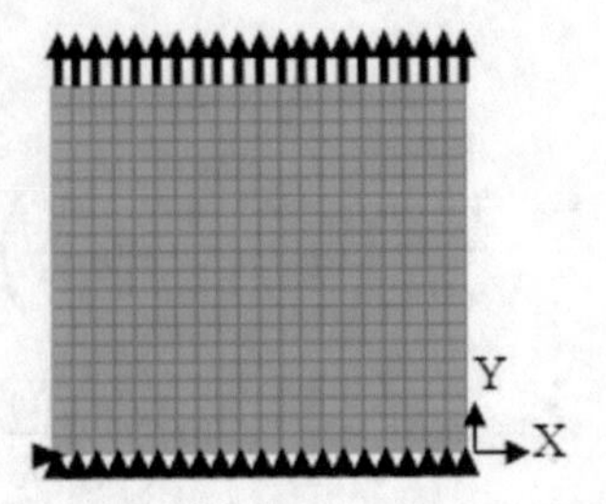

Fig. 4. Finite element mesh, boundary and loading conditions.

3.2.3 Results

Simulations for two shear band zones are conducted initially and an example of strain ε_{22} during the deformation process is shown in Fig. 5. The distribution of Von-Mises stress σ_e, strain ε_{22} and free volume concentration ξ are listed in Fig. 6. From each shear band zones (as denoted by A and B), two shear bands nucleate and then propagate simultaneously.

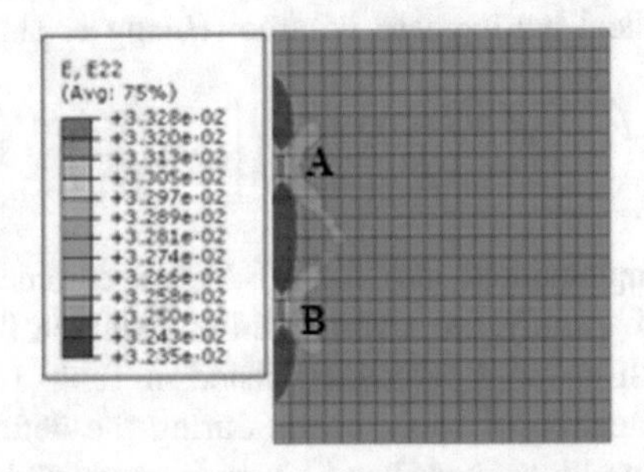

Fig. 5. Development of strain in the 2-D tensile specimen with two shear-band zones

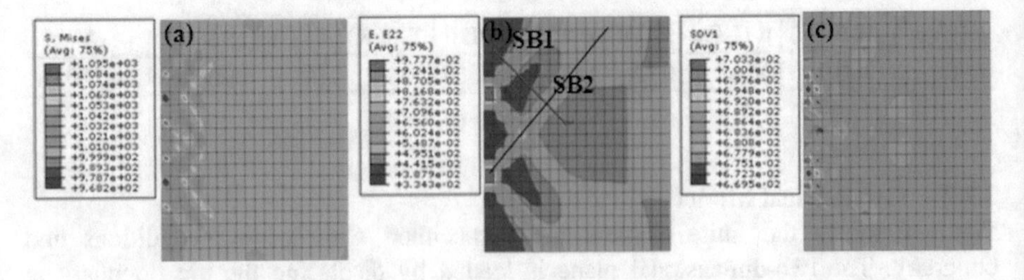

Fig. 6. Distributions of various parameters in the specimen with two shear band zones: (a) the Von Mises stress σ_e, (b) the strain, ε_{22}, and (c) the free volume concentration ξ

Figure 7 shows the distribution of Von-Mises σ_e and strain ε_{22}, along path1 denoted by the red discontinued line in Fig. 6b, inside and near the shear bands for $Cu_{60}Zr_{30}Ti_{10}$ alloy. Path1 crosses two shear bands, SB1 and SB2. From Fig. 7b, it is found that the strain magnitude inside the shear band increases significantly compared with the matrix (peak II and I correspond respectively to SB1 and SB2). This describes the highly localized deformation in $Cu_{60}Zr_{30}Ti_{10}$ alloy. Furthermore, Fig. 7a shows that the Von-Mises stress σ_e decreases significantly within the SB1 (see black arrow) and slowly decreases in SB2 (see green arrow). This corresponds to softening phenomena. Figure 8 shows the distribution of Von-Mises stress σ_e and strain ε_{22}, which is located within the shear band along path 2 (see black line in Fig. 6b). It is clear that, when two shear-bands propagate across each other, their directions remain inactive. Moreover, as shown in Fig. 8b, when one shear band intersects an existing shear band, a decrease of the strain magnitude in this shear band is observed.

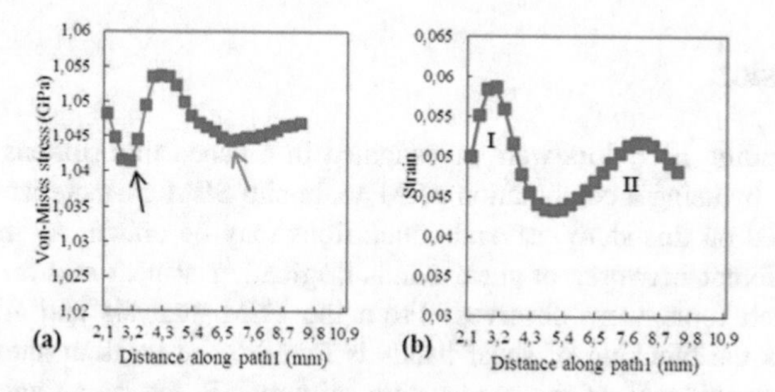

Fig. 7. Distributions of parameters inside and outside the shear band (along path 1): (a) Von-Mises stress, σ_e and (b) strain, ε_{22} for $Cu_{60}Zr_{30}Ti_{10}$

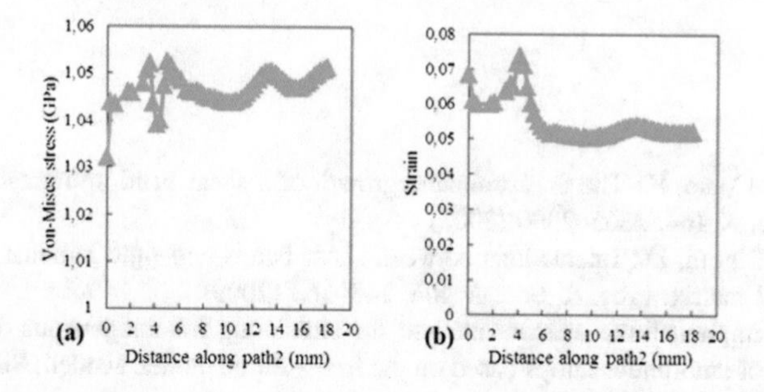

Fig. 8. Distributions of parameters in the shear band (along path 2): (a) Von-Mises stress, σ_e and (b) strain, ε_{22} respectively, for $Cu_{60}Zr_{30}Ti_{10}$

Figure 9 shows the results with the perturbation of the initial free-volume field at four elements, as denoted by 'A', 'B', 'C' and 'D' in Fig. 8a. By comparing Fig. 6 with Fig. 8, it is observed that the shear band density increase with the increase of shear band zones. Regular shear bands are displayed in symmetrical distribution through the sample. These shear bands propagate rectilinearly in the specimen and when two shear bands across each other, their directions grow rectilinearly. Moreover, we observe that the strain ε_{22} is localized into the shear bands.

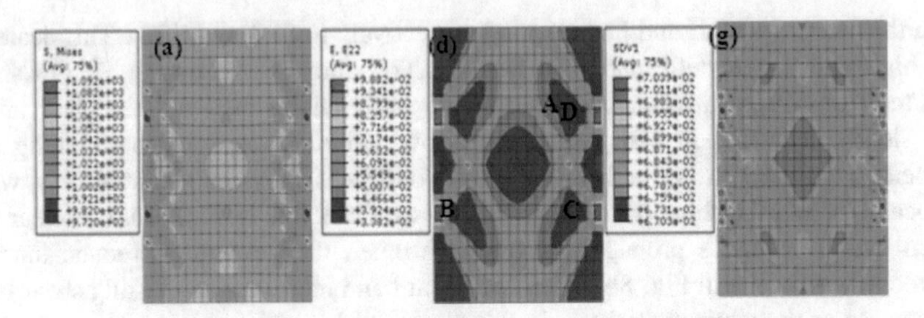

Fig. 9. Distributions of various parameters in the specimen with four shear band zones: (a) Von-Mises stress σ_e, (d) strain ε_{22}, and (g) free volume concentration ξ

4 Conclusion

The shear banding behaviors were investigated in notched thin ribbons $Cu_{60}Zr_{30}Ti_{10}$ metallic glass by using a combination of AFM, in-situ SEM characterization and FEM analysis. Based on this study, several conclusions may be drawn. As the tensile load increases, different networks of shear bands originating at near and far plastic zones from the notch roots were observed. From the FEM analysis and AFM frictional measurements the blocking of shear bands is found to be weak at sites of the intersection and no evidence of the pre-existing primary SB acts as an energy barrier to influence the propagation of the secondary SBs due to almost similar magnitude of frictional properties. The finite element simulation results demonstrated that shear bands nucleate and propagate from the shear band zones.

References

Bigoni, D., Dal Corso, F.: The unrestrainable growth of a shear band in a prestressed material. Proc. R. Soc. A **464**, 2365–2390 (2008)

Dal Corso, F., Bigoni, D.: Interactions between shear bands and rigid lamellar inclusions in a ductile metal matrix. Proc. R. Soc. A **465**, 143–163 (2009)

Gao, Y.F.: An implicit finite element method for simulating inhomogeneous deformation and shear bands of amorphous alloys based on the free-volume model. Modell. Simul. Mater. Sci. Eng. **14**, 1329–1345 (2006)

Greer, A.L., Cheng, Y.Q., Ma, E.: Shear bands in metallic glasses. Mat. Sci. Eng. R **74**, 71–132 (2013)

Huang, R., Suo, Z., Prevost, J.H., Nix, W.D.J.: Inhomogeneous deformation in metallic glasses. Mech. Phys. Solids **50**, 1011–1027 (2002)

Inoue, A.: Stabilization of metallic supercooled liquid and bulk amorphous alloys. Acta Mater. **48**, 279 (2000)

Inoue, A., Zhang, W., Zhang, T., Kurosaka, K.: High-strength Cu-based bulk glassy alloys in Cu-Zr-Ti and Cu-Hf-Ti ternary systems. Acta Mater. **49**, 2645–2652 (2001)

Johnson, W.L.: Bulk glass-forming metallic alloys: science and technology. Mater. Res. Bull. **24**(10), 42–56 (1999)

Sha, Z.D., et al.: Necking and notch strengthening in metallic glass with symmetric sharp-and-deep notches. Sci. Rep. **5**, 10797 (2015)

Spaepen, F.: A microscopic mechanism for steady state inhomogeneous flow in metallic glasses. Acta Metall. **25**, 407–415 (1977)

Steif, P.S., Spaepen, F., Hutchinson, J.W.: Strain localization in amorphous metals. Acta Metall. **30**, 447–455 (1982)

Wang, D.P., Sun, B.A., Gao, M., Yang, Y., Liu, C.T.: The mechanism of shear-band blocking in monolithic metallic glasses. Mater. Sci. Eng. A **703**, 162–166 (2017)

Wang, Z., Pan, J., Li, Y., Schuh, C.A.: Densification and strain hardening of a metallic glass under tension at room temperature. Phys. Rev. Lett. **111**, 135504 (2013)

Yavari, A.R., et al.: Crystallization during bending of a Pd-based metallic glass detected by X-Ray microscopy. Phys. Rev. Lett. **109**, 085501 (2012)

Zhao, J.X., Wu, F.F., Qu, R.T., Li, S.X., Zhang, Z.F.: Plastic deformability of metallic glass by artificial macroscopic notches. Acta Mater. **58**, 5420–5432 (2010)

Analytical Study of Curvature Radius Effect on the Bending Stress and Fatigue Life of Parabolic Leaf Spring

Rabï Ben Sghaier[1,2](✉), Akram Atig[1,2,3], and Raouf Fathallah[1,3]

[1] Unité de Génie de Production Mécanique et Matériaux (UGPM2), Ecole Nationale d'Ingénieurs de Sfax (ENIS), Université de Sfax, Sfax, Tunisia
rabibensghaier@gmail.com

[2] Institut Supérieur des Sciences Appliquées et de Technologie de Sousse, Université de Sousse, Rue Tahar Ben Achour Sousse, Sousse, Tunisia

[3] Ecole Nationale d'ingénieur de Sousse, Université de Sousse, Sousse Erriadh, BP 264, Sousse, Tunisia

Abstract. An accurate analytical approach is usually preferred in engineering design for the reason that it does not involve additional software and can be useful for different studied cases. In this work, an analytical approach to evaluate bending stress distribution and to generate fatigue life diagrams has been carried out for Single Asymmetric Parabolic Leaf Spring (SAPLS). This approach is based on an analytical model which considers the SAPLS as an initially curved beam. Bending stress distribution results obtained from straight beam, Finite element and proposed models are compared for a case study. It is observed that the proposed model is more precise than the straight beam model (SBM) compared to FEM of SAPLS. To predict the fatigue life of the parabolic leaf spring, the Morrow criterion was used. This criterion takes into account the effect of the average stress and the residual stresses introduced near the inner surface of the spring. The proposed model has been employed to implement the iso-line diagram of fatigue life regarding various design parameters.

Keywords: Single Asymmetric Parabolic Leaf Spring · Bending stress · Fatigue life diagram

1 Introduction

Leaf spring presents an essential component of vehicle suspension system. It is employed to absorb and to smooth out shocks caused by irregularities in road surface. Furthermore, leaf springs are utilized as a connecting element and link the axle to the vehicle chassis. In order to decrease the weight of leaf springs, Multi Leaf Springs (MLS) are substituted in light commercial vehicles by a single leaf spring. Single Parabolic Leaf spring (SPLS) becomes more useful than the multi-leaf spring because its ability of storing more energy per kilogram than MLS (McEvily 2002). Therefore, for the same maximum loadings, the SPLS is less heavy than the MLS (SAE Spring Committee 1996; Atig et al. 2017; Atig et al. 2018a, b). In this paper, analytical model of SPLS based on curved beam theory is proposed. The suggested analytical bending

A. Benamara et al. (Eds.): CoTuMe 2018, LNME, pp. 204–211, 2019.
https://doi.org/10.1007/978-3-030-19781-0_25

stress model has been used to implement the fatigue life diagrams in function of many SPLS design parameters. Such a diagram is exploited as a graphical solution, to choice the permissible values of design variables for a required number of cycles to failure or to predict the fatigue life for a couple of design factors.

2 Analytical Approach

In this approach, the Single Asymmetric Parabolic Leaf Spring (SAPLS) is considered as simply supported initially curved beam with rectangular cross section. The load applied was considered as a concentrated load in the middle of the seat length. We neglect the shear stress in this study to simplify the analytical model and we take into account only the normal bending stress. Figure 1 illustrates a schematic presentation of loading conditions applied on the SAPLS.

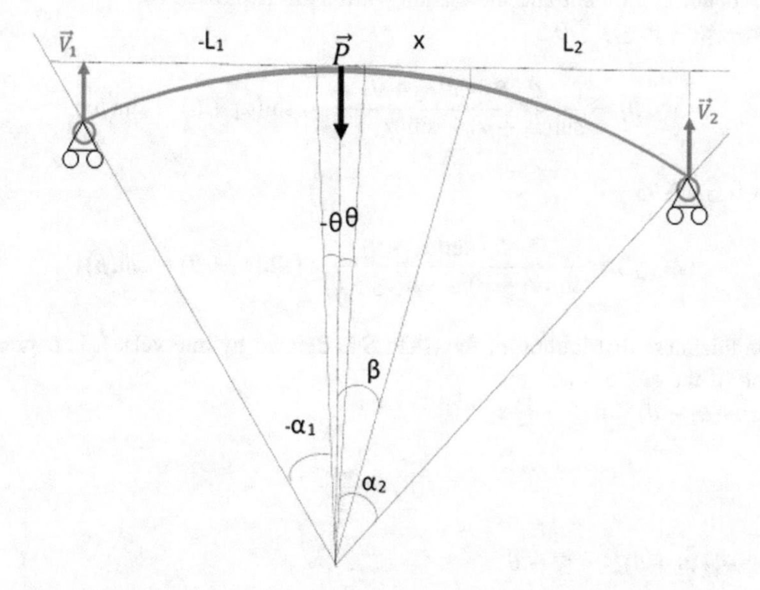

Fig. 1. Schematic presentation of loading conditions applied on the SAPLS

The bending stress along the SAPLS length is given by:

$$\sigma_b(\beta) = \frac{M_b(\beta) \cdot t(\beta)}{2 \cdot I_{GZ}(\beta)} \tag{1}$$

The moment of inertia of a rectangular cross section has the form:

$$I_{GZ}(\beta) = \frac{w \cdot t^3(\beta)}{12} \tag{2}$$

The bending stress becomes:

$$\sigma_b(\beta) = \frac{6 \cdot M_b(\beta)}{w \cdot t^2(\beta)} \tag{3}$$

The reactions at the two eyes of SAPLS are assumed vertical and they are expressed as:

$$V_1 = \frac{P \cdot L_2}{L_1 + L_2} = \frac{P \cdot \sin(\alpha_2 + \theta)}{\sin(\alpha_1 + \theta) + \sin(\alpha_2 + \theta)} \tag{4}$$

$$V_2 = \frac{P \cdot L_1}{L_1 + L_2} = \frac{P \cdot \sin(\alpha_1 + \theta)}{\sin(\alpha_1 + \theta) + \sin(\alpha_2 + \theta)} \tag{5}$$

The bending moment and the bending stress are expressed as:
For $-(\alpha_1 + \theta) \leq \beta \leq 0$

$$M_b(\beta) = \frac{P \cdot R \cdot \sin(\alpha_2 + \theta)}{\sin(\alpha_1 + \theta) + \sin(\alpha_2 + \theta)} \cdot (\sin(\alpha_1 + \theta) + \sin(\beta)) \tag{6}$$

For $0 \leq \beta \leq \alpha_2 + \theta$

$$M_b(\beta) = \frac{P \cdot R \cdot \sin(\alpha_1 + \theta)}{\sin(\alpha_1 + \theta) + \sin(\alpha_2 + \theta)} \cdot (\sin(\alpha_2 + \theta) - \sin(\beta)) \tag{7}$$

The thickness distribution of the SAPLS is defined by intervals. It is expressed in function of the angle β as:
For $-(\alpha_1 + \theta) \leq \beta \leq -\left(\frac{3}{4}\alpha_1 + \theta\right)$

$$t(\beta) = \frac{t_{\max}}{2} \tag{8}$$

For $-\left(\frac{3}{4}\alpha_1 + \theta\right) \leq \beta \leq -\theta$

$$t(\beta) = \sqrt{\frac{(\alpha_1 + \theta) + \beta}{\alpha_1}} \cdot t_{\max} \tag{9}$$

For $-\theta \leq \beta \leq +\theta$

$$t(\beta) = t_{\max} \tag{10}$$

For $\theta \leq \beta \leq \frac{3}{4}\alpha_2 + \theta$

$$t(\beta) = \sqrt{\frac{(\alpha_2 + \theta) - \beta}{\alpha_2}} \cdot t_{\max} \tag{11}$$

For $\frac{3}{4}\alpha_2 + \theta \leq \beta \leq \alpha_2 + \theta$

$$t(\beta) = \frac{t_{\max}}{2} \tag{12}$$

The description of bending stress analytical model over the length of the SAPLS can be divided in sex intervals:

For $-(\alpha_1 + \theta) \leq \beta \leq -(\frac{3}{4}\alpha_1 + \theta)$

$$\sigma_b(\beta) = \frac{12 \cdot P \cdot R \cdot \sin(\alpha_2 + \theta)}{w \cdot t_{\max}^2 \cdot (\sin(\alpha_1 + \theta) + \sin(\alpha_2 + \theta))} \cdot (\sin(\alpha_1 + \theta) + \sin(\beta)) \tag{13}$$

For $-(\frac{3}{4}\alpha_1 + \theta) \leq \beta \leq -\theta$

$$\sigma_b(\beta) = \frac{6 \cdot P \cdot R \cdot \sin(\alpha_2 + \theta)}{w \cdot (\frac{\alpha_1 + \theta + \beta}{\alpha_1}) \cdot t_{\max}^2 \cdot (\sin(\alpha_1 + \theta) + \sin(\alpha_2 + \theta))} \cdot (\sin(\alpha_1 + \theta) + \sin(\beta)) \tag{14}$$

For $-\theta \leq \beta \leq 0$

$$\sigma_b(\beta) = \frac{6 \cdot P \cdot R \cdot \sin(\alpha_2 + \theta)}{w \cdot t_{\max}^2 \cdot (\sin(\alpha_1 + \theta) + \sin(\alpha_2 + \theta))} \cdot (\sin(\alpha_1 + \theta) + \sin(\beta)) \tag{15}$$

For $0 \leq \beta \leq +\theta$

$$\sigma_b(\beta) = \frac{6 \cdot P \cdot R \cdot \sin(\alpha_1 + \theta)}{w \cdot t_{\max}^2 \cdot (\sin(\alpha_1 + \theta) + \sin(\alpha_2 + \theta))} \cdot (\sin(\alpha_2 + \theta) - \sin(\beta)) \tag{16}$$

For $\theta \leq \beta \leq \frac{3}{4}\alpha_2 + \theta$

$$\sigma_b(\beta) = \frac{6 \cdot P \cdot R \cdot \sin(\alpha_1 + \theta)}{w \cdot (\frac{\alpha_2 + \theta - \beta}{\alpha_2}) \cdot t_{\max}^2 \cdot (\sin(\alpha_1 + \theta) + \sin(\alpha_2 + \theta))} \cdot (\sin(\alpha_2 + \theta) - \sin(\beta)) \tag{17}$$

For $\frac{3}{4}\alpha_2 + \theta \leq \beta \leq \alpha_2 + \theta$

$$\sigma_b(\beta) = \frac{12 \cdot P \cdot R \cdot \sin(\alpha_1 + \theta)}{w \cdot t_{\max}^2 \cdot (\sin(\alpha_1 + \theta) + \sin(\alpha_2 + \theta))} \cdot (\sin(\alpha_2 + \theta) - \sin(\beta)) \tag{18}$$

3 FEM Approach

To validate the analytical model, FEA of a SAPLS case has been carried out in order to compare the bending stress distribution along the SAPLS span. Three Dimensional model of the SAPLS has been implemented using commercial finite element software ANSYS® Workbench®. Table 1 illustrates the geometric parameters of the SAPLS.

Table 1. Geometrical parameters of SAPLS

Geometrical parameters	Values (mm)
Maximum thickness	13,2
Width	63
Active length of rear cantilever	850
Active length of front cantilever	600
Seat length	100
Radius of curvature	1680

For material assignation, linear elastic material model was considered. The material considered in this study is the SAE 5160 steel, typically employed in the leaf spring manufacturing (Yamada 2007; Atig et al. 2018a, b). The mechanical proprieties of SAE 5160 are indicated in Table 2.

Table 2. Mechanical proprieties of SAE 5160 (Kong et al. 2014)

Mechanical properties of SAE 5160 steel	Mean value
Ultimate tensile strength (MPa)	1584
Yield strength (MPa)	1487
Young modulus (GPa)	207
Fatigue strength coefficient (MPa)	2063
Fatigue strength exponent	−0.08

For solid meshing, 8-nodes hexahedra elements were used with a maximal length of 3 mm. this meshing was selected to guaranteed a good accuracy for the maximum principal stress value and avoid additional computational time. 25662 hexahedra elements with 146202 nodes were generated.

4 Fatigue Life Prediction

The Morrow fatigue criterion is frequently employed to predict the fatigue life of leaf springs (Kong et al. 2014; Ghuku and Saha 2018). In fact, for high cycle fatigue region where the behaviour is assumed to be elastic represented by the Hook's law, the morrow criterion equation can be written as a stress-based form (Landgraf and Francis 1979):

$$(2N_f)^b = \frac{\sigma_a}{(\sigma'_f - \sigma_m)} \tag{19}$$

The effect of residual stresses is commonly superimposed with the mean stress correction (Webster and Ezeilo 2001). Therefore, to take into account the influence of the residual stress σ_r, the high cycle fatigue Morrow criterion is described as follows (Landgraf and Francis 1979):

$$(2N_f)^b = \frac{\sigma_a}{(\sigma'_f - \sigma_m - \sigma_r)} \tag{20}$$

5 Results and Discussion

As shown in Fig. 2, the maximum bending stress value of 773 MPa is located at the loading application point. It is observed that a high stress level is uniformly distributed along the leaf spring span. The maximum bending stress values is located at the seat length center for the three SAPLS models.

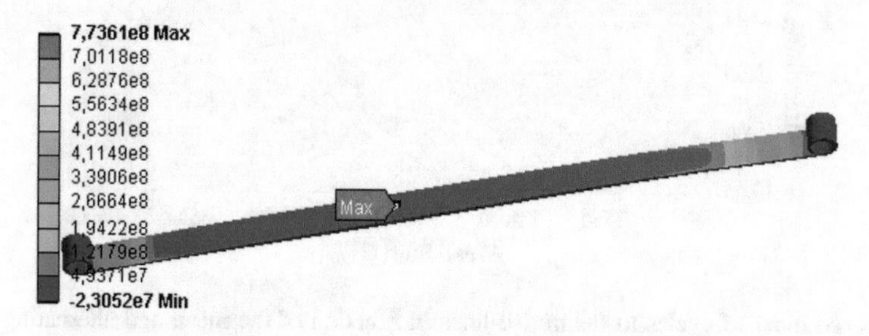

Fig. 2. Maximum principal stress distribution (MPa) of SAPLS

As depicted in Fig. 3, the variation of the maximum thickness has a significant effect on the maximum normal stress. It is observed that the increase of the maximum thickness decreases the normal stress peak value and reduces the margin error of the analytical models compared with FEA. It is clearly noted that, compared with the finite elements model, the proposed model is more precise than the classical model. As a result of the application of a design load of 4000 N, the proposed model gives an error of 3% compared with the FEM. Nevertheless, the margin error between the straight beam model and FEM is about 6.7%.

Next, the fatigue life diagram, that depicts the iso-lines of the estimated fatigue life as a function of the mean and range applied loadings has been shown in Fig. 4. Moreover, as shown in Fig. 5, the fatigue life diagram can be exposed in function of the maximum thickness and the curvature radius. As an example, it is observed that, to reach a 10^6 cycles to failure and for a given value of curvature radius of 1600 mm, a value of 13.5 mm of maximum thickness should be selected. In fact, fatigue life

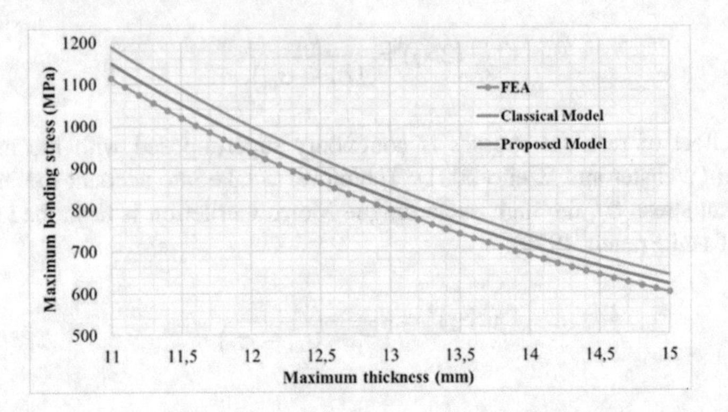

Fig. 3. Bending stress distribution regarding the SAPLS length

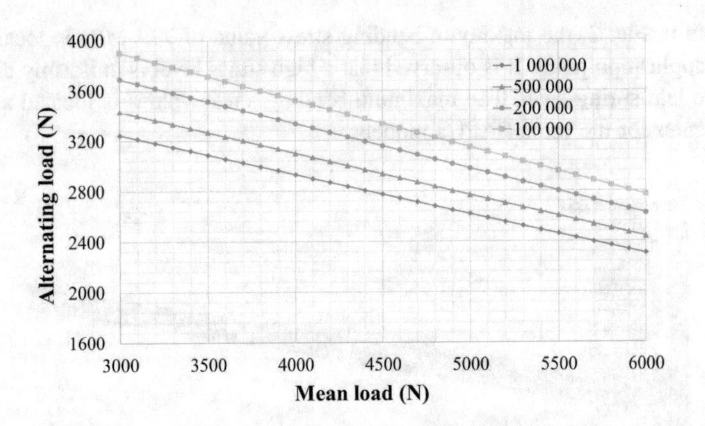

Fig. 4. Number of cycles to failure iso-lines in function of the mean and alternating loads

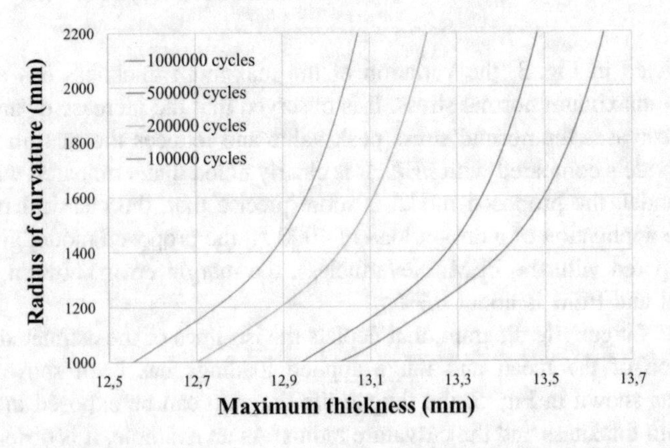

Fig. 5. Number of cycles to failure iso-lines in function of the maximum thickness and curvature radius

diagrams are very useful and indispensable for the leaf spring industrials (i) to select the permissible loadings field for a required number of cycles to failure or (ii) to predict the fatigue life for a couple of design parameters.

6 Conclusion

In this study, analytical bending stress model of SAPLS has been suggested. This model takes into account the initial curvature of the SAPLS. For validation, FEM of SAPLS has been implemented and bending stress results have been compared. It is observed that an acceptable accuracy of the proposed model and FEM has been obtained. Next, the iso-life fatigue diagrams have been established according to several design parameters. These diagrams indicate, based on the Morrow criterion, the iso-lines of the number of cycles to failure for several design parameters. It is noted that, based on the proposed analytical bending stress model, the fatigue diagrams have been generated with a low computational time and an acceptable level of accuracy compared with FEM.

References

Atig, A., Ben Sghaier, R., Seddik, R., Fathallah, R.: Reliability-based high cycle fatigue design approach of parabolic leaf spring. Proc. Inst. Mech. Eng. Part L J. Mater. Des. Appl. **233**, 588–602 (2017). 1464420716680499

Atig, A., Ben Sghaier, R., Seddik, R., Fathallah, R.: Probabilistic methodology for predicting the dispersion of residual stresses and Almen intensity considering shot peening process uncertainties. Int. J. Adv. Manuf. Technol. **94**(5–8), 2125–2136 (2018a)

Atig, A., Ben Sghaier, R., Seddik, R., Fathallah, R.: A simple analytical bending stress model of parabolic leaf spring. Proc. Inst. Mech. Eng. Part C J. Mech. Eng. Sci. **232**(10), 1838–1850 (2018b)

Ghuku, S., Saha, K.N.: Large deflection analysis of curved beam problem with varying curvature and moving boundaries. Eng. Sci. Technol. Int. J. **21**(3), 408–420 (2018)

Kong, Y.S., Omar, M.Z., Chua, L.B., Abdullah, S.: Fatigue life prediction of parabolic leaf spring under various road conditions. Eng. Fail. Anal. **46**, 92–103 (2014)

Landgraf, R., Francis, R.: Material and processing effects on fatigue performance of leaf springs. SAE Technical paper 790407 (1979)

McEvily, A.J.: Metal Failures: Mechanisms, Analysis, Prevention. Wiley, New York (2002)

SAE Spring Committee. Spring Design Manual, Warrendale, USA (1996)

Webster, G.A., Ezeilo, A.N.: Residual stress distributions and their influence on fatigue lifetimes. Int. J. Fatigue **23**, 375–383 (2001)

Yamada, Y.: Materials for Springs. Springer, Heidelberg (2007)

Improvement of the Predictive Ability of Polycyclic Fatigue Criteria for 42CrMo4 Nitrided Steels

Rafik Bechouel[1,2(✉)], Naoufel Ben Moussa[2], and Mohamed Ali Terres[2]

[1] National Engineering School of Sousse ENISO, University of Sousse, Sousse, Tunisia
bechouel.rafik@gmail.com

[2] Mechanical, Material and Processes Laboratory LMMP, Higher National Engineering School of Tunis, ENSIT, University of Tunis, Montfleury, 1008 Tunis, Tunisia
emailnaoufel@gmail.com, mohamedali.terres@gmail.com

Abstract. This work aims to estimate the residual stress relaxation during cyclic loading and the fatigue strength for the case of the 42CrMo4 nitrided steel. An original numerical methodology with Abaqus software has been established to determine the residual stress redistribution under fatigue loading. An important reduction of the residual stress is observed after the first cycles (45 to 60%) caused by the plastic deformation resulting from the superposition of residual stress and applied stresses. An experimental investigation was conducted by caring out 3-points bending fatigue tests on nitrided steel samples. The 42CrMo4 steel behaviour was described using the Chaboche model coupling isotropic and nonlinear kinematic hardening where the coefficients are identified experimentally. Residual stresses profiles measured by XRD methods where used to validate the proposed numerical methodology. It has been established that considering of the stabilised residual stress enhances the predictive ability of the polycyclic fatigue criteria.

Keywords: Modelling approach · Nitrided steel · Residual stress relaxation

1 Introduction

The favorable consequence of compressive residual stress resulting from machining (Ben Moussa et al. 2014a; Yahyaoui et al. 2015; Terres et al. 2001) or surface treatments (Sidhom et al. 2014b) on components fatigue life is well established. The improvement of the 42CrMo4 steel fatigue life due to compressive residual stress and hardening induced by ionic nitriding was previously proved experimentally (Terres et al. 2017; Chaouch et al. 2012; Terres and Sidhom 2012). Nevertheless, neglecting the cyclic relaxation of residual stresses, leads to inaccurate prediction of the resistance of structures and components. In order to determine relaxed residual stresses, many

A. Benamara et al. (Eds.): CoTuMe 2018, LNME, pp. 212–220, 2019.
https://doi.org/10.1007/978-3-030-19781-0_26

analytical models were developed from experimental results (Depouhon et al. 2014; Xie et al. 2017; Zaroog et al. 2011). These models can't take into account in their analytical equations of all the influence parameters with an adequate precision. In the current work an original numerical methodology is suggested to determine the residual stress redistribution caused by cyclic loading of 42CrMo4 steel treated by ionic nitriding. In the numerical simulation the material behaviour was described using the Chaboche model coupling isotropic and nonlinear kinematic hardening where the coefficients are identified from hysteresis loops established in this study. The classical method studied in several research works is limited to the considering of residual stresses in the tensor of maximal stresses by summing the applied stresses and the tensor of the relaxed residual stresses determined experimentally by XRD (Bernasconi and Papadopoulos 2005; Sidhom et al. 2014a; Wang 2004). The multiaxial residual stress induced by nitriding treatment is introduced in the FE model. The objective of this paper is to identify of effect of residual stress redistribution on cyclic bending limit of 42CrMo4 nitrided steel.

2 Material, Treatment and Experimental Procedure

The investigated material is the 42CrMo4 steel, usually used in mechanical industry for the manufacturing of the automotive transmission components and for dies of plastic injection molding. Its chemical composition is presented in Table 1. The heat-treatment processes used for 42CrMo4 steel were the followings (IN30): (i) solution treatment at 850 °C for 0.5 H plus oil cooling; (ii) tempering at 580 °C for 1 H plus air cooling and (iii) ion-nitriding at 520 °C for 30H under the gas mixture of 20% N_2 and 80% H_2. Three-point bending fatigue tests are conducted on notched specimens Kt = 1.6 to investigate the effect of initial and stabilized residual stress on the fatigue strength of nitrided specimens. Fatigue tests are carried out using an MTS-810 machine, with a frequency of 15 Hz and a loading ratio R = 0.1 and at room temperature. Initial and relaxed residual stresses are measured using X-ray diffraction methods.

3 Numerical Procedure of Relaxed Residual Stress Determination

3.1 Numerical Procedure

In this study, a numerical procedure, based on a 2D finite element model, is developed using ABAQUS standard software for simulating the 3-points bending loading tests. This procedure is summarized in Fig. 1.

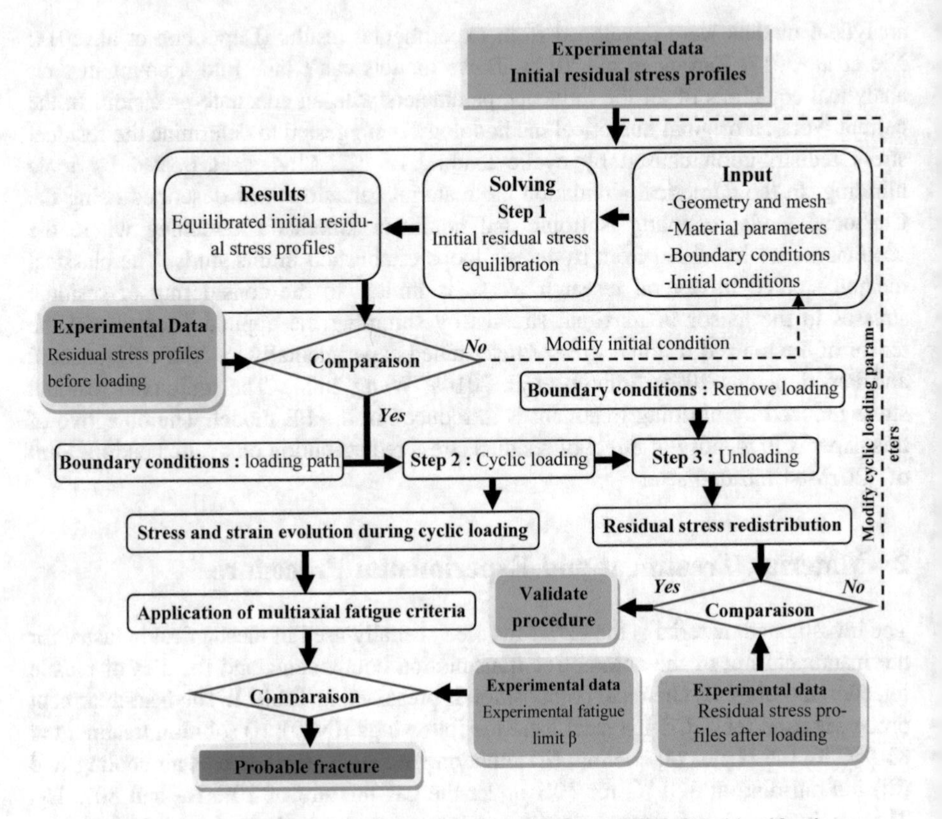

Fig. 1. Numerical procedure for residual stress relaxation profiles identification

3.2 Geometry, Mesh, Loading and Boundary Conditions

In the finite element model, illustrated in Fig. 2, only one-half of the nitrided specimens is considered due to its symmetry. The geometry was meshed by CPE4 type elements available in ABAQUS element library. The mesh in regions around the notch and the rollers was extremely refined until a length of 15 μm (Fig. 2) to enhances the precision the FE solutions, This value is retained after inspecting stress and strain evolutions for several mesh sizes. An appropriate 3-point bending boundary conditions are imposed on the half of the notched bending specimen as follow: On the cross-section of bending specimen we applied an X-symmetry boundary condition (U1 = UR2 = UR3 = 0). The analytical rigid part (top roller) controlled by a reference point (Rp2), where boundary conditions (U1 = U2 = U3 = UR1 = UR2 = UR3 = 0) are applied. The analytical rigid part (bottom roller) considered as a rigid part is controlled by a reference point (Rp1) with boundary conditions (U1 = UR3 = 0) and forces (F1 = 0, F3 = F(t)) are applied to describe bending loading.

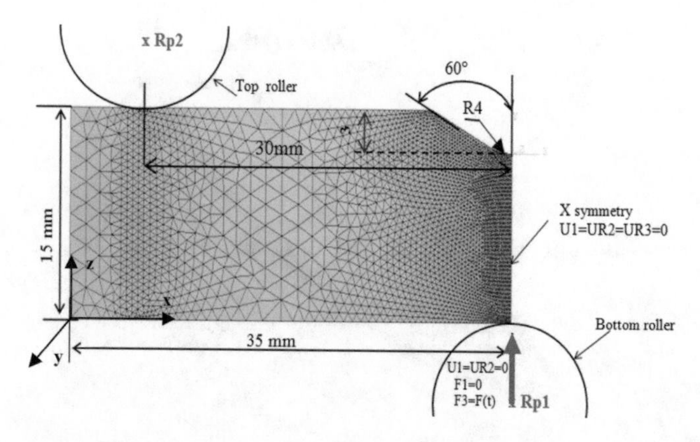

Fig. 2. Geometry and mesh of finite element model

3.3 Cyclic Hardening Model for the Base Material

On the objective to study the behavior of 42CrMo4 material under cyclic loading in this paper, Lemaitre and Chaboche model (isotropic and nonlinear kinematic hardening model) is considered. Hardening variables, the yield criterion of Von Mises and the plastic flow rule are given by the equations below.

$$\text{- Yield criterion } f = J_2(\sigma - X) - R - k = |\sigma - X| - R - k \leq 0 \quad (1)$$

$$\text{- Flow rule } \dot{\varepsilon}_p = \dot{p}\frac{\partial f}{\partial \sigma} = (\frac{2}{3}\dot{\varepsilon}_p)^{1/2}\frac{(\sigma - X)}{|\sigma - X|} \quad (2)$$

$$\text{- Isotropic hardening} \begin{cases} (\dot{R}) = b(Q - R)\dot{p} \\ (R) = Q(1 - e^{-bp}) \end{cases} \quad (3)$$

$$\text{- Kinematic hardening} \begin{cases} (\dot{X}) = \frac{2}{3}C\,\dot{\varepsilon}_p - \delta X\dot{p} \\ X_M = \frac{\Delta\sigma}{2} - k = \frac{C}{\delta}th(\delta\frac{\Delta\varepsilon_p}{2}) \end{cases} \quad (4)$$

With: X is the back stress indicating the center of yield stress surface, R is the drag stress, k is the initial size of the yield surface, $\dot{p}$ is the cumulated plastic strain rate.

The base material constants k, C, δ, Q and b of this model for the 42CrMo4 steel are identified from graphical methods with the experimental stress-strain hysteresis loops corresponding to imposed strain ε = 0.85% using the Abaqus program in accordance with CHABOCHE method. The obtained values are summarized in Table 1 and the result of simulation and verification parameters are illustrated in Fig. 3. This model was employed in the FE model to describe material behavior.

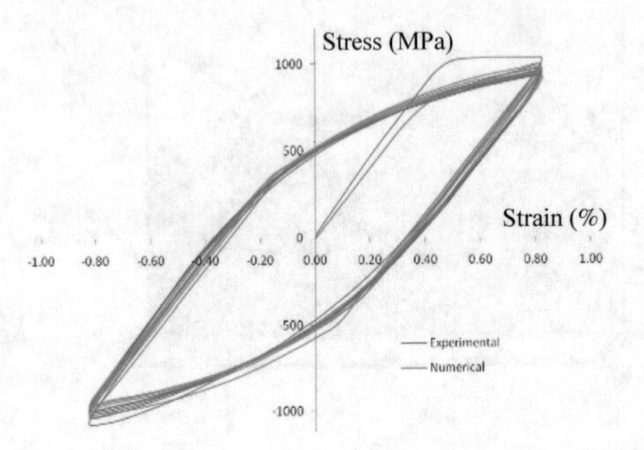

Fig. 3. Hysteresis loop response for experimental and numerical methods

Table 1. Cyclic behavior law coefficients of 42CrMo4 steel

E(GPa)	ν	K(MPa)	C (MPa)	δ	Q (MPa)	b
201	0.3	775	126228	100	−137	33.4

3.4 Application of SINES Criterion

This work aims to check the ability of the SINES criteria to determine the nitrided steel fatigue strength. This criterion is largely used in mechanical engineering (Laamouri et al. 2013; Sidhom et al. 2005; Terres et al. 2012; Terres et al. 2010; Sidhom et al. 2014a). In this case of ion nitriding, the redistributed residual stresses are introduced. The total cyclic stress tensor can be obtained by superimposing the relaxed residual stress tensor with the applied stress tensor. Therefore, the residual stresses effect is taken into account in these criteria by the change of the mean value of the hydrostatic pressure in Sines (Eq. 5) and Dang Van (Eq. 6) criterion.

$$\sqrt{J_{2,a}} + \alpha_S P_m \leq \beta_S \tag{5}$$

$$\tau_{a,D}(t) + \alpha_D P(t) \leq \beta_D \tag{6}$$

The total stress tensor is given by Eq. (7):

$$\underline{\underline{\sigma}}_t(t) = \underline{\underline{\sigma}}(t) + \underline{\underline{\sigma}}_R \tag{7}$$

The measured residual stresses using X-ray diffraction method was introduced in the code ABAQUS as initial conditions. Residual stresses were equilibrated in a first step of Abaqus calculation and compared with experimental measurements (Ben Moussa et al. 2014b). The FE simulation of nitrided residual stress relaxation described below was performed by applying a cyclic loading corresponding to the fatigue limit under 3-pointbending tests. The predicted aptitude for this criterion was improved by accounting for the stress state in the notched point. This multiaxial tensor stress state includes applied and redistributed residual stresses during cyclic loading.

4 Finite Elements Analysis Results and Discussion

The 3 point-bending fatigue test of nitrided steel highlight an accommodation of the material illustrated by a softening phenomenon consisting in a gradual decrease of the maximum stress value during cyclic loading (Fig. 4). The softening phenomenon much more marked for the first cycles, continues until 50 cycles loading. The isotropic hardening or softening level is determined by the parameter b of the CHABOCHE Model. During cyclic loading, new plastic deformations occur due to exceeding the local yield stress by stress field resulting from applied force and residual stress resulting from nitriding treatment. These plastic deformations lead to a redistribution of initial residual stress (Fig. 5). Residual stress (σ_{xx}) evolves rapidly during cycles and stabilizes at about 20loading cycles (Fig. 4). This phenomenon proves the effect of cyclic loading on the reorganization of dislocation induced by nitrided treatment.

The advantage of the accounting for residual stress field induced by nitriding treatment and its redistribution during cyclic 3-point bending can be visualized in the endurance diagram of Sines criterion (Fig. 6). It is demonstrated from the diagram of Sines that the nitriding residual stresses move the representative points of the loadings in the direction of the negative pressures and improve predictive ability of the criterion. Consequently, this effect leads to an increase of the material resistance and a retardation of the cracks nucleation in nitrided layers.

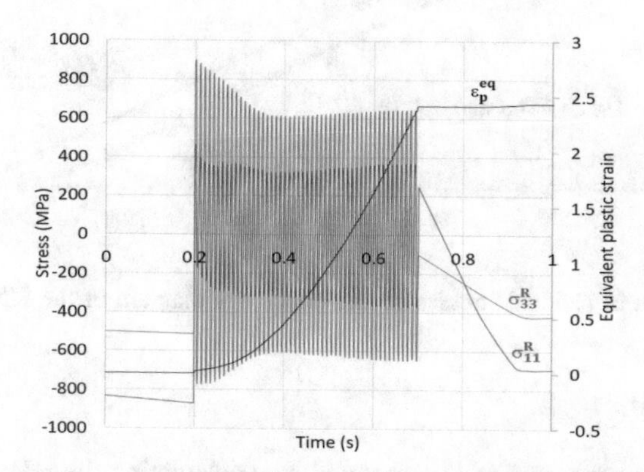

Fig. 4. Nitrided cyclic stress and strain evolutions at applied stress 980 MPa (fatigue limit) and stress ratio R = 0.1

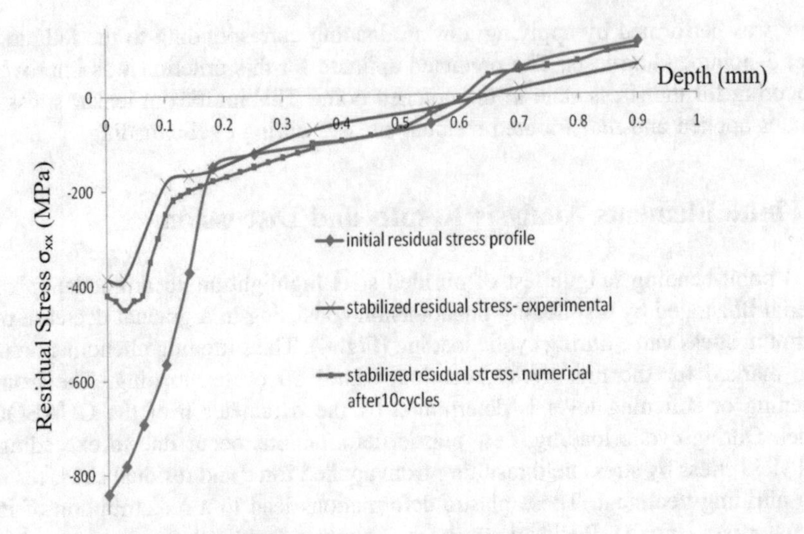

Fig. 5. Residual stress redistribution after cyclic loading at fatigue limit and stress ratio R = 0.1

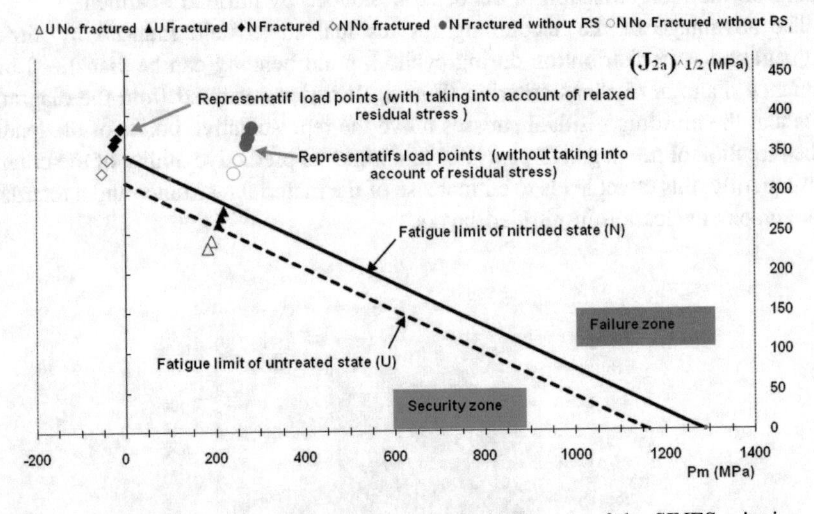

Fig. 6. Effect of residual stresses on the endurance diagram of the SINES criterion

5 Conclusion

The methodology suggested in this study allows calculating the relaxation of residual stresses induced by ionic nitriding after cyclic loading. Comparing the results obtained from the measurements and from the chosen numerical model shows that the simulation of the material response to a low-cycle loading to destruction is very reliable. A satisfactory capability of the proposed methodology is proved by taking into account relaxed residual stresses induced by ionic nitriding.

References

Terres, M.A., Bechouel, R., Ben Mohamed, S.: Low cycle fatigue behaviour of nitrided layer of 42CrMo4 steel. Int. J. Mater. Sci. Appl. **6**(1), 18 (2017). https://doi.org/10.11648/j.ijmsa.20170601.13

Ben Moussa, N., Al-Adel, Z., Sidhom, H., Braham, C.: Numerical assessment of residual stress induced by machining of aluminum alloy. Adv. Mater. Res. **996**, 628–633 (2014a). http://doi.org/10.4028/www.scientific.net/AMR.996.628

Ben Moussa, N., Sidhom, N., Sidhom, H., Braham, C.: Prediction of cyclic residual stress relaxation by modeling approach. Adv. Mater. Res. **996**, 743–748 (2014b). http://doi.org/10.4028/www.scientific.net/AMR.996.743

Bernasconi, A., Papadopoulos, I.V.: Efficiency of algorithms for shear stress amplitude calculation in critical plane class fatigue criteria. Comput. Mater. Sci. **34**(4), 355–368 (2005). https://doi.org/10.1016/j.commatsci.2005.01.005

Chaouch, D., Guessasma, S., Sadok, A.: Finite element simulation coupled to optimisation stochastic process to assess the effect of heat treatment on the mechanical properties of 42CrMo4 steel. Mater. Des. **34**, 679–684 (2012). https://doi.org/10.1016/j.matdes.2011.05.026

Depouhon, P., Sprauel, J.M., Mailhé, M., Mermoz, E.: Mathematical modeling of residual stresses and distortions induced by gas nitriding of 32CrMoV13 steel. Comput. Mater. Sci. **82**, 178–190 (2014). https://doi.org/10.1016/j.commatsci.2013.09.043

Laamouri, A., Sidhom, H., Braham, C.: Evaluation of residual stress relaxation and its effect on fatigue strength of AISI 316L stainless steel ground surfaces: Experimental and numerical approaches. Int. J. Fatigue **48**, 109–121 (2013). https://doi.org/10.1016/j.ijfatigue.2012.10.008

Sidhom, H., Ben Moussa, N., Ben Fathallah, B., Sidhom, N., Braham, C.: Effect of surface properties on the fatigue life of manufactured parts: experimental analysis and multi-axial criteria. Adv. Mater. Res. **996**, 715–721 (2014a). http://doi.org/10.4028/www.scientific.net/AMR.996.715

Sidhom, N., Laamouri, A., Fathallah, R., Braham, C., Lieurade, H.: Fatigue strength improvement of 5083 H11 Al-alloy T-welded joints by shot peening: experimental characterization and predictive approach. Int. J. Fatigue **27**(7), 729–745 (2005). https://doi.org/10.1016/j.ijfatigue.2005.02.001

Sidhom, N., Ben Moussa, N., Janeb, S., Braham, C., Sidhom, H.: Potential fatigue strength improvement of AA 5083-H111 notched parts by wire brush hammering: experimental analysis and numerical simulation. Mater. Des. **64**, 503–519 (2014b). http://doi.org/10.1016/j.matdes.2014.08.002

Terres, M.A., Laalai, N., Sidhom, H.: Effect of nitriding and shot-peening on the fatigue behavior of 42CrMo4 steel: experimental analysis and predictive approach. Mater. Des. **35**, 741–748 (2012). https://doi.org/10.1016/j.matdes.2011.09.055

Terres, M.A., Mohamed, S.B., Sidhom, H.: Influence of ion nitriding on fatigue strength of low-alloy (42CrMo4) steel: experimental characterization and predictive approach. Int. J. Fatigue **32**(11), 1795–1804 (2010). https://doi.org/10.1016/j.ijfatigue.2010.04.004

Terres, M.A., Sidhom, H.: Fatigue life evaluation of 42CrMo4 nitrided steel by local approach: equivalent strain-life-time. Mater. Des. **33**, 444–450 (2012). https://doi.org/10.1016/j.matdes.2011.04.047

Terres, M.A., Sidhom, H., Ben Cheikh Larbi, A., Ouali, S., Lieurade, H.P.: Influence de la résistance à la fissuration de la couche de combinaison sur la tenue en fatigue des composants nitrurés. Matériaux Techniques **89**(9–10), 23–36 (2001)

Wang, Y.: Evaluation and comparison of several multiaxial fatigue criteria. Int. J. Fatigue **26**(1), 17–25 (2004). https://doi.org/10.1016/s0142-1123(03)00110-5

Xie, X-f., Jiang, W., Luo, Y., Xu, S., Gong, J.-M., Tu, S.-T.: A model to predict the relaxation of weld residual stress by cyclic load: experimental and finite element modeling. Int. J. Fatigue **95**, 293–301 (2017). https://doi.org/10.1016/j.ijfatigue.2016.11.011

Yahyaoui, H., Ben Moussa, N., Braham, C., Ben Fredj, N., Sidhom, H.: Role of machining defects and residual stress on the AISI 304 fatigue crack nucleation. Fatigue Fract. Eng. Mater. Struct. **38**(4), 420–433 (2015). https://doi.org/10.1111/ffe.12243

Zaroog, O.S., Ali, A., Sahari, B.B., Zahari, R.: Modeling of residual stress relaxation of fatigue in 2024-T351 aluminium alloy. Int. J. Fatigue **33**(2), 279–285 (2011). https://doi.org/10.1016/j.ijfatigue.2010.08.012

Analysis of Surfaces Characteristics Stability in Grinding Process

Naoufel Ben Moussa(✉), Nasreddine Touati, and Nabil Ben Fredj

Laboratoire de Mécanique, Matériaux et Procédés LR99ES05,
Ecole Nationale Supérieure d'Ingénieurs de Tunis, Université de Tunis,
5 Avenue Taha Hussein, Montfleury, 1008 Tunis, Tunisia
emailnaoufel@gmail.com, benfredjnabil@gmail.com,
touatinasr@yahoo.com

Abstract. In this work, we investigate the effect of the grinding wheel morphology at grain scale on the stochastic aspects of the process. This morphology was controlled through dressing conditions in order to optimize the ground surface characteristics while minimizing their variance. It has been found that under the same grinding conditions, the surface characteristics and their dispersion vary significantly according to the dressing parameters. For the case of the grinding wheel (95A46M6V) dressed by a single-point diamond, a good choice of dressing conditions allows to reduce the roughness from 2 to 0.76 µm with a scatter less than 10% and to increase the hardening from 220Hv with a scattering of 40% to 382Hv with a dispersion of 23%.

Keywords: Grinding · Stainless steel · Surface properties · Wheel topography · Scatter

1 Introduction

The grinding process, and despite its efficiency in terms of dimensional and geometric accuracy of finished parts, is not sufficiently controlled owing to the large number of dispersion sources resulting mainly from the stochastic nature of the grinding wheel morphology. The removal of material is performed by a multitude of cutting edges with random sizes, shapes, orientations and spatial distribution leading to different material removal mechanisms, which consequently affects dispersions on mechanical surface characteristics of ground parts (Liu and Yang 2001). The accurate identification of ground surface characteristics is required to enhance the aptitude of component life predictive models (Ben Moussa et al. 2014a, b; Sidhom et al. 2014a, b; Yahyaoui et al. 2015; Zouhayar et al. 2013). In most research work, the effects of dressing conditions have been quantitatively evaluated through the finished surface characteristics and cutting forces without investigation of the real grinding wheel topography after dressing (Hadad and Sharbati 2016; Inasaki and Okamura 1985; Puerto et al. 2013). Last years, the technical evolution of measurement and visualization equipment has made it possible to evaluate the topography of the wheel (Badger and Torrance 2000; Woodin 2014; Coelho et al. 2001; Darafon 2013; Hadad and Sharbati 2016; Xie et al. 2008; Kapłonek and Nadolny 2013). In literature reviews, several indicators have been

A. Benamara et al. (Eds.): CoTuMe 2018, LNME, pp. 221–227, 2019.
https://doi.org/10.1007/978-3-030-19781-0_27

used to assess the grinding wheel topography at the microscopic scale. These indicators are commonly calculated from measurements made by contact and non-contact devices. The main obstacle often encountered in the investigation of the grinding wheel topography using the most of these devices is the requirement of the destruction of the wheel or its removal from the machine. In addition, in most research, the effects of grinding conditions on surface integrity have been established without studying the effects of dressing conditions on the stability of the ground surfaces characteristics. In this study, we propose an experimental methodology for investigating, on site, the effect of dressing conditions on the microscopic grinding wheel topography characteristics. This methodology is used to determine quantitative indicators of dispersions characterizing the grinding process. The analysis of ground surface characteristics for different dressing conditions allows improving the aptitude of the grinding process to remove material while reducing the scatter of ground surfaces characteristics by controlling the dressing parameters.

2 Experiments

2.1 Material

Grinding operations were carried out on samples of austenitic stainless steel AISI304 with dimensions 15 × 35 × 8 mm. The chemical composition and physical properties are given in Table 1. A stress-relief annealing treatment was applied to all samples to remove the anterior residual stresses (Fredj et al. 2006).

Table 1. Chemical composition and mechanical properties of the AISI 304 stainless steel.

C	Si	Mn	Cr	Ni	Mo	Cu	N	Fe
0.05	0.41	1.14	18.04	9	0.193	0.348	0.004	Balance

E (GPa)	Rm (MPa)	Rp0.2 (MPa)	A (%)	Hardness Hv0.1
193	670	315	53	270

2.2 Experimental Setup

The dressing and grinding operations are carried out in this work on a Teknoscuola RT600 grinding machine having two automated axes in vertical and longitudinal directions. Experiments were conducted using a Norton 95A46M6V vitrified aluminum oxide grinding wheel. A motorized device was designed and mounted on the grinding machine for controlling the dressing speed in the transverse direction Vd (Fig. 1). An electronic dimmer (Altivar11) is used to vary the dressing speed between 0.5 mm/s and 10 mm/s. In this work, a single-point diamond dresser was used for truing the wheel and generating fresh cutting edges (Fig. 1).

Fig. 1. Dressing feed variation device: dresser (1), trainer table (2), grinder magnet table (3)

The dressing conditions, summarized in Table 2, were choose according to the specifications recommended by the wheel manufacturer (Norton). The speed of the spindle is kept constant during the dressing operations at a value of 2700 rpm. After each dressing condition three grinding pass (Vw = 5.7 m/min, ae = 30 μm) are performed on the prismatic pieces of austenitic stainless steel AISI 304.

Table 2. Dressing conditions.

N°	1	2	3	4	5	6	7	8	9
Vd (mm/s)	1	1	1	5	5	5	10	10	10
ad (μm)	5	15	30	5	15	30	5	15	30

2.3 Examination of Grinding Wheel Surface

In this work, the characterization of the grinding wheel surface morphology is performed using a Carson MM-840 digital microscope type. For a magnification of 75x, ten zones of the grinding wheel with dimensions of 5.71 × 4.28 mm were observed after each dressing condition. In this study, the cutting-edges density A (%) and the spatial distribution of cutting edges R are chosen as evaluation indicators of the grinding wheel topography. The differentiation of peaks and hollows in acquired micrographs requires a good positioning of the light source (Fig. 2a).

For the observed zone, about 100 high quality micrographs with a resolution of 3200 × 2400 pixels were performed varying the angle of capture. These micrographs were processed using a freeware SFM (structure from motion) photogrammetry program (Wu 2013) to create a three-dimensional model of the observed zone. The intersection between the 3D model and a virtual plane distant from the highest point by the depth of pass ae allows determining the active edges surface area (Fig. 2b). Thereafter, the image processing program ImageJ (Schneider et al. 2012) is used to determine the cutting-edge density by analyzing micrographs of 10 spaced zones of grinding wheel after each dressing or grinding operations. For this purpose, all colored micrographs are converted to gray scale picture where for each pixel a gray level between 0 and 255 is assigned. Thereafter, a binary segmentation process was applied

Fig. 2. Micro topography of grinding wheel

to all pixels to convert those with a gray level below a threshold to black and the others representing the distribution of the active edge to white. The threshold is determined after testing several thresholds and a value of 200 was chosen since it gives a value of active surface equal to that obtained from three-dimensional model.

2.3.1 Evaluation of Cutting-Edge Density and Spatial Distribution

Figure 4 shows the effect of dressing velocity Vd and dressing depth of pass ad on cutting edge density. Coarse dressing (high ad and Vd values) results in macro-fractures of the grit and complete dislocations of grain blocks and coolant, which explains the decrease in the density of the cutting edges. On the other hand, the fine dressing (low ad and Vd values) makes the surface of the grinding wheel more closed, which results in the appearance of a higher number of light areas in the micrographs, therefore an increase in the density of the active surfaces is obtained.

The spatial arrangement of active cutting edges is determined from their center coordinates using a code developed in this work and based on the nearest neighbor search method (Taylor 1977). Figure 3 shows a decrease of the spatial dispersion of the active edges R when dressing is carried out under coarse conditions. It is noted that the decrease of the dispersion index R depicts an accumulation of active surfaces due to grits fracture and stripping under high dressing velocity and depth of pass.

2.4 Evaluation of Workpiece Characteristics

2.4.1 Effect on the Stability of the Roughness Indicators

Figure 4 shows an increase of roughness (Ra = 2 µm) when grinding wheel dressing is performed under coarse conditions. This effect is in agreement with the results found by (Hadad and Sharbati 2016) and (Puerto et al. 2013) when grinding a St37 mild steel and F-5229 steel, respectively with conventional alumina-grinding wheels dressed using a single tip dresser. In addition, roughness covariance is the highest (about 30%) for coarse dressing condition. This result can be explained by the increase in the heterogeneity of the grinding wheel morphology resulting from coarse dressing. For fine

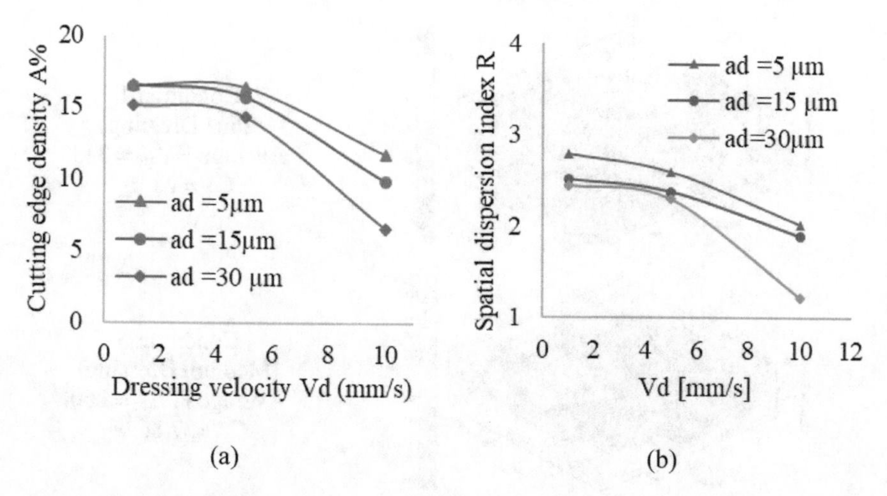

Fig. 3. Effects of dressing conditions on cutting edges: (a) density, (b) spatial dispersion

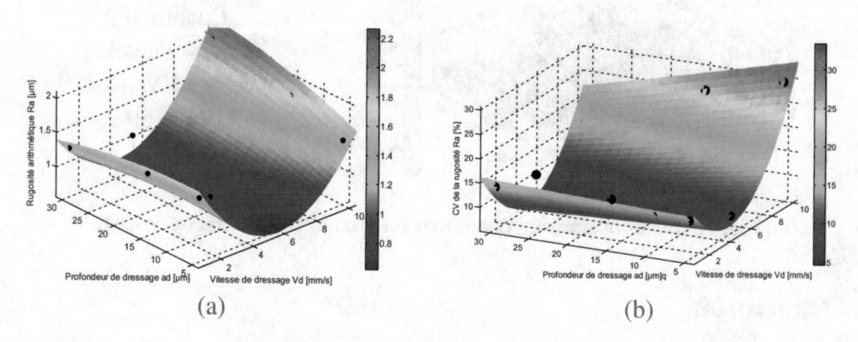

Fig. 4. Effect of dressing conditions on ground surface roughness values and dispersions

dressing condition, roughness value is decreased to 0.76 µm with a dispersion less than 15% due to the increase of the active edge density and the regularity of their spatial distribution on the grinding wheel surface.

2.4.2 Effect on the Stability of the Microhardness Distribution

The analysis of spatial distributions of microhardness on the ground surfaces is carried out by performing grids of 100 indentations spaced by 50 µm for all dressing conditions. The fine dressing conditions result in a smooth grinding wheel with dense active surfaces, which results in an increase of area interacting with the workpiece material. Figure 5 shows that fine dressing leads to an increase of averaged microhardness while minimizing its covariance. On the other hand, the coarse dressing condition leads to low microhardness average along with highest covariance.

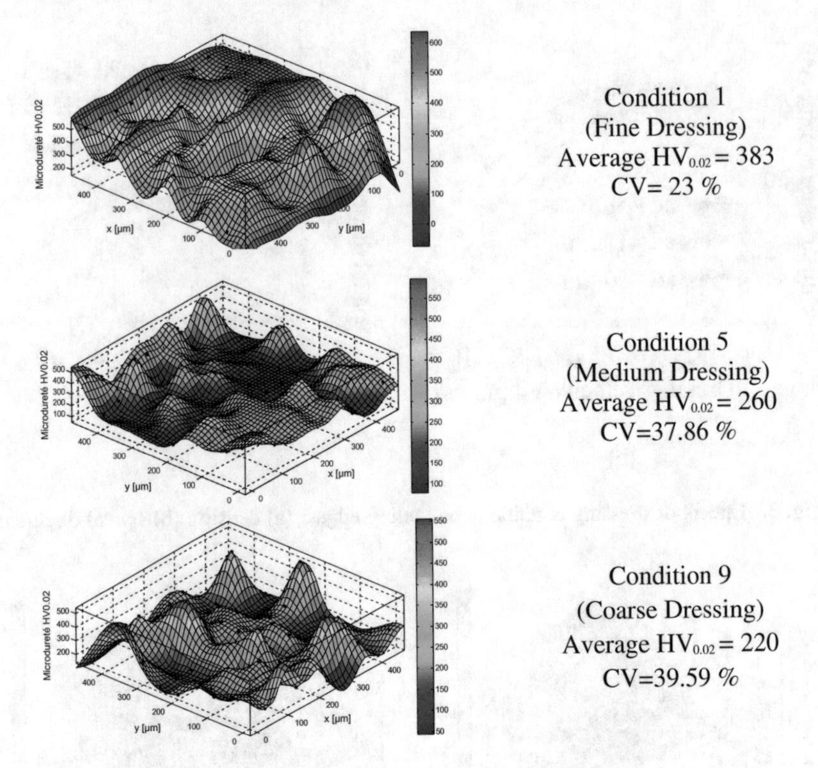

Fig. 5. Effect of dressing conditions on ground surface microhardness

3 Conclusion

The main objective of this study is to analyses the effect of the random aspect of the grinding process on ground surface characteristics and their stability.

A new method for characterization of the grinding wheel morphology and its ability to remove material has been proposed. This method is based on the acquisition and analysis of micrographs in real-time in different areas of the wheel that allowed determining the density of active surfaces and their spatial dispersion.

The grinding wheel morphology is successfully controlled by mean of dressing conditions allowing a significant enhancement of the distribution of surface roughness and hardness.

References

Badger, J., Torrance, A.: A comparison of two models to predict grinding forces from wheel surface topography. Int. J. Mach. Tools Manuf. **40**(8), 1099–1120 (2000)

Ben Moussa, N., Al-Adel, Z., Sidhom, H., Braham, C.: Numerical assessment of residual stress induced by machining of aluminum alloy. Adv. Mater. Res. **996**, 628–633 (2014a). http://doi.org/10.4028/www.scientific.net/AMR.996.628

Ben Moussa, N., Sidhom, N., Sidhom, H., Braham, C.: Prediction of cyclic residual stress relaxation by modeling approach. Adv. Mater. Res. **996**, 743–748 (2014b). http://doi.org/10.4028/www.scientific.net/AMR.996.743

Coelho, R.T., De Oliveira, J.F.G., Marinelli Filho, N.: The Application of Acoustic Emission (AE) Techniques on the Dressing Operation Using Synthetic Diamonds and Sintered Tools (2001)

Darafon, A.: Measuring and Modelling of Grinding Wheel Topography. Dalhousie University, Halifax (2013)

Fredj, N.B., Djemaiel, A., Rhouma, A.B., Sidhom, H., Braham, C.: Effects of the cryogenic wire brushing on the surface integrity and the fatigue life improvements of the AISI 304 stainless steel ground components. In: Youtsos, A.G. (ed.) Residual Stress and its Effects on Fatigue and Fracture, Dordrecht, 2006, pp. 77–86. Springer, Netherlands (2006)

Hadad, M., Sharbati, A.: Analysis of the effects of dressing and wheel topography on grinding process under different coolant-lubricant conditions. Int. J. Adv. Manuf. Technol. **90**, 1–12 (2016)

Inasaki, I., Okamura, K.: Monitoring of dressing and grinding processes with acoustic emission signals. CIRP Ann. Manuf. Technol. **34**(1), 277–280 (1985)

Kapłonek, W., Nadolny, K.: Assessment of the grinding wheel active surface condition using SEM and image analysis techniques. J. Braz. Soc. Mech. Sci. Eng. **35**(3), 207–215 (2013)

Liu, C.R., Yang, X.: The scatter of surface residual stresses produced by face-turning and grinding. Mach. Sci. Technol. **5**(1), 1–21 (2001)

Taylor, P.J.: Quantitative Methods in Geography, pp. 133–172. Waveland Press Inc., Prospect Heights (1977)

Puerto, P., Fernández, R., Madariaga, J., Arana, J., Gallego, I.: Evolution of surface roughness in grinding and its relationship with the dressing parameters and the radial wear. Procedia Eng. **63**, 174–182 (2013)

Schneider, C.A., Rasband, W.S., Eliceiri, K.W.: NIH image to ImageJ: 25 years of image analysis. Nat. Methods **9**, 671 (2012). https://doi.org/10.1038/nmeth.2089

Sidhom, H., Ben Moussa, N., Ben Fathallah, B., Sidhom, N., Braham, C.: Effect of surface properties on the fatigue life of manufactured parts: experimental analysis and multi-axial criteria. Adv. Mater. Res. **996**, 715–721 (2014a). http://doi.org/10.4028/www.scientific.net/AMR.996.715

Sidhom, N., Moussa, N.B., Janeb, S., Braham, C., Sidhom, H.: Potential fatigue strength improvement of AA 5083-H111 notched parts by wire brush hammering: Experimental analysis and numerical simulation. Mater. Des. **64**, 503–519 (2014b). https://doi.org/10.1016/j.matdes.2014.08.002

Woodin, C.T.: Effects of Dressing Parameters on Grinding Wheel Surface Topography. Georgia Institute of Technology (2014)

Wu, C.: Towards linear-time incremental structure from motion. In: 2013 International Conference on 3D Vision - 3DV 2013, 29 June–1 July 2013, pp 127–134 (2013). https://doi.org/10.1109/3dv.2013.25

Xie, J., Xu, J., Tang, Y., Tamaki, J.: 3D graphical evaluation of micron-scale protrusion topography of diamond grinding wheel. Int. J. Mach. Tools Manuf. **48**(11), 1254–1260 (2008). https://doi.org/10.1016/j.ijmachtools.2008.03.003

Yahyaoui, H., Ben Moussa, N., Braham, C., Ben Fredj, N., Sidhom, H.: Role of machining defects and residual stress on the AISI 304 fatigue crack nucleation. Fatigue Fract. Eng. Mater. Struct. **38**(4), 420–433 (2015). https://doi.org/10.1111/ffe.12243

Zouhayar, A.-A., Naoufel, B.M., Houda, Y., Habib, S.: Surface integrity after orthogonal cutting of aeronautical aluminum alloy 7075-T651. In: Design and Modeling of Mechanical Systems, pp. 485–492. Springer, Heidelberg (2013)

Fluid Mechanics and Energy, Mass and Heat Transfer

Numerical Modelling of Cavitating Flows in Venturi

Aicha Abbassi[1(✉)], Rabeb Badoui[1(✉)], Lassaad Sahli[1(✉)], and Ridha Zgolli[2(✉)]

[1] Laboratory of Applied Mechanical Research and Engineering, National School of Engineers of Tunis, University of Tunis EL MANAR, 1002 Tunis, Tunisia
aicha.abbassi@enit.utm.tn, rabeb.bedoui@gmail.com, sahli.lassaad555@gmail.com

[2] Laboratory of Hydraulic and Environmental Modeling, National School of Engineers of Tunis, University of Tunis EL MANAR, 1002 Tunis, Tunisia
ridha.zgolli@enit.utm.tn

Abstract. The cavitation is one of the most binding physical phenomena influencing the performances of the hydraulic machines. This paper presents a theoretical study modeling of cavitation in cavitating flows and a numerical study of a cavitation pocket developing in a Venturi flow. The study of the flows by the digital present simulations CFD also has big interest. For the numerical simulation we used the free software to access the code and the algorithms. OpenFOAM is a toolbox for computational fluid dynamics.

Keywords: Cavitation flows · Numerical simulation · Venturi 8° · RANS · Open FOAM

1 Introduction

This study joins in the continuity of the works on the modelling and the simulation of cavitants flows. The present paper concentrates on the digital simulation of cavitating flows. These flows are studied in particular be-cause of the negative impact led by the appearance of the cavitation in various systems. The cavitation is a change of phase which allows crossing from the liquid state to the state vapor by a decrease of the pressure without the contribution of heat. A large number of studies and developments in research are conducted on cavitation. The unsteady character of the cavity behavior of Venturi profile is responsible for many issues like erosion, noise and vibrations. The study of the flows surging by the digital present simulations CFD also has big interest. The modelling of the cavitation has to base itself on the algorithms coupling the equations of the dynamics of the fluids as the equations of Navier-Stokes and model one, said of cavitation, often empirical, which has to predict correctly the way the phase vapor appears, disappears, and interacts with the liquid phase in a process of evaporation and condensation. The models of cavitation thus check the appearance and the disappearance of the vapor in the liquid flow. They often rest on the homogeneous

A. Benamara et al. (Eds.): CoTuMe 2018, LNME, pp. 231–238, 2019.
https://doi.org/10.1007/978-3-030-19781-0_28

approach defined previously both phases liquid and vapor establish a diphasic mixture represented by a unique fluid defined by the average physical properties. OpenFOAM (Moukalled et al. 2016) is used for numerical simulation in fields as fluid dynamics, cavitating flow, turbulent flow, heat transfer, solid mechanics and other fields of engineering. Kubota et al. (1992) proposed the premise of homogeneous flows, he presents a model based on Rayleigh's equation and neglects the effects of viscous damping, and surface tensions. The mixture of liquid vapor is treated as a single fluid for Navier-Stokes equations. On the basis of this premise, Coutier-Delgosha in 2003 proposed that the barotropic state law allows controlling the condensation and vaporization, and he elaborated the unstable cavitation flow in two types of geometries Venturi nozzles (Coutier-Delgosha et al. 2003); (Barre et al. 2009). Also, we find other models based on transport equations whose multiphase flow formed by a liquid fraction and a vapor fraction (Singhal et al. 2002); (Charrière et al. 2015).

2 Mathematical and Numerical Model

2.1 Governing Equations

In the case of a homogeneous mixture fluid we write the following hydrodynamic equations: the continuity equation (1), the momentum equation (2) for vapor-liquid mixture considered homogeneous and incompressible and the volume fraction equation for the liquid phase (3).

$$\nabla.U = \dot{m}\left(\frac{1}{\rho_l} - \frac{1}{\rho_v}\right) \tag{1}$$

$$\frac{\partial \rho}{\partial t} + \nabla.(\rho U U) = -\nabla P + \nabla.\tau + S_M \tag{2}$$

$$\frac{\partial \gamma}{\partial t} + \nabla.(\gamma U) = \frac{\dot{m}}{\rho_l} \tag{3}$$

Where $\dot{m}$ represents the inter-phase mass transfer rate due to cavitation, P the time averaged pressure, U represents the time averaged mixture velocity, ρl the liquid density, ρ_v the vapor density, S_M are momentum sources, τ is stress tensor.

The vapor volume fraction α and the water volume fraction γ are defined as follows:

$$\alpha = \frac{Volume\ of\ vapor}{Total\ volume} \quad \gamma = \frac{Volume\ of\ liquid}{Total\ volume} \tag{4}$$

The vapor volume fraction is related to liquid volume fraction as:

$$\alpha + \gamma = 1 \tag{5}$$

A last the effective density ρ and the dynamic viscosity μ of the vapor-water mixture are given by "(6)," and "(7)," respectively:

$$\rho = \rho_v \alpha + (1 - \alpha)\rho_l \tag{6}$$

$$\mu = \mu_v \alpha + (1 - \alpha)\mu_l \tag{7}$$

2.2 Mass Transfer Models

The modeling of cavitating flow must be based on the algorithms coupling the equations of the dynamics of the fluids like the Navier-Stokes equations and a model, known as of cavitation. In the present study, this cavitation model based on the Rayleigh-Plesset equation has been applied.

2.2.1 Kunz Model

The Kunz transport model is heuristic model based on the work by Merkle et al. (1998) and one of the mass transfer models implanted in Open-FOAM CFD. The model of Kunz consists in subdividing it source term in a term related to vaporization m.+ and another related to condensation as shown in "(8)".

$$\dot{m} = \begin{cases} \frac{C_{prod}\ \rho_v(\gamma^2 - \gamma^3)}{t_\infty} & \text{if p} < \text{pv} \\ \frac{C_{dest}\ \rho_v \gamma \min(0, P - Pv)}{\left(\frac{1}{2}\rho_l U_\infty^2\right) t_\infty} & \text{if p} > \text{pv} \end{cases} \tag{8}$$

In the above equations U_∞(m/s) is the free-stream velocity, $t_\infty = L/U_\infty$ is the mean flow time scale, where L is the characteristic length scale. Cdest and Cprod are two empirical coefficients. In the original formulation Cdest = 100, Cprod = 100.

2.3 Turbulence Models

The models of turbulence have for objective to couple the performance of the methods RANS. Indeed, the model RANS allows the simulations, large number of Reynolds but are incapable to feign correctly the unsteady zones of flow three-dimensional as the trails or the coats of the mixture.

3 Results and Discussions

3.1 Test Case: Geometry Venturi 8°

The numerical simulations were implemented for a Venturi profile with a divergence angle just downstream of the leading edge of 8° and a convergence angle of 18° (Fig. 1). The length of the vein is 1,272 m and its entrance section is 0.044 m wide and 0.05 m high. This type of Venturi was tested in test at LEGI (Laboratory of Geophysical and Industrial Flows) by Stutz and Reboud (1997).

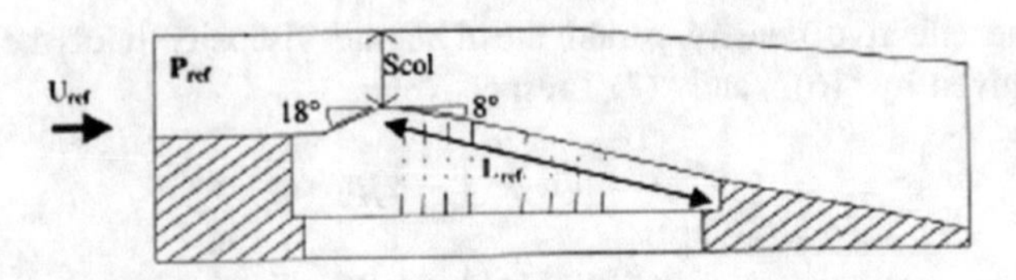

Fig. 1. Cavitation pocket: Schematic view of Venturi 8°

3.2 Computation Domain and Mesh Generation

This geometry is an unsteady reentrant cavity with quasi-periodic fluctuations of the attached sheet and vapor clouds shedding has been obtained.

3.2.1 Meshing

The blockMesh integrated into OpenFOAM. It contains 300 nodes in the flow direction and 65 in the orthogonal direction. The y+ values of the mesh, at the center of the first cell, vary between 10 and 30 for a non-cavitating computation (Fig. 2).

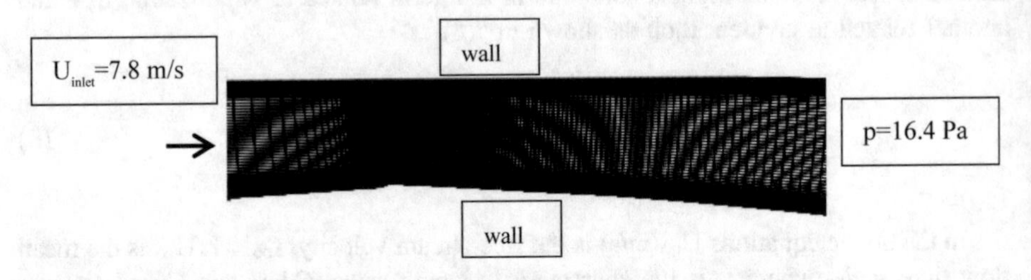

Fig. 2. Mesh venturi 2D

3.2.2 Open Foam: InterPhaseChangeFoam

Solver to simulate the cavitation in OpenFOAM is interphaseChangeFoam. It provides a large variety of RANS turbulence models and cavitation models. For cavitation modelling, the way is available to use a transport equation for the volume fraction of liquid. The last one is retained for the present simulation.

3.2.3 Boundary and Operation Conditions

The boundary conditions related to turbulent quantities are represented as a turbulent intensity imposed at the entry of the calculation domain, which is 1% in our case. The numerical simulation starts at 0 s and lasts 2×10^{-1} s with time steps of 1×10^{-4} s. The boundary conditions are described in Table 1.

Table 1. Boundary conditions

Venturi	Condition
Inlet	Velocity in x axis U = 7.8 m/s
Outlet	Pressure p = 16.4 Pa
Top and bottom	Wall
Front and back	Symmetry planes
Venturi wall	Wall

3.3 Study 2D: Global Validation

Turbulent viscosity models are based on a constitutive equation that linearly links the Reynolds tensor (turbulent stresses) to the strain tensor. For models with two equations, this viscosity is often expressed as a function of two quantities such as turbulent kinetic energy and dissipation in models k-ε and k-ω;

3.3.1 K-ε RNG Model

It should be noted that this model was developed for the study of the monophasic incompressible flows and it is also known for its dissipative character. The flow studied here is strongly unsteady and compressible; because of the local presence and quasi-perms both phases settle and vapor. This strong compressibility is not taken into account in this type of model. It turns out that the model k-ε RNG is not capable of reproducing the instationnarité strong of the flow observed in the experiments (Fig. 3).

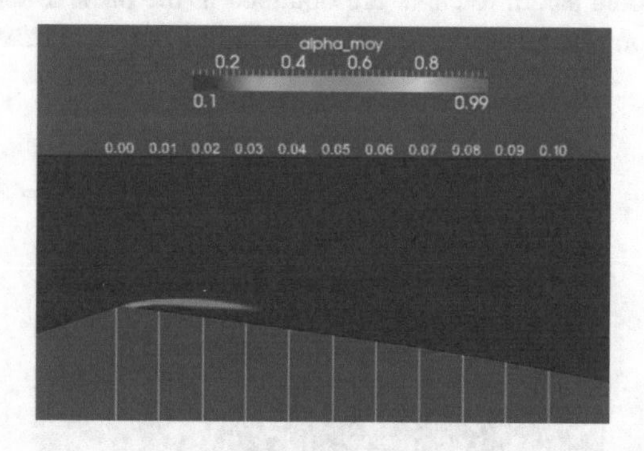

Fig. 3. Numerical simulation: Venturi 8° case. Conditions of calculation: Velocity inlet = 7.8 m/s, cavitation number = 2.8.

Thus the obtained behavior by applying the model k-ε RNG is not physical essentially because it does not reproduce the instability of the flow.

3.3.2 K-ω SST Model

The second model tested in this study is the model Shear Stress Transport) proposed by Menter. It is an approach which aims at combining the respective advantages of both models k-ε and k-ω. The second is activated near walls while the first one is applied to the rest of the flow. The model k-ω is indeed based on the hypothesis that the quantity is proportional in the normal component of the turbulent kinetic energy in the viscous zone of sublayers what allows to have a very good agreement with the experience and the results DNS in this zone (Fig. 4).

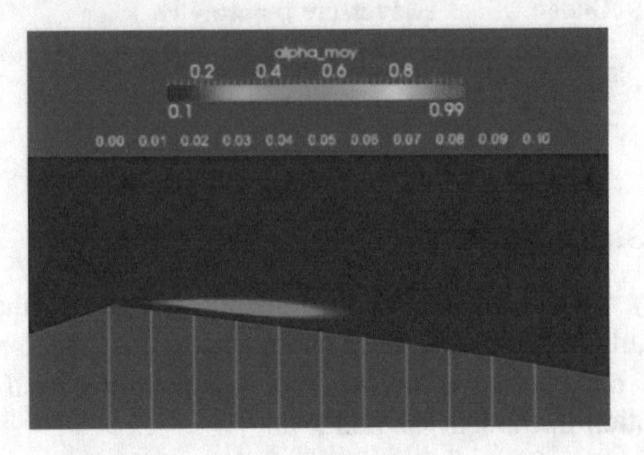

Fig. 4. Numerical simulation: Venturi 8° case. Conditions of calculation: Velocity inlet = 7.8 m/s, cavitation number = 2.8.

This model k-ω SST gives very interesting results in terms of overall behavior and average pocket length on this flow pattern.

3.3.3 Modified RNG K-ε Model

The unsteadiness of the flow is now well reproduced: the incoming jet is correctly predicted and periodic steam releases are obtained at the back of the pocket. Figure 5 shows the cavitation pocket at a given time, including steam releases entrained by the main flow.

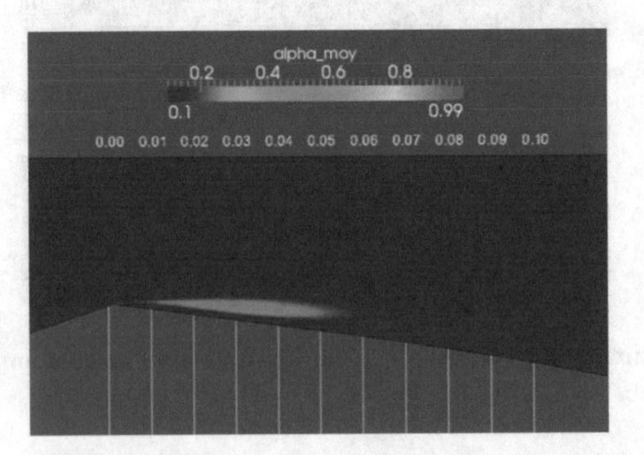

Fig. 5. Numerical simulation: Venturi 8° case. Evolution of the pocket and appearance of the vapor cloud. Conditions of calculation: Velocity inlet = 7.8 m/s, cavitation number = 2.8

The average pocket length in this case is estimated at 60 mm and the characteristic frequency obtained from the input pressure signal is 37.5 Hz, which gives a Strouhal number St = 0.31 very close to the experimental value. Observation of the evolution of the cavitation pocket to the over a period of time provides a better understanding of what these pressure peaks correspond to.

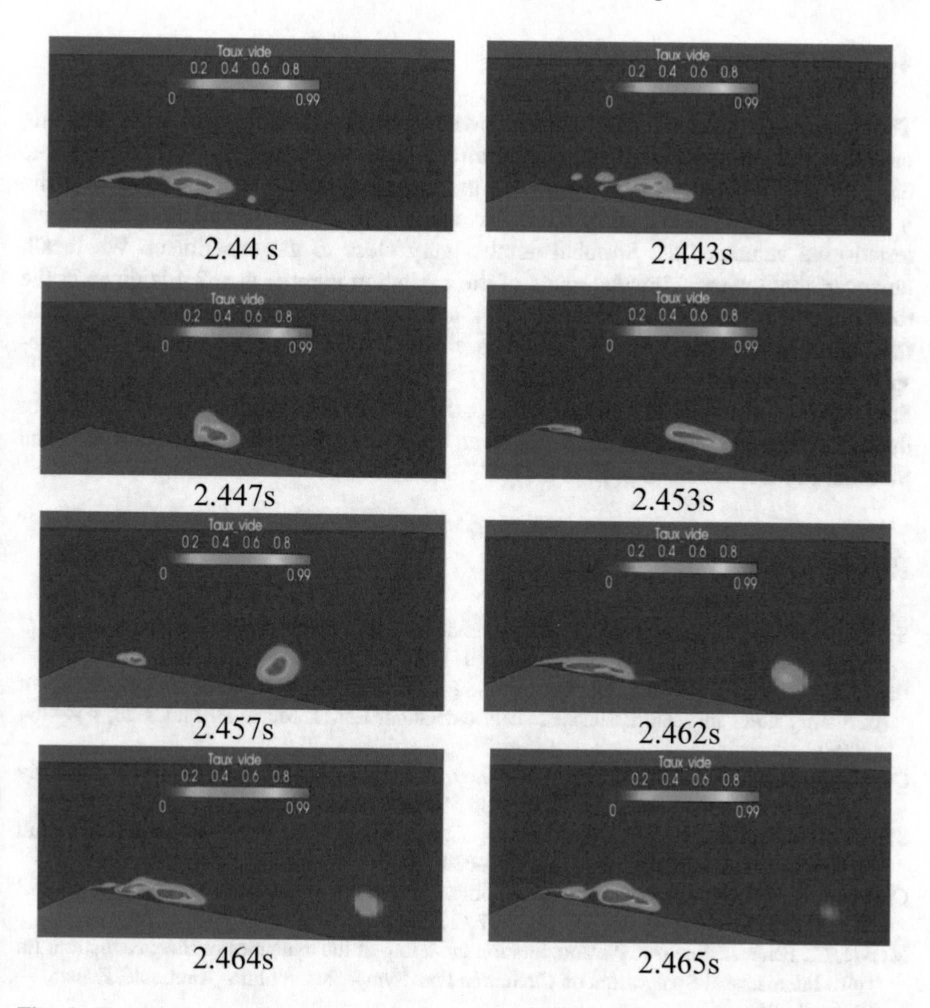

Fig. 6. Numerical simulation: Venturi 8° case. Analysis of unsteady behavior: the evolution of the pocket during a cavitation cycle. Conditions of calculation: Velocity inlet = 7.8 m/s, cavitation number = 2.8

At t = 2.44 s the pocket reaches its maximum size and its interface is very disturbed by the rise of the incoming jet. A first cloud of steam then begins to form (t = 2,443 s). The cavitation pocket under the effect of pressure waves emitted by these implosions disappears at t 2,447 s. The cloud formed is then convected downstream as the pocket reappears a second time at the throat (t = 2,453 s) and a second cloud of vapor is formed accompanied by the temporary disappearance of the pocket near the cervix at t = 2,457 s which generates the second peak obtained during the same period. The pocket allows to reappear quickly it develops until reaching the second cloud to form a pocket attached at the instant 2.462 s. The main cloud always continues its journey until it meets the areas of highest pressure at t = 2,465 s, this is the boundary between the end of this cycle and the beginning of another cavitation cycle (Fig. 6).

4 Conclusions

The different turbulence models applied have resulted in different behaviors depending on the model used. Standard versions of turbulent viscosity models k-ε RNG and k-ω SST do not allow the instationarity of the cavitation flows to be reproduced; the incoming jet is well predicted by numerical simulations. In the Venturi case, the models tested give values of the Strouhal number very close to the experiment. We recall, however, that the experimental value of the cavitation number $\sigma = 2.4$ is given at the inlet of the flow: In the case of the model k-ω SST, the number of cavitation is fixed at the output of the calculation domain, the value $\sigma = 2.8$ has been imposed, it corresponds to a number of input cavitation $\sigma = 2.4$, The Strouhal number obtained is $St = 0.3$. For the model Modified RNG k-ε, the number of cavitation $\sigma = 2.8$ is fixed at the output also, this value corresponds to an upstream cavitation number $\sigma = 2.5$. The Strouhal number obtained is $St = 0.31$.

References

Kubota, A., Kato, H., Yamaguchi, H.: A new modelling of cavitating flows: a numerical study of unsteady cavitation on a hydrofoil section. J. Fluid Mech. **240**, 59–96 (1992)

Barre, S., Rolland, J., Boitel, G., Goncalvés, E., Fortes, R.: Experiments and modelling of cavitating flows in venturi: attached sheet cavitation. Eur. J. Mech. B. Fluids **28**, 444–464 (2009)

Coutier-Delgosha, O., Reboud, J.L., Delannoy, Y.: Numerical simulation of the unsteady behavior of cavitating flows. Int. J. Numer. Methods Fluids **42**, 527–548 (2003)

Singhal, A.K., Athavale, M.M., Li, H., Jiang, Y.: Mathematical basis and validation of the full cavitation model. J. Fluids Eng. **124**, 617–624 (2002)

Charrière, B., Decaix, J., Goncalvès, E.: A comparative study of cavitation models in a Venturi flow. Eur. J. Mech. B. Fluids **49**, 287–297 (2015)

Merkle, C., Feng, J., Buelow, P.: Computation modeling of the dynamics of sheet cavitation. In: Third International Symposium on Cavitation Post Symposium Volume, Grenoble, France, 7–10 April 1998

Stutz, B., Reboud, J.L.: Experiments on unsteady cavitation. Exp. Fluids **22**, 191–198 (1997)

Moukalled, F., Mangani, L., Darwish, M.: The Finite Volume Method in Computational Fluid Dynamics: An Advanced Introduction with OpenFOAM® and Matlab, Switzerland (2016)

Numerical Simulation of a Water Jet Impacting a Titanium Target

Ikram Ben Belgacem[1(✉)], Lotfi Cheikh[1,2], El Manaa Barhoumi[3], Waqar Khan[4], and Wacef Ben Salem[1,5]

[1] Laboratoire de Génie Mécanique, Ecole Nationale d'Ingénieurs de Monastir, Université de Monastir, Monastir, Tunisia
ikrambenbelgacem@gmail.com, wacef.bensalem@gmail.com
[2] Institut Préparatoire aux Etudes d'Ingénieur de Monastir, Université de Monastir, Monastir, Tunisia
lotfi.cheikh@ipeim.rnu.tn
[3] Department of Electrical and Computer Engineering, College of Engineering, Dhofar University, Salalah, Oman
[4] University of Waterloo, Waterloo, ON N2L 3G1, Canada
w.khan@mu.edu.sa
[5] Institut Supérieur des Arts et Métiers de Mahdia, Université de Monastir, Monastir, Tunisia

Abstract. This paper is dedicated to the numerical simulation of a round water jet impinging on a plate (1.5 × 6 × 3.5) made of titanium alloy Ti555-03. Two configurations which differ from each other by the position (angle of inclination) of the plate relatively to the axis of revolution of the jet inlet are investigated in this study. This study aims to predict the behavior of the plate Solicited by the effect of the fluid produced under an initial high speed 500 m/s. The state of stress and displacement of the plate will be investigated. The modeling of this coupling fluid structure problem is hybrid: the water by the smoothed particle hydrodynamics (SPH) method and the target by the finite element method. The titanium alloy is modeled by the Johnson Cook law. Subsequently the results of this study will be useful for the modeling of a problem of orthogonal cutting assisted by a jet of water at high pressure. The numerical simulations are carried out under "ABAQUS".

Keywords: Titanium alloy (Ti555-03) · ABAQUS · Smoothed Particle Hydrodynamics (SPH) · Finite Element Method (FEM) · Fluid Structure Interaction (FSI)

1 Introduction

The impact of a high-velocity jet on an obstacle is widely studied because of its large applications from household appliances to space technology. The high-velocity water jet is characterized by its many advantages and ease of industrialization (Kaushik et al. 2015). The technology of water jets is now applied In many industrial fields, such as welding, cleaning (Turbine, engine parts), decorating process, cutting (metals sheet,

A. Benamara et al. (Eds.): CoTuMe 2018, LNME, pp. 239–247, 2019.
https://doi.org/10.1007/978-3-030-19781-0_29

plastic), removing materials by Abrasive water jet, high-pressure water jet assisted machining (Chizaris et al. 2008, 2009; Mabrouk et al. 2000; Mabrouk and Raissi 2002; Saxena and Paul 2007; Ayed et al. 2016). The most used way to produce a high-velocity water jet is forcing a quantity of water through a converging nozzle. By using this method, the water is accelerated and the jet can attain the speed up to 4000 m/s (Hsu et al. 2013) Due to the strong multi-physics coupling, the numerical modeling of Fluid-Structure Interaction (FSI) is considered very difficult (Kaushik et al. 2015). Recently, Simulations start to take into consideration the capability of the water jet technique in different domains. Only a few studies have been focused experimentally and numerically on high pressure water-jet taking into account the FSI problems. This study is a part of a whole project about High-speed water jet-assisted machining of a new Titanium alloy: Ti555-03. Within this framework we propose a numerical study in aim to investigate the behaviour of a pure water jet by continues impact on a metal sheet. In this paper we propose a numerical model using ABQUS: it is a meshfree smoothed particle hydrodynamics based on Lagrangian formulation developed to handle and analyze in detail the material behavior under the water jet impingement. By using the above method; we propose in this study tow configurations which differ from each other by the position (angle of inclination) of the target relatively to the axis of revolution of the jet nozzle. A convergence study was indispensable to validate the stability of the numerical results for each model.

2 Problem Formulation

We are going to impact a horizontal plate of Titanium alloy: TI55503 with a pure water jet which has an initial velocity of V_e = 500 m/s through a nozzle having a diameter D of 0.3 mm. The Target is a sheet of titanium alloy metal Ti555-03 having 6 mm as a length, 1 mm as width and 3.5 mm as thickness as presented in Fig. 1. These geometrical parameters values correspond to a high-pressure water jet assisted machining applications (Braham-Bouchnak et al. 2010). In addition, in order to minimize the computation time CPU. The domain of the jet flow is fixed to h = 1.5 mm far from the target. The axis of the nozzle is coaxial with the normal of the target. We propose to study tow configurations of impacting a target. For configuration 2 the only difference is the incline of the target by α = 45° from the horizontal level.

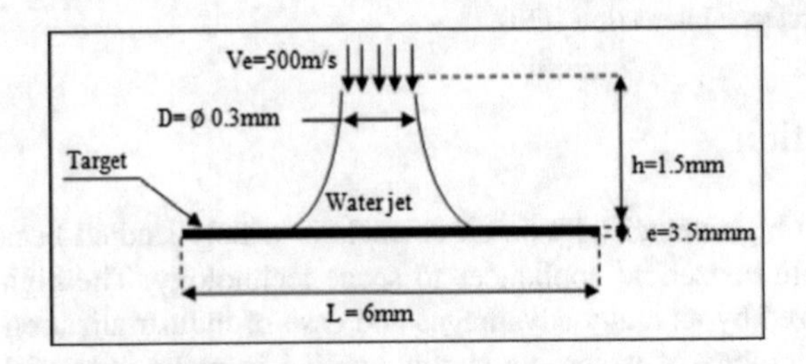

Fig. 1. Geometrical parameters for the impacting problem

3 Numerical Models

A water column acting as a water source having a diameter D = 0.3 mm and a height of 6 mm was taken. The length of the water jet was fixed to be long enough to obtain a stable flow during the period of cooling. The Target is a sheet of titanium alloy metal Ti555-03 having 6 mm as a length 1 mm as width and 3.5 mm as thickness.

3.1 Material Modeling

For the hydrodynamic behavior law, the water is modeled using the linear Hugoniot form of the Mie-Greisen equation of state. ABAQUS/Explicit provides a linear U_s–U_p equation of state model that can simulate incompressible viscous and in-viscid laminar flow governed by the Navier-Stokes Equation of Motion (ABAQUS 6.14 documentation). Equation parameters are shown in Table 1.

Table 1. Parameters for the US-UP equation of water

C_0 (m/s)	s	Γ_0
1450	0	0

Where C_0 is reference speed of sound in water; s is slope of U_s–U_p curve; Γ_0 is Grüneisen ratio of water. For the Sheet modeling, Titanium is often used as alloys. The Ti-555-has excellent properties (Braham-Bouchnak et al. 2010). In Table 2 mechanical and thermal proprieties are presented.

Table 2. The mechanical propriety of the Titanium alloy Ti-5553 (Braham-Bouchnak et al. 2010)

Parameter	Density ρ (kg/m^3)	Hardness Hv	Yield σ_e (MPa)	Tensile strength σ_R (MPa)	Elongation A (%)	Young's modulus E (GPa)	Poisson coefficient υ	Phase change temperature T_β (°K)	Melting temperature T_m (°K)
Value	4650	345	1174	1236	6	112	0.32	1118	1943

The constitutive law of Johnson–Cook and the Johnson–Cook damage model have been chosen to take into account the dynamic behavior of the material. Parameters of this law are presented in Tables 3 and 4.

Table 3. Parameters of the Johnson Cook Law for the Ti555-03 (Braham-Bouchnak et al. 2010)

Parameter	A (MPa)	B (MPa)	C	n	m	$\dot{\varepsilon}_0$	T_f (°K)	T_{ref} (°K)
Value	1175	728	0.035	0.26	0.55	0.1	1934	293

Table 4. Parameters of Johnson–Cook damage law for the Ti555-03 (Braham-Bouchnak et al. 2010)

Parameter	D_1	D_2	D_3	D_4	D_5
Value	–0.09	0.27	0.48	0.014	3.870

3.2 Mesh and Conversion to Particles

In this modeling part, a hybrid discretization technique has been adopted. The SPH method is adopted for the water source. Starting by meshing the column of water with C3D8R (8-node linear brick, reduced integration, hourglass control) to generate "parent" elements. These "parent" elements will be after that converted to particles (PC3D elements). Since we have selected a C3D8R element types and after a convergence studies, we have fixed the total number of elements in the column of water at 515967. After conversion to particles, we have a total number of particles equal to 13931109 particles. For the Titanium target, it is presented as a shell membrane. Hence the elements type is a 4-node general-purpose shell, reduced integration with hourglass control, finite membrane strains (S4R). The total number of elements is fixed at 515967.

3.3 Boundary Conditions and Predefined Field

An initial velocity field of 500 m/s is applied on the water column. The four edges of the target were fixed as shown in Fig. 2 Moreover, general contact was used, and the type of contact domain was chosen.

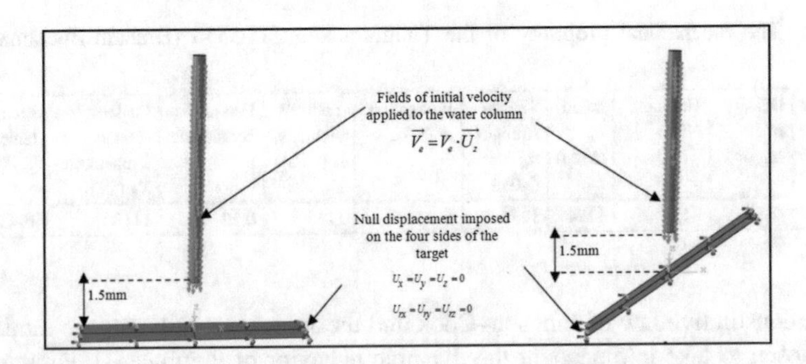

Fig. 2. Initial conditions and boundary conditions

4 Numerical Results

Analyzing the impact of the high-speed water jet on a solid target requires considering connected solid mechanics and fluid mechanics theories. This study proposes a method using ABAQUS to simulate the dynamic process of impact for tow configurations already shown in Table 1. Simulations are presented in Figs. 3a and b.

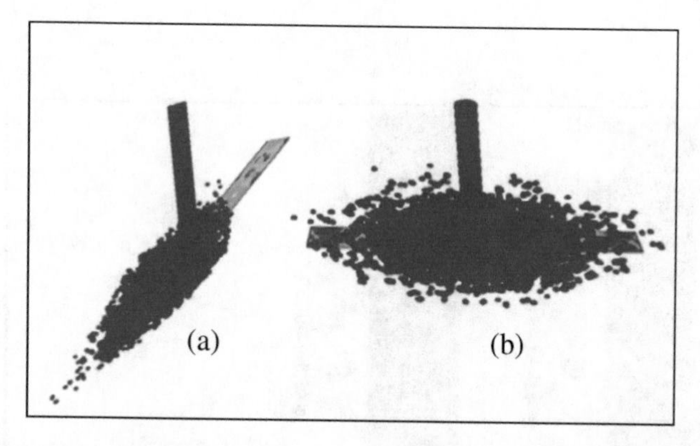

Fig. 3. Simulation of the high-speed water jet impact on (a) the HT and (b) the IT after 8e-006 s

4.1 Pressure on the Target

For the HT, the simulation was extended to a period of 8e–006 s from the beginning of the fall of the water column. The distribution of the impact pressure on the surface of the titanium sheet at different moments of the process was investigated as shown in Fig. 4. The color contour represents the magnitude of the pressure.

Figure 4a presents the first moment of impact. In earlier stage, the pressure has a symmetric distribution (Figs. 4b, c and d) along the *Y* and *X* axis. In Fig. 4e, the pressure does not present a symmetrical distribution any more. There was a chaotic distribution at points closer to the sheet limits. The pressure fluctuates as far as the water hit the target. The particles of the fluid are more dispersed on the entire target in a random and unpredictable way. The distribution of the pressure on the target shows zones of pressure and depression in alternately way. At approximately 4.5×10^{-6} s, a peak value of pressure approached just over 0.36 MPa and a peak value of depression approached to –0.54 MPa. This high fluctuation then propagated strongly out to the surrounding areas as a wave. The depression is more approached at the limit of the target as shown in Fig. 4e. This fluctuating response of the target under the water impact could be in consistency with the vibration of the metal sheet. For the IT, the simulation was extended to a period of 8×10^{-6} s from the beginning of the fall of the water column. The distribution of the impact, pressure on the surface of the inclined plate at different intervals is presented in Fig. 5. During the impact maximum pressure keeps one position located in the center of the metal plate unlike the first configuration. Figure 5a shows the state at 2.8×10^{-6} s, which is when the water jet began hitting the structure. At approximately 4.5×10^{-6} s (Fig. 5b), the pressure approached a peak value of just over 0.7 MPa more approached at the limit of the target. This high pressure then propagated out to the surrounding areas as a wave. In the next period, there was a chaotic distribution of the pressure (Fig. 5c), which may be due to the changing state of the flow before the stability occurred. The distribution of the stagnation pressure is shown in (Fig. 5e) at 8×10^{-6} s. Furthermore, pressure distribution presents alternatively zones of pressure and depression as far as the jet hit the target. The jet head is not perfectly flat. It could be considered as punctual relatively to the

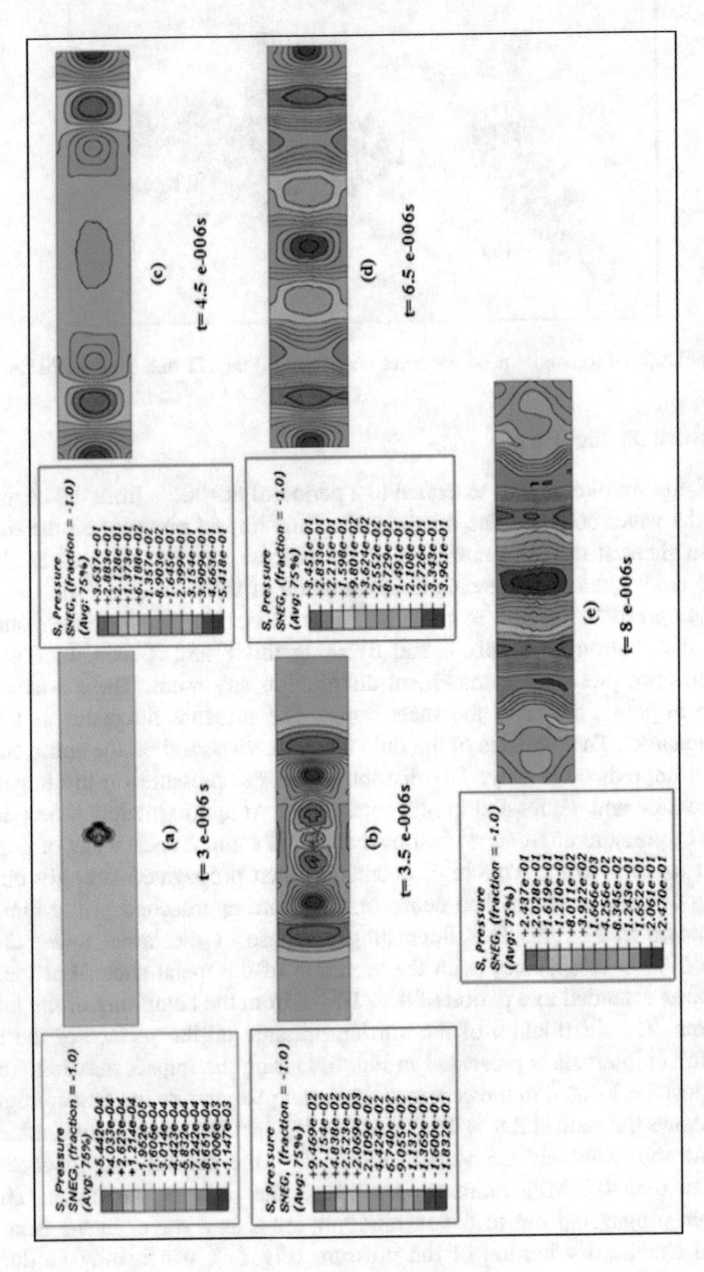

Fig. 4. Distribution of pressure on the impact area at different time steps on the HT

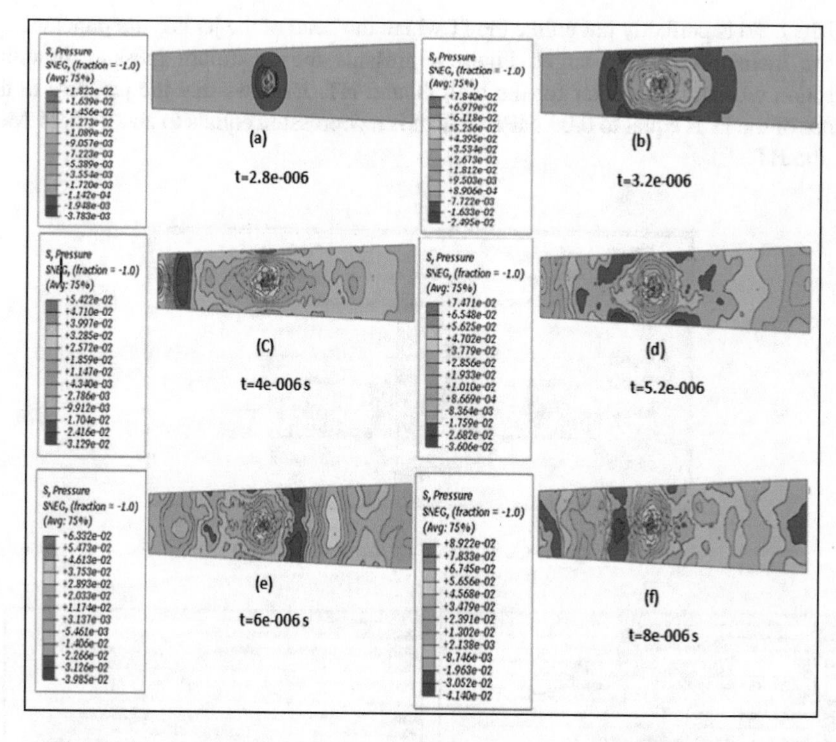

Fig. 5. Distribution of pressure on the impact area of IT at different time steps

length of the plate due to the 45° angle position of the plate, which means that the points localized in the part of the plate under the Y axis reach their maximum values earlier than other points.

5 Comparison Between the Inclined Target and the Horizontal Target

Figure 6 presents the magnitude displacement through a radial direction crossing the center of the target for both of configuration. It shows that the horizontal target presents the haier displacement and stronger fluctuations compared to the second configuration which presents lower magnitude and smoother fluctuation. Indeed water in the first configuration keeps contact with the target but for the second configuration water finds a way for evacuation faster thanks to the incline of the target. Figure 7 presents the maximum displacement versus the time history. The IT shows a smoother shape compared to the HT. The IT presents a maximum value at the end of the impact equal to 0.008 mm while the HT presents a maximum at about 6e–006 s. At the first moment of the impact (3e–006 s) a pic of 0.004 mm is shown in the IT. For the HT, the first pic of displacement 0.008 mm is observed at 4e–006 s. This pic is less acute. This could be explained by the shape of the head of the jet at the first moment of impact. In fact for the

HT the head is perfectly flat unlike the IT where the head of the jet is quite punctual due to the inclination of the target. Figure 8 presents the maximum pressure in radial direction crossing the center for the both IT and HT. It shows that the pressure in the centre of the IT is equal to 0.08 MPa while it is a depression equals to about –0.24 MPa for the HT.

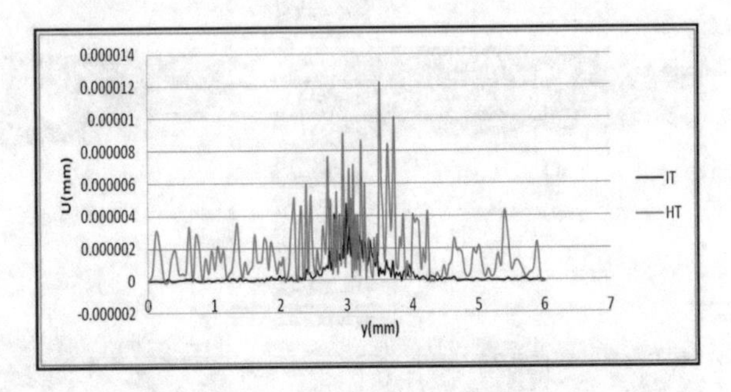

Fig. 6. Comparison of the displacement

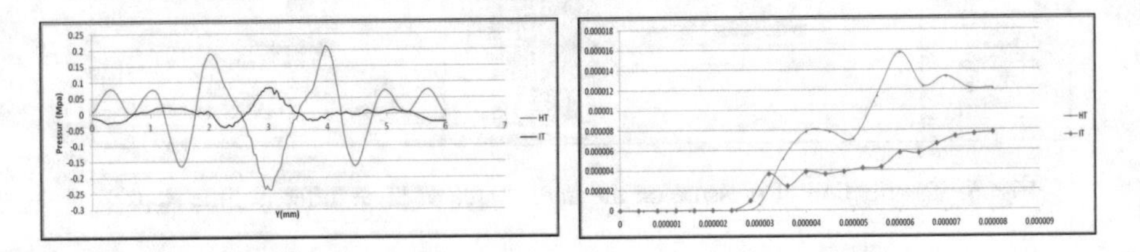

Fig. 7. Maximum pressure in radial direction crossing the center for IT and HT

Fig. 8. The maximum displacement of the IT and HT

6 Conclusions

In this paper, numerical simulations of a high velocity water jet impacting a horizontal plate and an inclined plate were carried out under ABAQUS. The plate is a metal sheet of Ti555-03. Its mechanical behavior is handled by the Johnson cook law. Two points of view are presented for this study. The first one is a CFD study of the jet impacting where it was possible to demonstrate the structure of a free jet and impacting for the two cases of positioning the plate (0° and 45°). A second part was devoted to study the FSI by a hybrid method a finite element modeling for the plate and free mesh method called SPH method to investigate the Von Mises stress; the pressure and the displacement of the target. This result will be useful to investigate and to understand the behavior of a water assistance jet impact during high-speed machining of a Ti555-03 titanium material in terms of the interaction between the water of assistance and the work piece and the tool. Furthermore these results will be useful in a study of tool/workpiece/cutter system.

References

Kaushik, M., Kumar, R., Humrutha, G.: Review of computational fluid dynamics studies on jets. Am. J. Fluid Dyna. **5**(3), 1–11 (2015)

Chizari, M., Al-Hassani, T.S., Barrett, L.M.: Experimental and numerical study of water jet spot welding. J. Mater. Process. Technol. **198**(1–3), 213–219 (2008)

Chizari, M., Barrett, L.M., Al-Hassani, A.: An explicit numerical modeling of the water jet tube forming. Comput. Mater. Sci. **45**(2), 378–384 (2009)

Mabrouki, T., Raissi, K., Cornier, A.: Numerical simulation and experimental study of the interaction between a pure high-velocity waterjet and targets: contribution to investigate the decoating process. Wear **239**(2), 260–273 (2000)

Mabrouki, T., Raissi, K.: Stripping process modelling: interaction between a moving waterjet and coated target. Int. J. Mach. Tools Manuf. **42**(11), 1247–1258 (2002)

Saxena, A., Paul, O.: Numerical modelling of kerf geometry in abrasive water jet machining. Int. J. Abras. Technol. **1**(2), 208–230 (2007). https://doi.org/10.1504/ijat.2007.015385

Ayed, Y., Roberta, C., Germaina, G., Ammarab, A.: Development of a numerical model for the understanding of the chip formation in high-pressure water-jet assisted machining. Finite Elem. Anal. Des. **108**, 1–8 (2016)

Hsu, C.-Y., Liang, C.-C., Teng, T.-L., Nguyen, A.-T.: A numerical study on high-speed water jet impact. Ocean Eng. **72**, 98–106 (2013)

Braham-Bouchnak, T., Germain, G., Robert, P., Lebrun, J.L.: High pressure water jet assisted machining of duplex steel: machinability and tool life. Int. J. Mater. Form. **3**(1), 507–510 (2010)

Documentation of ABAQUS 6.14

Effect of Co-flow Stream on a Plane Turbulent Heated Jet: Concept of Entropy Generation

Amel Elkaroui[1](✉), Amani Amamou[1], Mohamed Hichem Gazzah[2], Nejla Mahjoub Saïd[3], and Georges Le Palec[4]

[1] LGM, National Engineering School of Monastir, University of Monastir, Monastir, Tunisia
amel.karoui@hotmail.fr, amani.amamou@yahoo.fr

[2] URPQ, Faculty of Science of Monastir, University of Monastir, Monastir, Tunisia
hichem.gazzah@fsm.rnu.tn

[3] LGM, Preparatory Institute for Engineering Studies, University of Monastir, Monastir, Tunisia
nejla.mahjoub@fsm.rnu.tn

[4] Aix-Marseille University, CNRS, IUSTI, Marseille, France
georges.lepalec@univ-amu.fr

Abstract. The present paper numerically investigates the effect of a co-flowing stream on the mean and turbulent flow properties, air entrainment and entropy generation rate of a heated turbulent plane jet emerging in a co-flowing stream. The first order k-ε turbulence model is used and compared to the existing experimental data. The Finite Volume Method (FVM) is used to discretize the governing equations. The predicted results were consistent with the experimental data. It is found that a jet in a co-flowing stream is known to be a quicker mixer than a jet in a quiescent ambient, and it is proved that the presence of a co-flow enhances mixing. Therefore, the mixing of a jet in co-flow is more efficient.

Keywords: Numerical modeling · k-ε model · Co-flow · Entropy generation concept

1 Introduction

Free turbulent jets in a stagnant ambient stream as well as in a moving external stream have been the interest of numerous investigations. This interest is due to their practical applications as turbulent diffusion flames in combustion chambers when a fuel jet flow is commonly injected into a co-flowing stream. The important parameters that influence the mixing characteristics of a jet are the presence of density difference and a co-flowing between the jet and its surroundings. The entropy generation concept, which is based on the second law of thermodynamics, has recently appeared in many applications in engineering, such as combustion engines and convective heat transfer system flows. Much more attention is given to round jets emerging in a co-flowing stream. Indeed, some experimental investigations were conducted on co-flowing jets. Nickels and Perry [1], Antonia and Bilger [2], Smith and Hughes [3] made measurements in the

A. Benamara et al. (Eds.): CoTuMe 2018, LNME, pp. 248–256, 2019.
https://doi.org/10.1007/978-3-030-19781-0_30

strong region of round jets as well as in the strong-to-weak jet-transition region. However, the measurements of Davidson and Wang [4] extend into the weak region. Furthermore, Gazzah et al. [5, 6] have numerically investigated the co-flow effect on heated turbulent round jets. Less works, however, deal with co-flowing plane jets. Bradbury [7] and Bradbury and Riley [8] have experimentally studied a turbulent plane jet in a slow moving air stream. They found that the spread of jets could be merged, with varying co-flow velocity ratio, when accounting for the effective origins. Elkaroui et al. [9] have investigated the dynamic and thermal behavior of a free turbulent plane jet, including the influence of temperature on entropy generation. They found that high rates of entropy generation correspond to higher inlet jet temperatures. The present contribution extends this study by investigating the effect of a moving external stream on the jet development and entropy generation. The presence of co-flow stream is known to improve the mixing process for a non-reactive jet and to generate flame stability. Free and co-flowing round jets, which can be used in chambers and the mixer, have received much more attention than plane jets. That is why this paper aims to investigate the configuration of turbulent plane jets issuing in a co-flowing stream and applicable in air curtains. The effect of co-flow on entropy generation, which can be used as an effective tool for the optimal design of thermal systems, is studied through a numerical simulation using the first order k-ε model. The effect of the co-flow velocity ($U_{co} = 0.0$ m/s, $U_{co} = 1.20$ m/s and $U_{co} = 2.0$ m/s) on the dynamic and thermal behavior and local entropy generation for a turbulent plane jet are predicted.

2 Problem Formulation

The considered flow is a heated turbulent plane jet issuing in a co-flowing ambient stream with various velocity ratios, temperature ratios. The actual dimension of the slot width H is equal to 0.04 m. The flow is only weakly compressible in the sense that the

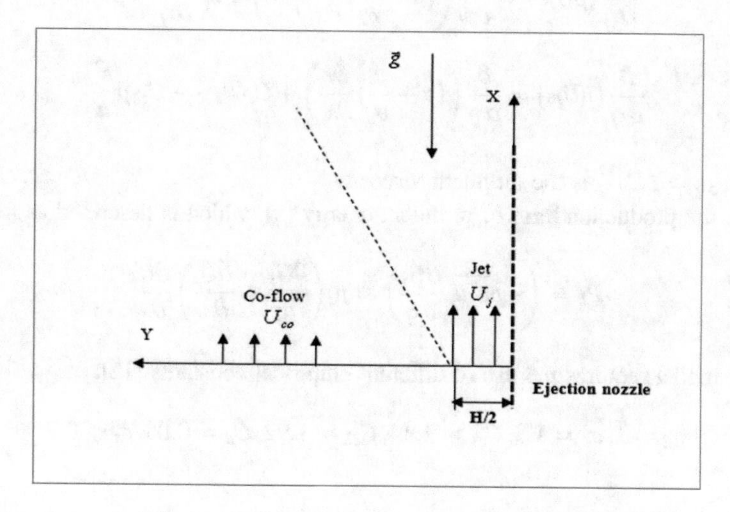

Fig. 1. Geometric configuration with co-flow

Mach number is low. Hence, the flow development is characterised by the Reynolds number $\mathrm{Re} = \frac{U_j H}{\upsilon}$, the densimetric Froude number $Fr = \frac{\rho_j U_j^2}{\left(gH\left(\rho_{co} - \rho_j\right)\right)}$ as defined by Chen and Rodi [10], where U_j = 18 m/s is the jet velocity (Fig. 1).

2.1 Governing Equations

Under these assumptions, the governing equations including continuity (1), momentum (2) and temperature (3) conservation equations are considered in the cartesian coordinates system as follows:

$$\frac{\partial \rho U_j}{\partial x_j} = 0 \tag{1}$$

$$\frac{\partial}{\partial x_j}\left(\rho U_i U_j\right) = -\frac{\partial P}{\partial x_i} - \frac{\partial}{\partial x_i}\left(\overline{\rho u_i'' u_j''}\right) + \frac{\partial}{\partial x_j}\left[\mu\left(\frac{\partial U_i}{\partial x_j} + \frac{\partial U_j}{\partial x_i}\right) - \frac{2}{3}\mu\frac{\partial U_k}{\partial x_k}\delta_{ij}\right] \tag{2}$$

$$\frac{\partial}{\partial x_j}\left(\rho T U_j\right) = \frac{\partial}{\partial x_j}\left(\frac{\lambda}{C_P}\frac{\partial T}{\partial x_j}\right) - \frac{\partial}{\partial x_j}\left(\overline{\rho t'' u_j''}\right) \tag{3}$$

With δ_{ij}: is the Kronecker symbol; δ_{ij} = 1 if i = j, 0 if not

$\overline{\rho u_i'' u_j''} = -\mu_t\left(\frac{\partial U_i}{\partial x_j} + \frac{\partial U_j}{\partial x_i}\right) + \frac{2}{3}\rho k \delta_{ij}$: is Reynolds stress tensor

$\overline{\rho t'' u_j''} = -\frac{\mu_t}{\sigma_t}\frac{\partial T}{\partial x_j}$: is the Reynolds heat flux vector

The $k-\varepsilon$ model is used for the closure of this system: the equations of the turbulent kinetic energy (k) and its dissipation rate (ε) can be written as [11]:

$$\frac{\partial}{\partial x_j}\left(\rho U_j k\right) = \frac{\partial}{\partial x_j}\left(\left(\mu + \frac{\mu_t}{\sigma_k}\right)\frac{\partial k}{\partial x_j}\right) - \overline{\rho u_i'' u_j''}\frac{\partial U_i}{\partial x_j} - \rho\varepsilon \tag{4}$$

$$\frac{\partial}{\partial x_j}\left(\rho U_j \varepsilon\right) = \frac{\partial}{\partial x_j}\left(\left(\mu + \frac{\mu_t}{\sigma_\varepsilon}\right)\frac{\partial \varepsilon}{\partial x_j}\right) + C_{\varepsilon 1} P_k \frac{\varepsilon}{k} - C_{\varepsilon 2}\rho\frac{\varepsilon^2}{k} \tag{5}$$

With $\mu_t = \rho C_\mu \frac{k^2}{\varepsilon}$ is the turbulent viscosity

P_k is the production term of turbulent energy (k), which is described as follows:

$$P_k = \left(-\overline{\rho u_i'' u_j''}\frac{\partial U_i}{\partial x_j}\right) \approx \mu_t\left(\frac{\partial U_i}{\partial x_j} + \frac{\partial U_j}{\partial x_i}\right)\frac{\partial U_i}{\partial x_j} \tag{6}$$

This model requires the use of different empirical constants [12]:

$$\sigma_k = 1, \sigma_\varepsilon = 1.3, C_{\varepsilon 1} = 1.44, C_{\varepsilon 2} = 1.92, C_\mu = 0.09, \sigma_t = 0.7$$

2.2 Local Entropy Generation Rate

In the non-reacting jet flow, when both temperature and velocity fields are known and based on the second law of thermodynamics, the volumetric entropy generation rate ($\dot{S}_{gen}$) as given by Bejan [13] at each point in the fluid, is calculated as follows:

$$\left(\dot{S}_{gen}\right) = \left(\dot{S}_{gen}\right)_{heat} + \left(\dot{S}_{gen}\right)_{fric} \tag{7}$$

Where $\left(\dot{S}_{gen}\right)_{heat}$ and $\left(\dot{S}_{gen}\right)_{fric}$ represent the volumetric entropy generation rates due to heat transfer and fluid friction, respectively with the following expressions [17]:

$$\left(\dot{S}_{gen}\right)_{heat} = \frac{K_{eff}}{T^2}\left(\frac{\partial T}{\partial x_j}\frac{\partial T}{\partial x_j}\right) \tag{8}$$

$$\left(\dot{S}_{gen}\right)_{fric} = 2\frac{\mu_{eff}}{T}\left(S_{ij}S_{ij}\right) \tag{9}$$

Where K_{eff} and μ_{eff} are the effective thermal conductivity and the effective viscosity respectively.

S_{ij} is the mean strain rate and: $S_{ij} = \frac{1}{2}\left(\frac{\partial U_j}{\partial x_i} + \frac{\partial U_i}{\partial x_j}\right)$.

3 Boundary Conditions

At the jet exit, a uniform velocity U_j, and a temperature T_j are imposed. The jet discharged into a co-flowing stream U_{co} is symmetric to the plane (y = 0) and the calculation is performed over half of the flow. The computational domain is rectangular with a domain size, in terms of the jet inlet height H, of $L_x/H = 60$ and $L_y/H = 25$ in the longitudinal and transversal directions, respectively. The foregoing system of equations is completed with the following boundary conditions:

- In the free boundary parallel to the axis, the considered conditions are as follows:

$$\left(\frac{\partial V}{\partial y}\right)_{y=wall} = 0; U = 0; T = T_{co}; k = 0; \varepsilon = 0$$

- On the symmetry axis, the lateral velocity and gradients of all variables are set to zero.

$$\left(\frac{\partial \Phi}{\partial y}\right)_{y=0} = 0;\ \Phi = U, T, k, \varepsilon;\ V = 0$$

– At the outflow boundary, the gradient of dependent variables in the axial direction and the lateral velocity are set to zero.

$$\left(\frac{\partial \Phi}{\partial x}\right)_{outlet} = 0;\ \Phi = U, T, k, \varepsilon;\ V = 0$$

– At the inlet, and in order to overcome the jet emission influence as much as possible, the axial velocity profile was calculated from the following relation (10):

$$x = 0: \begin{cases} 0 < y < H/2: & U = U_j; \quad V = 0; \quad T = T_j; \quad k = 10^{-3} U_j^2; \quad \varepsilon = \frac{C_\mu k^{3/2}}{0.03H} \\ y \geq H/2: & U = U_{co}; \quad V = 0; \quad T = T_{co}; \quad k = 0; \quad \varepsilon = 0 \end{cases}$$

4 Numerical Solution Method

The equations are solved by the Finite Volume Method (FVM) with a staggered grid as described by Patankar [15]. The transport equations of momentum, energy, turbulent kinetic energy and its dissipation rate are discretized through this equation:

$$A_P \Phi_P = \sum_{nb} A_{nb} \Phi_{nb} + \Phi_P. \tag{10}$$

Where, the subscript 'nb' designates neighbours which mean (i + 1, j), (i − 1, j), (i, j + 1) or (i, j − 1). In the present study, the diffusion and the convection coefficients are discretized using a hybrid scheme [16], which is first order or second order, depending on the local cell Reynolds number.

Thus, to find the numerical solution of these equations, a computer code was developed. The velocity-pressure coupling is solved with the SIMPLE algorithm (Semi-Implicit-Method for Pressure-Linked-Equations). The system of algebraic equations is solved line by line using the TDMA (Tri-Diagonal Matrix Algorithm). The TDMA is based on the Gaussian elimination procedure associated with the over-relaxation technique described by Patankar [15].

The used mesh is not uniform and gradually extends according to the longitudinal and transversal directions.

5 Results and Discussion

5.1 Mean Centerline Velocity Variation

Figure 2 shows the axial evolution of the centerline longitudinal velocity $\left((U_j - U_{co})/(U_c - U_{co})\right)^2$ as a function of the normalized distance x/H, where U_c is the jet centerline mean axial velocity. Indeed, the normalized centerline longitudinal velocity obeys the self-similar decay law as: $\left(\frac{U_j - U_{co}}{U_c - U_{co}}\right)^2 = K_1\left(\frac{x}{H} - C_1\right)$.

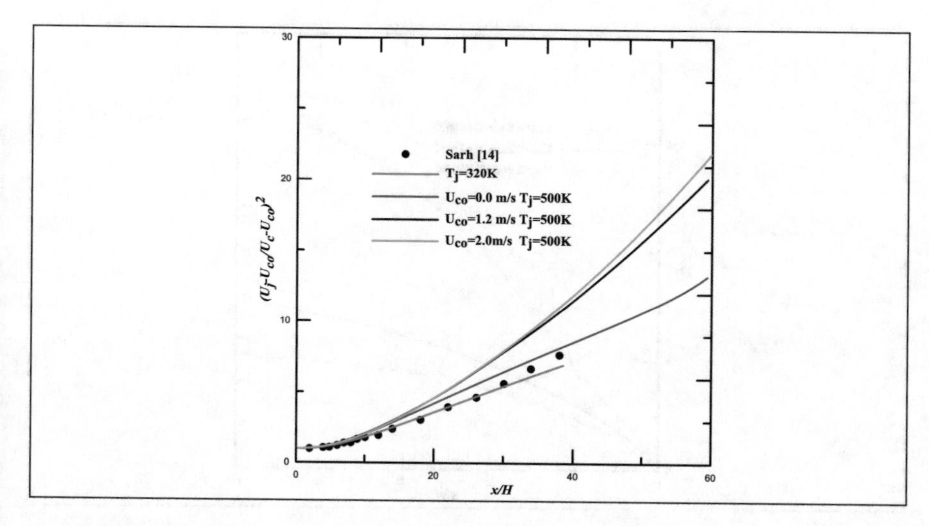

Fig. 2. Effect of co-flow variation on the axial evolution of the centerline longitudinal velocity

It appears clearly that the computed results agree well with the experimental values of Sarh [14]. For three different co-flow velocities (U_{co} = 0.0 m/s, U_{co} = 1.20 m/s and U_{co} = 2.0 m/s), the centerline longitudinal velocities are constant in the potential core area, near the jet exit, and are equal to the centerline exit velocity. In this region, the co-flow velocity has no effects on this parameter and the flow behaves as a jet in a stream at rest. However, in the similarity zone, away from the jet exit, the centerline velocity decreases as ($x^{1/2}$). It is found that the jet in a co-flowing stream is a quicker mixer than a jet in a quiescent ambient, which means that the presence of a co-flow enhances mixing. Therefore, the mixing of a jet in co-flow is more efficient.

5.2 Entrainment Variation

Figure 3 represents the axial evolution of air entrainment with T_j = 500 K and for three different co-flow velocities (U_{co} = 0.0 m/s, U_{co} = 1.20 m/s and U_{co} = 2.0 m/s). The amount of air entrainment by the jet can be determined by the time-average lateral profiles of velocity. This quantity, which relates the mass flow rate of the surrounding fluid entrained into the jet to the characteristic velocity difference between the jet and the co-flow, is given by:

$$E = 2 \int_{0}^{y(U=U_{co})} \rho(U - U_{co})dy \tag{11}$$

It is shown that the presence of co-flow stream decreases considerably the air entrainment. This is due to the reduction of the jet lateral expansion in the presence of co-flow. Qualitative analyses suggest that a co-flowing stream would restrict the lateral

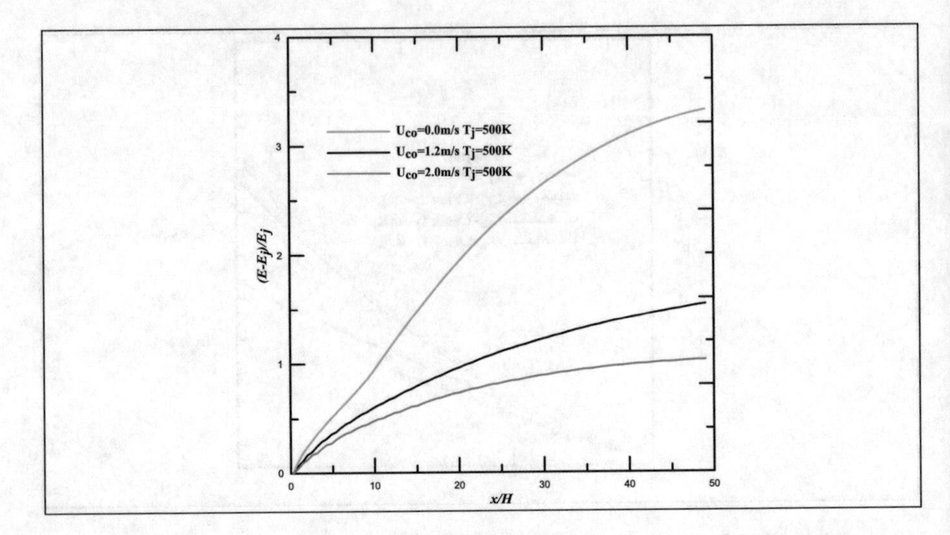

Fig. 3. Effect of co-flow on the axial evolution of the air entrainment

in flow of air into the jet. The free jet entrains from 30% to 75% more air than the co-flowing jet at any given axial location.

5.3 Entropy Generation Rate Variation

Figure 4 represents the axial evolution of the total entropy generation rate in the jet with a fixed inlet jet temperature $T_j = 500$ K and for three co-flow velocities ($U_{co} = 0.0$ m/s, $U_{co} = 1.20$ m/s and $U_{co} = 2.0$ m/s).

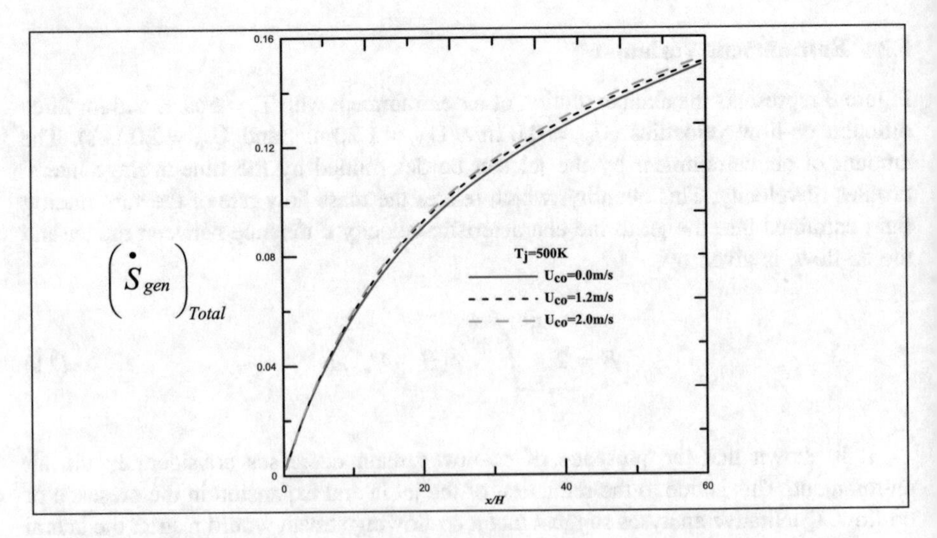

Fig. 4. Effect of co-flow on the axial evolution of the total entropy generation rate

In the potential cone region, it is seen that the entropy generation rate grows as x/H increases. Furthermore, there is a sudden increase in entropy generation in the flow region where x/H is less than 20. In this region, the jet is more unsteady and also has the most distorted profiles, right before, it reaches the self-similarity region. Moreover, in the self-similarity region, (x/H > 20) there is a smooth increase of entropy generation with respect to the observed longitudinal coordinate. The obvious conclusion from this Figure is that the co-flow velocity has very little effect.

6 Conclusion

A comprehensive analysis of the evolution of a heated turbulent plane jet emerging in a co-flowing stream is provided. The entropy generation rate in a turbulent plane jet is investigated, taking into account the effect of co-flow. Numerical simulations are carried out using the first order k-ε turbulence model. In particular, the numerical results for mean and turbulent quantities, air entrainment, entropy generation rate are studied. The main conclusions from the present study can be summarized as follows:

First of all, the predicted results agree reasonably well with the experimental data available in the literature for plane jets.

The increase of the co-flow velocity reduces the amount of air entrainment and the mixing efficiency of the jet. Moreover, the total entropy generation rate decreases with the increase of the co-flowing velocity.

The calculation results confirm that the entropy generation rate grows and attains an asymptotic value along the flow direction and depends directly on the entrainment with the still ambient fluid.

References

1. Nickels, T.B., Perry, A.E.: The turbulent co-flowing jet. J. Fluid Mech. **309**, 157–182 (1996)
2. Antonia, R.A., Bilger, R.W.: An experimental investigation of an axisymmetric jet in a co-flowing air stream. J. Fluid Mech. **61**, 805–822 (1973)
3. Smith, D.J., Hughes, T.: Some measurements in a turbulent circular jet in the presence of a co-flowing free stream. Aeronaut. Q. **28**, 185–196 (1977)
4. Davidson, M.J., Wang, H.J.: Strongly advected jet in a co-flow. J. Hydraul. Eng. **128**, 742–752 (2002)
5. Gazzah, M.H., Belmabrouk, H.: Local entropy generation in co-flowing turbulent jets with variable density. Int. J. Numer. Meth. Heat Fluid Flow **24**, 1679–1695 (2014)
6. Gazzah, M.H., Belmabrouk, H.: Directed co-flow effects on local entropy generation in turbulent heated round jets. Comput. Fluids **105**, 285–293 (2014)
7. Bradbury, L.J.S.: The structure of a self-preserving turbulent plane jet. J. Fluid Mech. **23**, 31–64 (1965)
8. Bradbury, L.J.S., Riley, J.: The spread of a turbulent plane jet issuing into a parallel moving air stream. J. Fluid Mech. **27**, 381–394 (1967)
9. Elkaroui, A., Gazzah, M.H., Mahjoub Saïd, N., Bournot, P., Le Palec, G.: Entropy generation concept for a turbulent plane jet with variable density. Comput. Fluids **168**, 328–341 (2018)

10. Chen, C.J., Rodi, W.: Vertical Turbulent Buoyant Jets-A Review of Experimental Data. Heat & Mass Transfer, vol. 4. Pergamon Press, Oxford (1980)
11. Argyropoulos, C.D., Markatos, N.C.: Recent advances on the numerical modelling of turbulent flows. Appl. Math. Model. **39**, 693–732 (2015)
12. Launder, B.E., Spalding, D.B.: The numerical computation of turbulent flows. Comput. Methods Appl. Mech. Eng. **3**(2), 269–289 (1974)
13. Bejan, A.: A study of entropy generation in fundamental convective heat transfer. J. Heat Transfer **101**, 718–725 (1979)
14. Sarh, B.: Contribution à l'étude des jets turbulents à masse volumique variable et des flammes turbulentes de diffusion. Thèse de doctorat à l'Université Pierre et Marie Curie Paris (1990)
15. Patankar, S.V.: Numerical Heat Transfer and Fluid Flow. Hemisphere Publishing, Washington, D.C. (1980)
16. Spalding, D.B.: A novel finite-difference formulation for differential expression involving both first and second derivatives. Int. J. Numer. Meth. Eng. **4**, 551–559 (1972)
17. Moore, J., Moore, J.G.: Entropy production rates from viscous flow calculations, Part I. A turbulent boundary layer flow. ASME Paper 83-GT-70, ASME Gas Turbine Conference, Phoenix, AZ (1983)

CFD Study of a Pulverized Coal Boiler

Lakhal Fatma Ezzahra[1(✉)], Agrebi Senda[1], and Mouldi Chrigui[1,2]

[1] Research Unit "Mechanical Modeling, Energy and Material", National School of Engineering of Gabes, Gabes, Tunisia
lakhal.fatma@gmail.com, aguerbi.sinda@hotmail.com, mouldi.chrigui@rnu.tn

[2] Institute of Energy and Power Plant Technology, Technical University of Darmstadt, Darmstadt, Germany
mchrigui@ekt.tu.darmstadt.de

Abstract. Numerical simulation of pulverized coal combustion is carried out in order to investigate the behavior of the dispersed phase. Combustion is done in a combined cycle unit (a coal-fired boiler) coupled with a gas/oil-fired turbine. Results are presented for laboratory scale burner SCO (Single Central Orifice) type. Combustion is performed in atmospheric air conditions. Coal particle trajectories are analyzed to capture Gas-particle interaction within a turbulent flow. An Eulerian-Lagrangian combustion approach is chosen. A reduced reaction mechanism is used to describe coal combustion in the furnace. Particle transport, heating and devolatilization, char combustion involving heterogeneous kinetic reaction and heterogeneous combustion are modeled. These models are combined with a finite rate/eddy dissipation model to describe turbulence chemistry interaction. Results are compared against previous data produced by researcher Godey et al. experimental data.

Keywords: Combustion · Discrete phase · Turbulence · SCO burner · Numerical simulation

1 Introduction

Pulverized Coal Combustion (PCC) is reported by the World Energy Council to be the major shared application (Hein 2013). As environmental regulations are getting more stringent, existing coal thermal power plants need to control their pollutants emissions such as CO, CO2, NOx, SOx and ash particles. It is therefore important to understand the pulverized-coal combustion mechanisms and develop clean coal technology for pulverized coal fired power plants. However, the combustion of pulverized coal is a complex phenomenon compared to that of gaseous or liquid fuels, since dispersion of coal particles, devolatilazation and oxidation reactions take place simultaneously (Hwang et al. 2006; Tsuji et al. 2002). The efficiency improvement and pollutant formation reduction of already built PCC and/or upcoming one, requires a good understanding of Gas particle interaction in turbulent reacting flows. Reynold Averaged Navier Stokes (RANS) has been the standard technique to model turbulence mixing in the flow (Harding et al. 1982; Michel and Payne 1980; Lockwood et al. 1984; Hassan et al. 1983; Xu et al. 2018 and 2017). RANS solves the time averaging equations to

A. Benamara et al. (Eds.): CoTuMe 2018, LNME, pp. 257–264, 2019.
https://doi.org/10.1007/978-3-030-19781-0_31

give an indication of the mean flow. The approach is useful to give order of magnitude estimations as explained by (Michel and Payne 1980). The present work combines usage of Eulerian-Lagrangian approach and a reduced reaction mechanism to numerically study Gas-particle of the reactive turbulent flow. The investigated furnace is developed in International Flame Research Foundation (IFRF) and experimentally investigated by (Godoy et al. 1988). The main goal of the present work is to analyze particle trajectories in a combusting flow to capture the interaction between Gas-particle, then study its effect on the evolution of multiphase air-coal combustion properties in the furnace. Focus is put on predicting the CO specie production in relation with particle trajectory and temperature rise.

The methodology consists in (1) validation of results, (2) Gas velocity study, (3) temperature analysis, (4) Particle trajectory interaction with gas axial velocity, (5) CO concentration prediction.

2 Numerical Configuration and Boundary Conditions

The configuration under study corresponds to a downward fired cylindrical combustion chamber equipped with a swirl burner. The combustor consists of a cylinder, with an internal diameter of 0.6 m as shown in Fig. 1. The burner (SCO) consists mainly of two coaxial inlets. Primary air is injected throw a central tube with pulverized coal Particle. A secondary swirling air is supplied to the furnace throw an annular duct. A full description of the experimental facility can be found in (Lockwood et al. 1984; Hassan et al. 1983; Hirji and Loockwood 1986; Hassan et al. 1985; Hirji 1985). The simplified computational domain is similar to (Shang and Zhang 2009) work. Geddling coal is utilized in experimental test. Its proximate analysis data are fixed carbon 53.7%, volatile 35.8%, moisture 6.3% and ash 4.2%. The simulation conditions are similar to (Godey's 1988) experimental conditions. In accordance with experiments Table 1 summarizes all flow parameters used in this work.

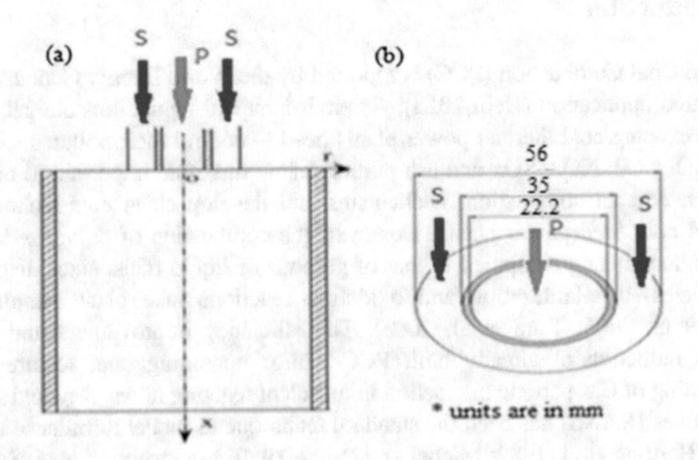

Fig. 1. Combustion chamber and burner dimensions (Shang and Zhang 2009)

Table 1. Operating conditions (Godey et al. 1988)

Parameters	Value	Units
Primary air axial velocity	19.7	[m/s]
Secondary air axial velocity	32.5	[m/s]
Secondary air tangential velocity	30.8	[m/s]
Primary preheat temperature	353	[K]
Secondary air preheat temperature	600	[K]
Wall temperature	1293	[K]
Coal flow rate	11.7	[Kg/h]

3 Numerical Inflame Measurements

3.1 Validation

Numerical results are compared with previous work of (Godey et al. 1988), for validation purpose. Experimental data of temperature mainly for three different cross sections X = 0.124, 0.204 and 0.284 m are used to validate flame temperature as in Fig. 3. For the aerodynamic behavior of the SCO burner especially in the near burner region (NBR) is confirmed using previous numerical results of (Shang and Zhang 2009). As depicted by Fig. 2, good agreement between the two numerical results of gas axial velocity is observed. The temperature results comparison shows simultaneous occurrence with previous experimental and numerical results. For X = 0.128 m and X = 0.204 m, a slight discrepancy at the center line of the combustion chamber is marked.

3.2 Velocity

Figure 2 shows a pattern of the axial velocity. At the core of the configuration, negative axial velocity zones are observed. They demonstrate Internal recirculation Zone, IRZ, witch are caused by the swirling flow at the secondary inlet. Mainly two recirculation zones are noticed. The first recirculation is at the axis, caused by the flow swirl, and the second one is close to the wall, caused by the sudden expansion of the combustor geometry, at the inlet. The results show that the devolatilization occurs immediately while particles are entering into the combustion chamber. Coal particle are rapidly heated then release their volatiles upstream the IRZ. As known the devolatilization kinetics affects the overall combustion characteristics, the combustion is controlled by the mixing process of these volatiles with the surrounding oxidizer as confirmed in Fig. 3. The combustion is mainly governed by the turbulence. The Dahmkoler number is large $\gg 1$.

3.3 Gas Temperature Study

The devolatilization kinetics impacts the overall combustion characteristics as the combustion is controlled by the mixing process of the released volatiles with the

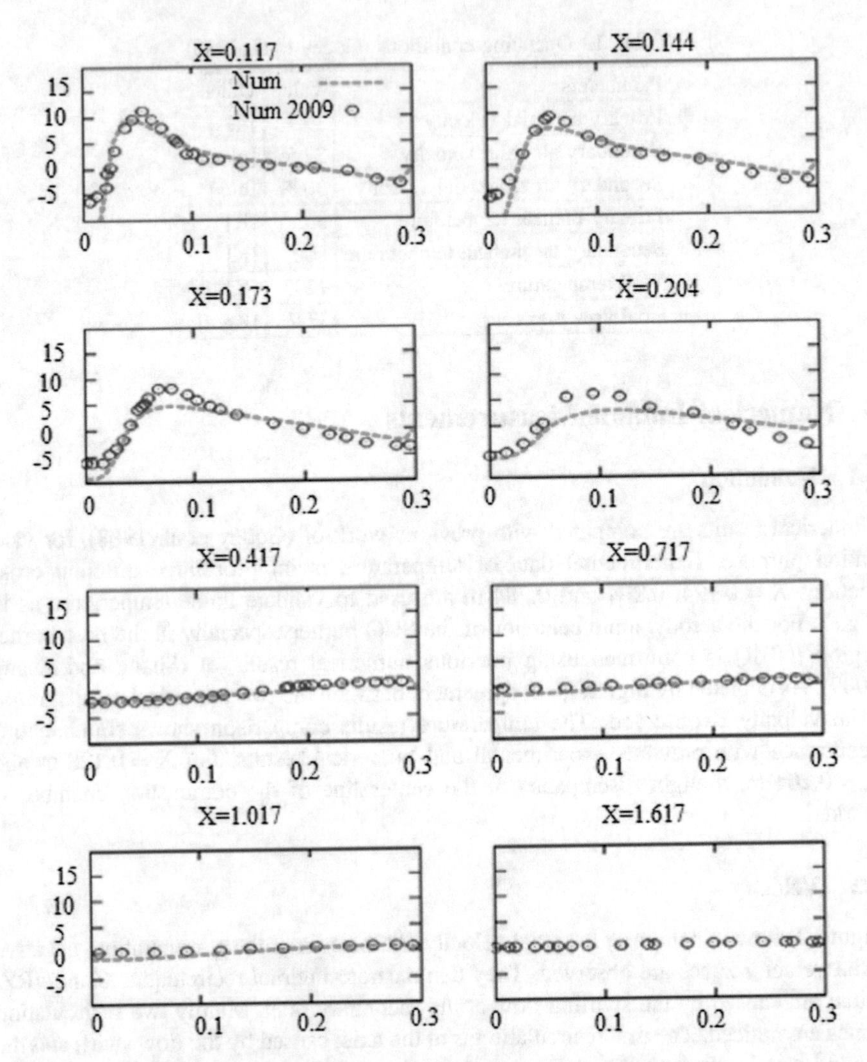

Fig. 2. Axial velocity in (m/s) for different radial cross sections

surrounding oxidizer. Figure 3 displays the temperature distribution at different line cutting the meridian of the combustion chamber. Contours of temperature highlight burned region in the configuration. The flame reaches a maximum temperature of 1500 K close to the inlet region. This maximum temperature which gives hints about the flame location wish is experimentally measured starting from a distance X = 0.124.

3.4 Discrete Phase Trajectory Study and CO Concentration

It is instructive to delineate the aerodynamic behavior of the SCO burner especially in the near burner region (NBR) because of its important role in establishing a stable flame and its influence on the formation of CO and overall the particle burnout.

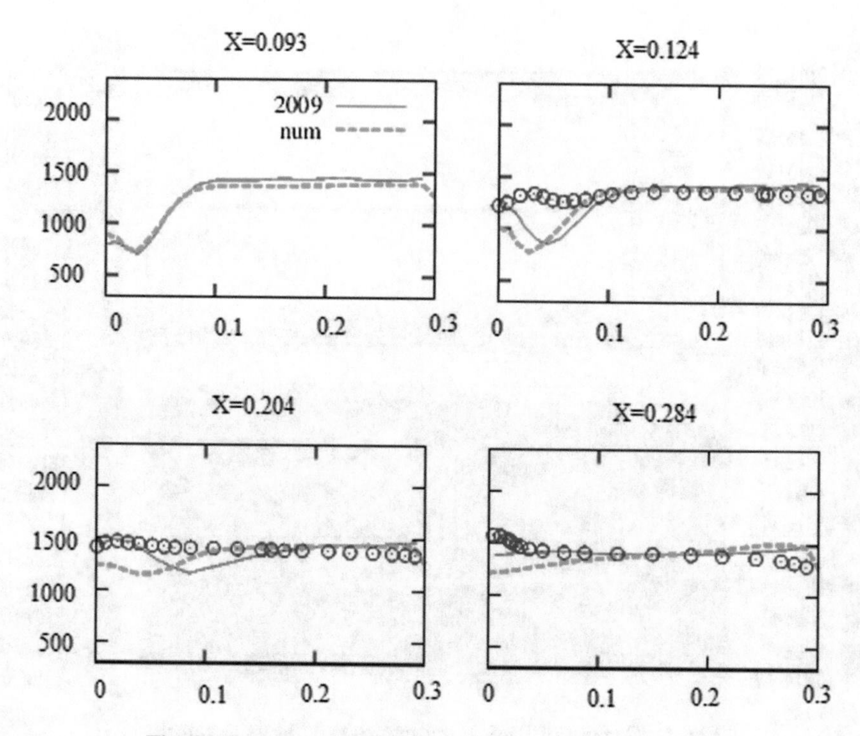

Fig. 3. Temperature in (K) for different radial cross sections

As soon as the coal particles are injected in the burner they penetrate into the internal recirculation zone (IRZ) as seen in Figs. 4 and 5. The secondary air flow influences their trajectory. Particles are then entrained into high shear zone between the IRZ boundary and the secondary air stream. CO generation is marked to be strongly dependent of the devolatilazation of the particles. Particles penetrate a certain distance into the internal recirculation zone before losing their initial momentum and entrained backward to the burner inlet. Therefore devolatilization takes place near the burner region. As a result the on axis gas temperature increases rapidly to its peak near the burner region were CO generation takes place as observed in Fig. 6.

Predicted results of CO generation did not present the same order of magnitude obtained by experimental results, although reached numerical results captured CO generation in the same region detected by experiment. CO generation increases as the temperature increases. Thus, the discrepancy of the flame temperature at the center line of the combustion chamber may be the raison behind the lack in order of magnitude of CO generation in the combustion chamber.

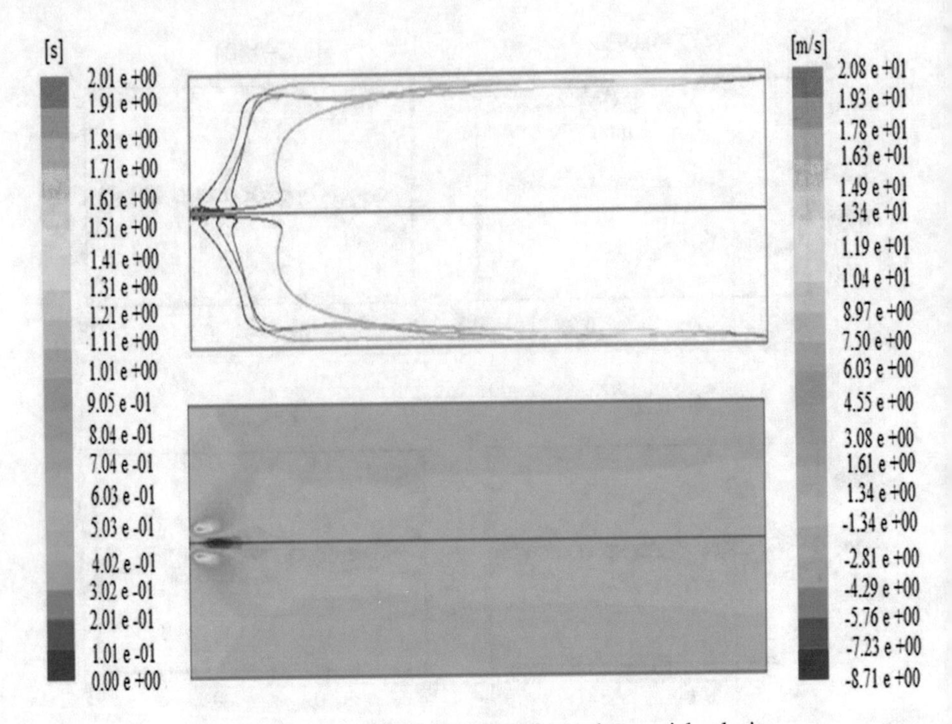

Fig. 4. Contour of Particle trajectories and gas axial velocity

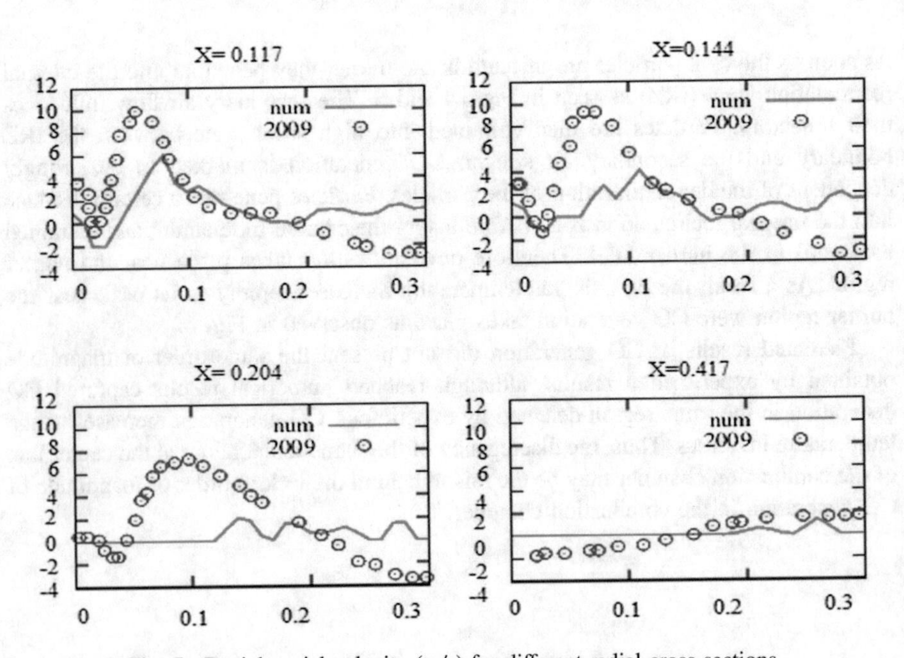

Fig. 5. Particle axial velocity (m/s) for different radial cross sections.

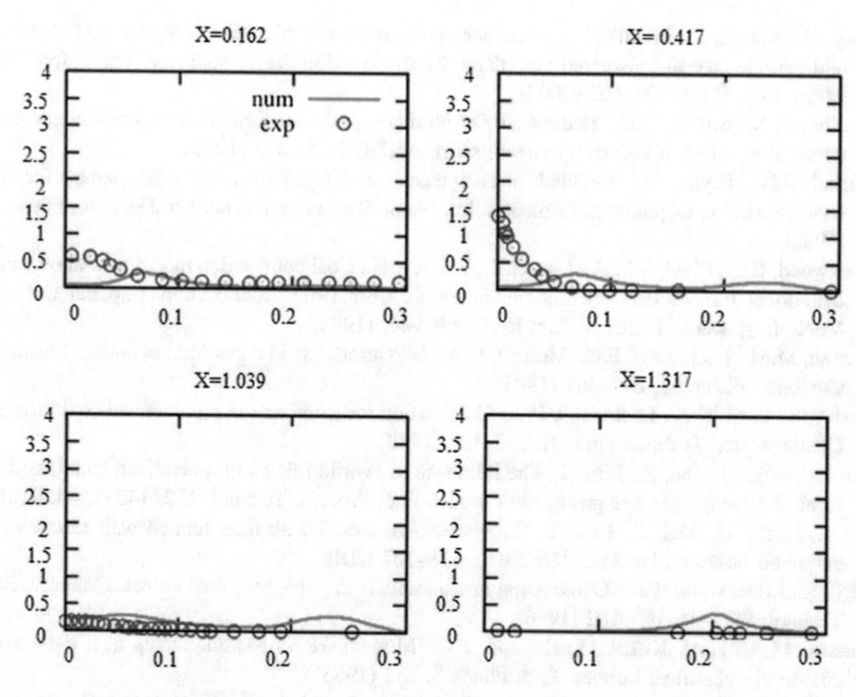

Fig. 6. Co concentration for different radial cross sections

4 Conclusion

An Eulerian-Lagrangian air-coal combustion module is applied to capture the behavior of the dispersed particle within a burning environment. RANS models are used to capture turbulence. A multi-reaction mechanism involving six reactions, three of them are homogeneous and three are heterogeneous. Turbulence chemistry interaction is modeled using the finite rate/eddy dissipation. Reasonable agreement between experimental data and numerical simulation of temperature are predicted. An under estimation of CO generation is observed. The reason behind is the coal distribution and volatilization which collocate into cluster producing important concentration. Further investigation will focus on the effects of higher turbulence model, LES, on the transport of coal and species combustion in the furnace.

References

Hein, K.: Behaviour of non-swirling pulverized coal flames. IFRF Doc. F32/a341 (2013)
Hwang, S., Kurose, R., Akamatsu, F., Tsuji, H., Makino, H., Katsuki, M.: Observation of detailed structure of turbulent pulverized-coal flame by optical measurement. JSME Int J., Ser. B **49**, 1316–1327 (2006)

Tsuji, H., Kurose, R., Makino, H.: Simultaneous measurement of particle velocity, particle shape and particle size in pulverized coal flame by shadow doppler velocimetry. Trans. Jpn. Soc. Mech. Eng. B **68**, 596–602 (2002)

Harding, S.N., Smoot, L.D., Hedman, P.O.: Nitrogen pollutant formation in a pulverized coal combustor: effect of secondary stream swirl. AIChE J. **28**, 573 (1982)

Michel, J.B., Payne, R.: Detailed measurements of long pulverized coal flames for the characterization of pollutant formation. Int. Flame Res. Found. Ijmuiden, Doc. No. F09/a/23 (1980)

Lockwood, F.C., Rizvi, S.M.A., Lee, G.K., Whaley, H.: Coal combustion model validation using cylindrical furnace data. In: 20th International Combustion Symposium, Imperial College, Mech. Eng. Dept., Fluids Section Rep. FS/83/41, (1984)

Hassan, M.M., Lockwood, F.C., Moneib, H.A.: Measurements in a gas-fired cylindrical furnace. Combust. Flame **51**, 249–261 (1983)

Godey, S., Hirji, K.A., Lockwood, F.C.: Combustion measurement in a pulverized coal furnace. Combust. Sci. Technol. **59**(1–3), 165–182 (1988)

Xu, J., Liang, Q., Dai, Z., Liu, H.: The influence of swirling flows on pulverized coal Gasifiers using the comprehensive gasification model. Fuel Process. Technol. **172**, 142–154 (2018)

Xu, J., Liang, Q., Dai, Z., Liu, H.: Comprehensive model with time limited wall reaction for entrained flow gasifier. Fuel **184**(2016), 118–127 (2016)

Hirji, K., Loockwood, F.C.: Combustion measurements in pulverized coal flames. Combust. Sci. Technol. **59**(1–3), 165–182 (1986)

Hassan, M.A., Hirji, K.A.A., Lockwood, F.C., Moneib, H.A.: Measurements in a pulverized coal-fired cylindrical furnace. Exp. Fluids **3**, 153 (1985)

Hirji, K.A.A.: Combustion measurements in pulverized coal flames. Ph.D. thesis Dissertation, Imperial College of Science and Technology, London (1986)

Shang, Q., Zhang, J.: Simulation of gas-particle turbulent combustion in a pulverized coal-fired swirl combustor. Fuel **88**, 31–39 (2009)

Numerical Investigation of Turbulent Swirling n-Heptane Spray Ignition Behavior: Cold Flow

Senda Agrebi[1(✉)], Mouldi Chrigui[1,2], and Amsini Sadiki[2]

[1] Research Unit of Mechanical Modelling, Energy and Materials, National Engineering School of Gabes, The University of Gabes, Omar Ibn Elkhattab Street, 6029 Gabes, Tunisia
aguerbi.sinda@hotmail.com

[2] Institute of Energy and Power Plant Technology, Department of Mechanical and Process Engineering, Technical University of Darmstadt, Jovanka Bontschits-Str. 2, 64847 Darmstadt, Germany
{mchrigui, sadiki}@ekt.tu-darmstadt.de

Abstract. A detailed numerical simulation of a cold flow without spray injection of Marchoine et al. (2009) configuration are carried out. The introduced swirling air were simulated with Large Eddy Simulation (LES) and the Reynolds-averaged Navier–Stokes equations (RANS). Focus is put on the turbulent kinetic energy level on the spark location. Simulations results are plotted against experimental data.

Keywords: Turbulence · RANS · LES · Swirl

1 Introduction

The ignition of a gas turbine combustor is an increasingly important issue for engine manufacturers due to the current trend towards lean operation that makes flame initiation more difficult. The initiation of a flame through a spark in a flammable mixture is one of the fundamental problems in combustion and has been studied very thoroughly (Spalding 1979; Lefebvre 1998). A large effort has also been devoted to the effects of flow, and of the turbulence in particular, on the success of ignition (see Ahmed et al. 2007; Mastorakos 2017).

Based on Marchoine et al. (2009) work, numerical simulations will be performed in order to deeply examine the ignition process of n-heptane turbulent. This paper as first part of whole work focuses on the cold flow without spray study. RANS and LES simulation are carried out to characterize the flow field and evaluating the kinetic turbulent energy as well as the velocities fluctuations in the sparks locations.

A. Benamara et al. (Eds.): CoTuMe 2018, LNME, pp. 265–272, 2019.
https://doi.org/10.1007/978-3-030-19781-0_32

2 Numerical Configuration and Boundary Conditions

Numerical simulations are performed on Marchoine et al. (2009) test facility experimentally investigated at Cambridge University, UK. It consisted a circular duct of D = 35 mm inner diameter, fitted with a conical bluff body of diameter D_b = 25 mm giving a blockage ratio of 50% (Fig. 1). The fuel injection system consisted of a pressure swirl hollow-cone atomizer with 0.5 MPa gauge pressure. The nozzle exit diameter was 0.15 mm and the spray cone angle was 60°. The fuel used was n-heptane due to its quick evaporation that hence allows spray flames to be stabilized at the burner without air preheat. The combustion chamber consisted of a 70 mm inner diameter tube 80 mm in length. The air entered from the annular open area between the outer duct wall and the bluff body. Swirl was imparted by static swirl vanes at 60° with respect to the flow axis along the flow passage between the inner and outer ducts (the geometrical swirl number is about 1.4). After the n-heptane injection with swirling air in combustion chamber, Marchoine et al. (2009) used sparks for the spray ignition. Ignition system consists of:

- Multiple spark: 100 to 500 sparks with 5 mm gap lasting for 8 ms separated by a break of 2 ms. The energy of the spark is 60 mJ and located at (R = 30 mm, Z = 5–80 mm)
- Single spark: spark of 2 mm gap lasting 500 μs, of 200 mJ of energy and located in all domain.

In this paper, only cold flow without fuel injection and spark ignition was investigated. The introduced air mass flow rate was 0.42 kg/min with bulk velocity, U_b = 12.7 m/s. The numerical configuration as shown by Fig. 1 has structured mesh of size 817046 of cells. Calculations were performed ANSYS Fluent.

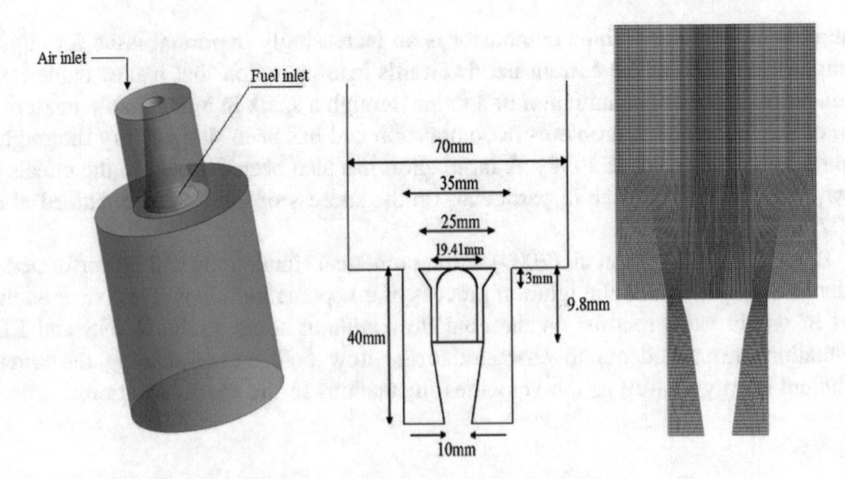

Fig. 1. Furnace geometry and its numerical grid of the studied configuration

3 Modeling Approach

3.1 LES

LES consists of directly resolving the three-dimensional time dependent turbulent motions associated with large eddies, while eddies on the order of the grid size and smaller are resolved using SGS model. The governing equations are obtained by applying a spatial filter to the evolution equations of mass and momentum, according to Chrigui et al. (2012). The resulting filtered continuity and momentum equations of the turbulent flow emerge as:

$$\frac{\partial \overline{\rho}}{\partial t} + \frac{\partial \left(\overline{\rho} \widetilde{U}_i \right)}{\partial x_j} = 0 \tag{1}$$

$$\frac{\partial \left(\overline{\rho} \widetilde{U}_i \right)}{\partial t} + \frac{\partial \left(\overline{\rho} \widetilde{U}_i \widetilde{U}_j \right)}{\partial x_j} = -\frac{\partial \overline{p}}{\partial x_i} + \overline{\rho} g_i + \frac{\partial}{\partial x_j} \left[\overline{\rho} \widetilde{\upsilon} \left(\frac{\partial \widetilde{U}_i}{\partial x_j} + \frac{\partial \widetilde{U}_j}{\partial x_i} \right) - \frac{2}{3} \overline{\rho} \widetilde{\upsilon} \frac{\partial \widetilde{U}_k}{\partial x_k} \delta_{ij} - \overline{\rho} \tau_{ij}^{sgs} \right] \tag{2}$$

Where the dependent filtered variables obtained from spatial filtering is defined as $\tilde{\psi} = \overline{\rho \psi} / \overline{\rho}$ with $\psi = \tilde{\psi} + \psi''$. Overbars and tildes express spatially filtered and density-weighted filtered values with a filter width Δ_{mesh}, respectively, while double prime represents sub-grid scale (SGS) fluctuations. The quantity ρ is the air density, U is the air velocity and x_i the Cartesian coordinate in i-direction.

In the flow, the effect of the small scales appears through the SGS stress tensor, expressed as:

$$\tau_{ij}^{sgs} = \rho \overline{U_i U_j} - \rho \overline{U}_i \overline{U}_j \tag{3}$$

The Wall-Adapting Local Eddy-Viscosity (WALE) model (Nicoud et al. (2008)) was used in this work, thus for the SGS stress tensor, expressed:

$$\tau_{ij}^{sgs} - \frac{1}{3} \tau_{kk}^{sgs} \delta_{ij} = -2 \mu_t \overline{S}_{ij} \tag{4}$$

Where δ is Kronecker delta. sub-grid scale turbulent viscosity, μ_t has the form:

$$\mu_t = \rho L_s^2 \frac{\left(S_{ij}^d S_{ij}^d \right)^{3/2}}{\left(\overline{S}_{ij} \overline{S}_{ij} \right)^{5/2} + \left(S_{ij}^d S_{ij}^d \right)^{5/4}} \tag{5}$$

The strain rate tensor S_{ij} is expressed by:

$$\overline{S}_{ij} = -\frac{1}{2}\left(\frac{\partial \overline{U}_i}{\partial x_j} + \frac{\partial \overline{U}_j}{\partial x_i}\right) \tag{6}$$

$$L_s = \min\left(\kappa d, C_w V^{1/3}\right) \tag{7}$$

$$S_{ij}^d = \frac{1}{2}\left(\overline{g}_{ij}^2 + \overline{g}_{ji}^2\right) - \frac{1}{3}\delta_{ij}\overline{g}_{kk}^2;\ \overline{g}_{ij} = \frac{\partial \overline{U}_i}{\partial x_j} \tag{8}$$

In Eq. (7) κ is the von Kármán constant, d the distance to the closest wall and V the volume of the computational cell. The default value of the WALE constant C_w is 0.325.

3.2 Rans

As mentioned above, a steady-state RANS simulation has been performed for comparison purposes. The same set of equations similar to Eqs. (1) and (2) now for Reynolds averaged quantities has been solved, in addition to the turbulent kinetic energy, k and its dissipation rate, ε. Thus, a steady-state RANS simulation has been carried out using the realizable k-ϵ model. It is well known that advanced RANS models have been suggested in the literature (see in Hanjalic (2005)). The realizable k-ϵ model is only used for commodity, as it is the most used RANS models in industrial applications.

4 Results and Discussion

The measurement of mean and R.M.S of the axial and swirl velocities can help after in interpreting the ignition probability results.

Figures 2 and 3 show the comparisons of the radial profile of the mean and the R.M.S of the axial velocities obtained from RANS and LES, and experiments in the absence of the fuel spray. Both models are able to predict the flow behavior in reasonable agreement. In Fig. 2, the highest mean axial velocities are found at a radius between 12 and 18 mm corresponding to the annular inlet of air at the burner face. The central area is characterized with low and negative velocity (see also Fig. 4), in contrast to the side region, that has positive values. The large recirculation zone created by the bluff body and the swirl is evident and this causes a rapid decrease in axial velocity and a wide angle of the flow coming out of the burner inlet. The highest R.M.S. of axial velocities as shown by Fig. 3 are about 1.0 to 3 m/s localized at radial position between 25 mm and 30 mm, which imply very high local relative turbulent intensities.

Figure 5 depicted the radial profile of the mean swirl velocity. The mean swirl velocity has a maximum value of about half of the bulk velocity, U_b close to the burner exit; it quickly decreases and the central part of the flow seems to be in solid body rotation. The maximum radius of the recirculation zone is about 25 mm and occurs between 30 and 35 mm from the bluff-body; these results when expressed in terms of

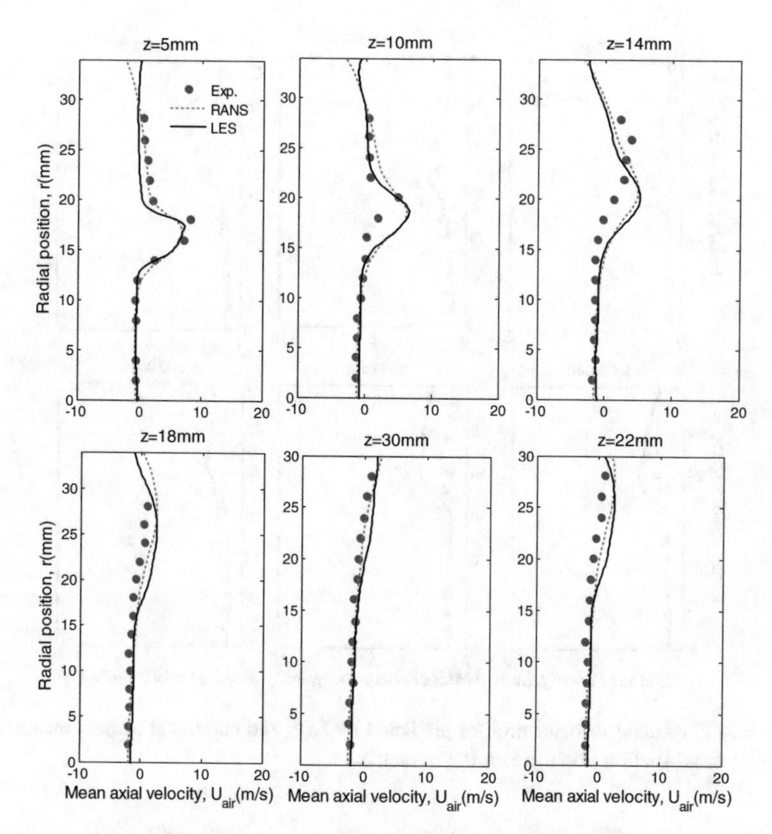

Fig. 2. Mean axial velocity profiles predicted by RANS and LES and compared against measurements at different cross-section downstream the burner inlet.

bluff-body diameter D_b imply that the width of the central recirculation zone (CRZ) is about $2D_b$ and occurs at about $1.4D_b$. The R.M.S of the swirl velocity, depicted by Fig. 6, are about 1.0 to 2 m/s

The flow results were extrapolated to the location of the multiple and single spark chosen by Marchoine et al. (2009) experiments. For multiple spark (r = 30 mm, z = 5–80), the mean swirl velocity is about 3 m/s and the mean axial velocity is negative up to about z = 10 mm and reaches its highest positive value at about r = 15 mm. The R.M.S fluctuating velocities are about 1.0 to 3 m/s. Thus, there will be strong local turbulent intensities at the spark position. Comparing the discharge duration of the multiple spark (which is 8 ms) to the large turbulent time scale which is 5 ms a large velocity and mixture fluctuations in the flow passing by the spark gap during the time of energy deposition by the spark. In contrast, for the single spark experiments, where the spark duration was 0.5 ms, smaller, but finite composition fluctuations throughout the spark.

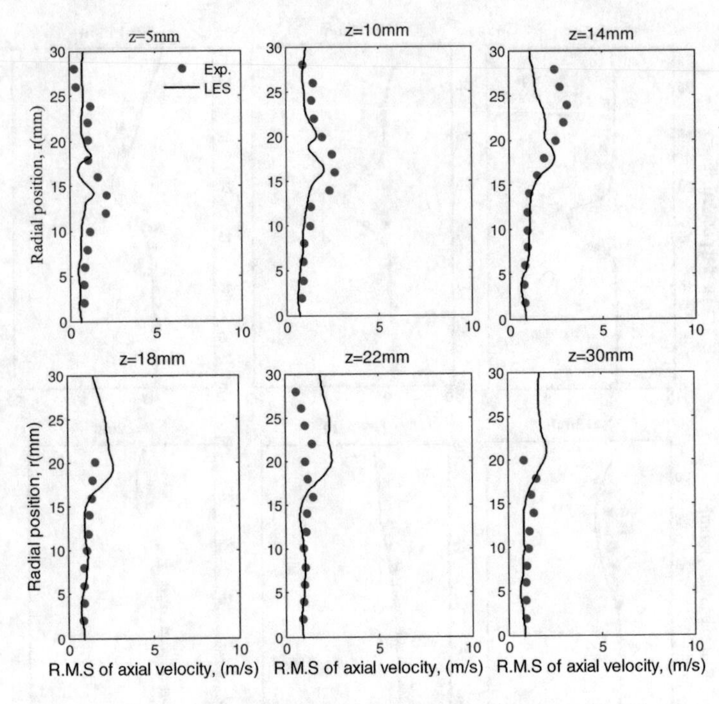

Fig. 3. R.M.S of axial velocity profiles predicted by LES and compared against measurements at different cross-section downstream the burner inlet.

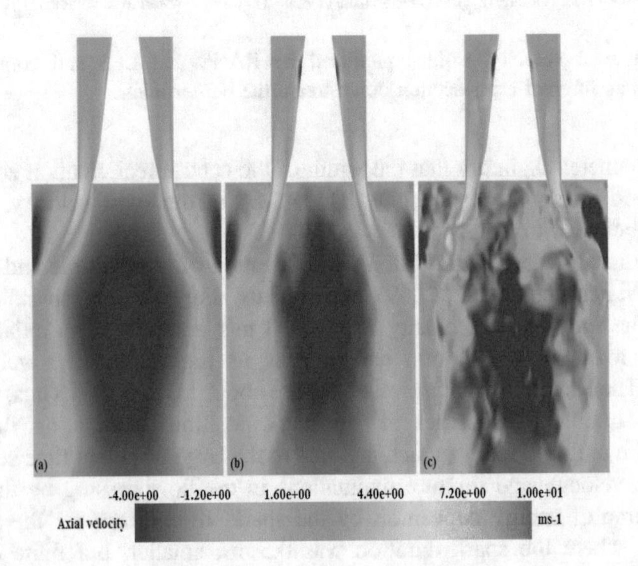

Fig. 4. Axial velocity contours (a) RANS (b) instantaneous LES (c) mean LES.

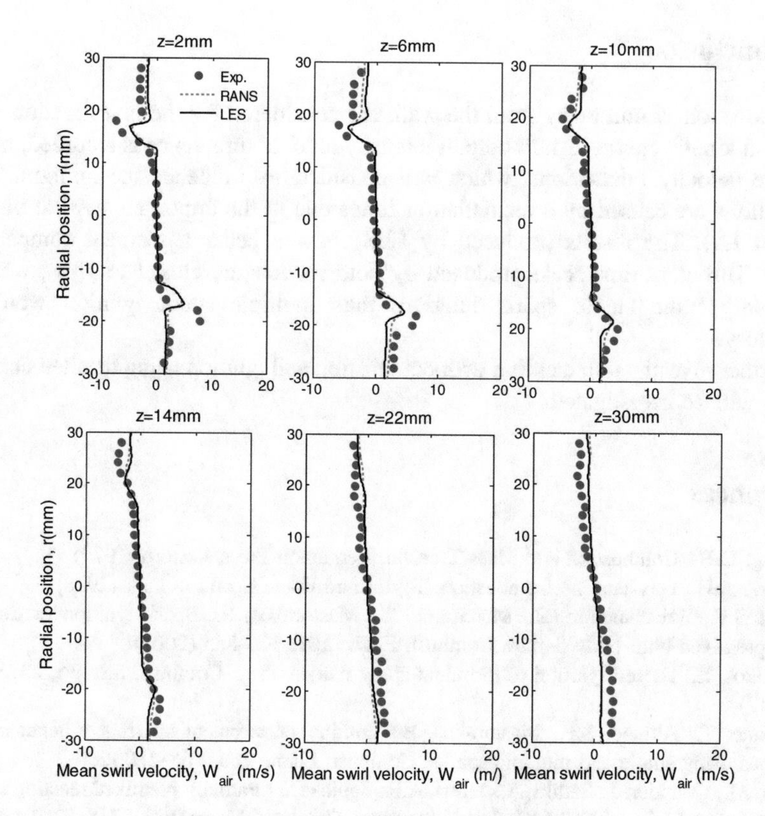

Fig. 5. Mean swirl velocity profiles predicted by RANS and LES and compared against measurements at different cross-section downstream the burner inlet.

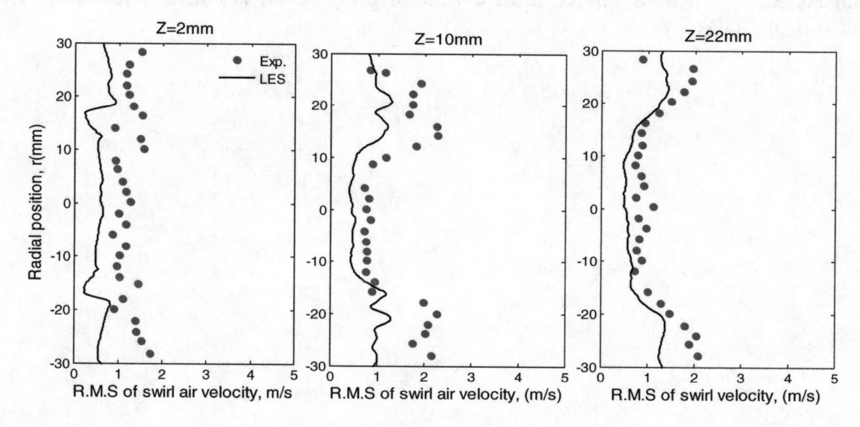

Fig. 6. R.M.S of swirl velocity profiles predicted by LES and compared against measurements at different cross-section downstream the burner inlet.

5 Conclusion

Spark location, 5 mm away from the wall, are considered for the investigation of the turbulent kinetic energy. An important intensity level of turbulence is recorded, hinting to large velocity fluctuations, which will considerably influence the ignition. These fluctuations are caused by a recirculation zones due to the important swirl at the inlet (swirl = 1.4). The results produced by LES show a better agreement compared to RANS. Turbulent time scale produced by both models are closed to 1 ms, which is much larger the single spark duration, thus multiple spark ignition would be mandatory.

Further investigation tackling with combustion and ignition using detailed chemical kinetic will be investigated.

References

Spalding, D.B.: Combustion and Mass Transfer. Pergamon Press, Oxford (1979)

Lefebvre, A.H.: Gas Turbine Combustion. Taylor and Francis, Milton Park (1998)

Ahmed, S.F., Balachandran, R., Marchione, T., Mastorakos, E.: Spark ignition of turbulent nonpremixed bluff-body flames. Combust. Flame **151**, 366–385 (2007)

Mastorakos, E.: Forced ignition of turbulent spray flames. Proc. Combust. Inst. **36**, 2375–2391 (2017)

Marchione, T., Ahmed, S.F., Mastorakos, E.: Ignition of turbulent swirling n-heptane spray flames using single and multiple sparks. Combust. Flame **156**, 166–180 (2009)

Chrigui, M., Gounder, J., Sadiki, A., Masri, A.R., Janicka, J.: Partially premixed reacting acetone spray using LES and FGM tabulated chemistry. Combust. Flame **159**, 2718–2741 (2012)

Nicoud, F., Ducros, F.: Subgrid-scale stress modelling based on the square of the velocity gradient tensor. Flow Turbul. Combust. **62**(3), 183–200 (1999)

Hanjalic, K.: Will RANS survive LES: a view of perspectives. ASME J. Fluids Eng. **127**, 831–839 (2005)

Author Index

A. Benamara et al. (Eds.): CoTuMe 2018, LNME, pp. 273–274, 2019.
https://doi.org/10.1007/978-3-030-19781-0

Printed in the United States
By Bookmasters